A Hidden Simplicity

of NUCLEAR PARTICLES

A SYSTEMATIC FULL RE-CLASSIFICATION OF THE STANDARD MODEL PARTICLES

that even UNDERGRADS WILL UNDERSTAND

by Fred Howard

**+- CHARGE SPIN INITIATES MASS
SINGLE SOURCE UNIFIES ALL FORCES**

(THE CAUSAL CORE OF WHAT HAPPENS IN QUANTUM MECHANICS)

Copyright © 2018 by Fred Howard.

Library of Congress Control Number: 2018908028
ISBN: Hardcover   978-1-9845-3976-2
                Softcover   978-1-9845-3975-5
                eBook   978-1-9845-3974-8

All rights reserved. No part of this book may be reproduced or transmitted in any form or by any means, electronic or mechanical, including photocopying, recording, or by any information storage and retrieval system, without permission in writing from the copyright owner.

Any people depicted in stock imagery provided by Getty Images are models, and such images are being used for illustrative purposes only.
Certain stock imagery © Getty Images.

Print information available on the last page.

Rev. date: 08/07/2018

To order additional copies of this book, contact:
Xlibris
1-888-795-4274
www.Xlibris.com
Orders@Xlibris.com
781722

A 2005 Particle Mass Power Law based on PDG Data
Anticipated the QM 2010
Lightest Up Quark Mass (m_p)
Within QM Error Limits
Using a New Conic Vortex Microquantum (m_c)

$$m_p = N^5 \sum m_c \left[\frac{n_\pm}{n} + \frac{n_0}{an} \right]$$

- N = Number of microquanta components c, $N = 2n$
- m = mass of particle p or c in eV mass energy
- n = number of microquantal pairs (in regular LQ particles)
- $n_\pm$ = number of charged pairs of microquanta
- n_0 = number of neutral pairs
- m_c = 10.9525... eV, $\Sigma\, m_c = m_c N$ in all LQ particles
- $a = 3^x$, integral *exponent* $x \geq 1$, = 1 in the <u>U</u>sual range
- **<u>Forceful Gyre Charge Is a Microquantal Mass Interactor</u>**

A Hidden Simplicity

of NUCLEAR PARTICLES

that even UNDERGRADS WILL UNDERSTAND

*Fresh Insight from
a Century's hard accredited PDG data
Re-Structures All Sub-Atomic Particles*
(Both Nuclear and Nuclear-Related)
as **Necessarily ONLY**
Interacting Sets of Energetic Micro-Quanta Pairs

with ALL the FORCES of Nature
- **Exposing its Unified Hidden Variables**
the CAUSE of MASS & LIGHT

(A Clarifying Auxiliary to Quantum Mechanics --
What's In Those Probabilistic Black Boxes?)

by
Fred Howard

(Einstein - - "It would be enough to understand the Electron;"
Feynman, a Founder of QM-SM - - "Nobody understands Quantum Mechanics.")

[The world's Particle Data Group (PDG) has not labored a Half Century in vain;
there IS a Solid New Basis of Physical Reality in their Mountains of Data.]

To my deceased wife,
Rietta Winn Bailey Howard

And to other members of the Howard Particle Physics Group without whose long-term assistance and encouragement the originals and this rewriting of the similar physics papers and briefings would never have been accomplished, our children living nearby:

Fred Ervin Howard, III, Mathematical Physicist
Henry Blevins Howard, Computer/Research Assistant
Katherine Marie Douglas, Research Analyst/Program Director

...............this book provides a coherency of concept
for fundamental particles........................
Fred III

To be Copyrighted 2014 by Fred E. Howard, Jr. with credit to The Florida Academy of Sciences publisher for most Figures of Chapter 1 which was rewritten and condensed from its two first edition papers in Florida Scientist (2005 & 2006). Copyright assigned to Mr. & Mrs. Fred E. Howard, Jr., Family Trust.

Published by Howard Particle Physics Group
306 Gardner Drive NE
Fort Walton Beach, FL 32548 USA
www.particlephysics.info
www.electron-particlephysics.org
(850) 244-5465 and 974-7546
fredhoward1@cox.net

Table of Contents

A Hidden Simplicity of NUCLEAR PARTICLES

Vol. 1

Foreword - - Validations, Single & General	8
Beating the Complex Nuclear Quantum Mechanics Game with SIMPLICITY	8
How Casually it All Began - - & Now How Intensively Proven !!	9
The Many Striking Correspondences with Quantum Mechanics' Standard Model	17
Many Matched Paradigm/SM Features are Especially Related to NUCLEAR QUARKS.	18
Other Matched Paradigm-QM Features are Especially Related to ELECTRONS.	20
There Are also Matched Features Especially Related to NEUTRINOS.	21

SECTION I   The Charge-Mass Power Tangle   24
(A Simple Twist that Multiplies Nature's Cross-Links)

Chapter

1 - A Sub-Structure Power Law of the Masses & Charges of "Elementary" Particles	25
SUMMARY and Overview of Discovery of New Natural Laws	25
A Sub-Structural Equation for MASSES of Composite Nuclear Baryons and Mesons	28
Search for Micro-Components in Extending the MASS Equation to Lepton-Quark Particles	31
Broadly Clarifying Consequences of the CHARGE-MASS Power Law	38
Some More Immediate Consequences of the CHARGE-MASS Power Law.	40

SECTION II   The Necessary Particle Sub-Structure & Quantum Formations   48

2 - Inherently Required Sub-Structures for General Particle Functions & Micro-Quanta	49

SECTION III   Quantal Vortex Formation Data & Structural Equations for a Particle   64

3 - Turbulent Conic Vortices - Lab Structure Data, Flow Ratios, Macro-Scaling Equations	65
4 - Turbulent Conic Vortices - Lab Forces Source Data, Equations, Single//Dual Macro-Scaling	78
5 - A Self-Consistent Structure for the Electron/Positron - the Prototype Micro-Quantal Particle	112
Appendix 5-A, Entanglement - - Another Aspect of Vortices in Electron Structure	133
Appendix 5-B, Necessary Space Medium & Energy Storage in Gyre Cone Drivers	134

SECTION IV   Re-Classification of Constrained & Expanded Proliferations of Large Particles   138

6 - A Consequential Step Into Quarks, Protons, Neutrons, Neutrinos, Muons & Tauons - An Overview	139
Appendices 6-A,B,C Neutrino Structures -- Chemistry Basis—Quark Structures	149-151
7 - Effects of Interference Constraints on Quark, Charged Lepton, & Baryon Proliferations	154
8 - The Mesons as Broken Baryons (with Quarklet Debris)	201
9 - The Baryon "Decay" Necessities for Further Neutrinos - - Dark Matter	228
10 - Expanded Neutrinos - - Pre-Quarks in Pre-Baryons & "Higgs" to Photons/Cosmos/Black Holes	235
Appendix 10-A Is there expansion of the Universe?	254
11 - Entanglement throughout the Cosmos	258
12 - The Micro-Quantal Particle (MQP) Paradigm, An inherent Unifier of All Forces (Full EarlyBriefing)	262-380

After-word and Acknowledgements	381
Primary Chapter References (for full references see References & Tables Vol.2)	382
CD Title	389

Vol. 2 APPENDICES, REFERENCES, AND TABLES (IN PREP, SEE WEBSITE)

Illustrations

Figures & Tables by Chapters of Volume 1	Page
1-1 Standard Model/Particle Data Group Nuclear Particle Data--2000 ±	29
1-2 Systematic Variation of Exponent y	30
1-3 Extension to LQ Particles of the Exponent y at Asymptotic Limit 5	32
Table 1-1: Equn. 1-2 Computed LQ Particle/Baryon Masses versus 2004 PDG SM Values	35
1-4 New Paradigm Resolves Quark Masses	36
1-5 New Empirical Paradigm Scale to Atomic Nature	37
1-6. Log Graph of Factor F in Usual Range of Equ'n 1-3 Vs LQ Particle Mass	43-4
2-1 to 5 Basic Prototype Orbital Sphere Structure for Electron/Usual Particles	51-55
2-6 The Sum Equator Quasi-elliptical or Circular Linking Orbit, & Orbit Close Approaches	56
2-7 The Basic Baryon Prototype Structural Form in a Proton	57
2-8 Sphere Orbit Internal Sync Start Sites and Conflicts If SEq on Sphere	58
2-9 Structural Form of a Six-Quark Meson	59
2-10 A Centrally Driven Turbulent Conic Vortex with Toroid Rings	61
3-1 Step-by-Step Lab Vortex Drive Process	69,71,72
3-2 Spiral Wave Planform	74
3-3 Full Spheric Conic Vortex – Immersed Elevation	75
4-1 Turbulent Gyre Diameter Versus Peripheral Velocity of Two Cones	82
4-2 Electric Drive Power & Stored Energy of Turbulent Conic Vortices.	84
4-3 Point Thrust Force Versus Peripheral Velocity of Drive Cones	86
4-4a-b-c Relative Axial Angles & Initial Mutual Force Examples of Symmetric Gyres	90-1-3
4-5a to f. Empirical Equn. 4-5 to -24 Curves for Coefficients in Equn. 4-4 per Table 4-1	94-8
4-5f Basic Master Equation 4-15a For $f_R(R)$ in Equn. 4-4 (plus Coefficient Equns)	98
Table 4-1 Symmetric Mutual Force Equation 4-4 Factors, Schedule of Usage	101
4-6 Grams Radial Mutual Force for Two Symmetric Near-Surface Turbulent PI Vortices	102
4-7 Grams Radial Mutual Force between 2 Symmetric Immersed Turbulent PI Vortices	103
4-8 Grams Radial Mutual Force for 30° Conic Symmetric Turbulent BI Vortices.	104
4-9a Symmetric Coaxial Fully Immersed Mutual Forces Orthogonal to Radial Force	105
4-9b-c Crawl and Lateral Force Data	106-7
4-10 Gram Mutual Forces vs Cm Separation of Asymmetric Gyres	108
5-1 (per Fig. 2-10) Quanta Require Centrally Driven Turbulent Conic Vortices	114
5-2 Steps of Orbital Rotations of Three Pairs of Quanta in an Electron	115-118
Table 6-1 Overview of Lepton-Quark (LQ) and Hadron Particles. (Photocopy with Tbl 1-1)	141
6-1 Electron 3D Schematic of Gyre Drive Cones At Their Synchronization Points	143
Table 6-2 Electron and Muon Neutrino Pair Orbit Sites Compared	144
Table 6-3 Empirical Sequence Priorities in Filling LQ Particle Orbit & Static Spin Sites	145
Table 6-4 Charged Lepton Pair Orbit & Static Spin Sites with Some Neutrinos & Quarks	146
Table 6-5 Actual Spherical Angle Degree Clearances between All Orbit & Static Sites	148
6-2 Extreme Neutrino Pair-Structure in Static Spin Sites & Cylinder	150
Table 6-6a-b Sphere Sites for Dual Quark Masses (& Neutrinos)	152-3
7-1a-b-c-d Proton, Neutron, & Omega⁻ Orbit Figures With Inter-Quark Conflicts	157-161
Table 7-1 Prototype Orbits and Single Spin Site for the Basic Baryons	160
Table 7-2 Key Concepts of Quark and Baryon-Linked Orbiting Spheres of Vortex Quanta	162
Table 7-3 MQP Paradigm Anticipated QM Small Quark Masses At Least Five Years	163
Tables 7-4a-to-l Summarize the Small Quarks & Their Baryon Effects	166-178, 183-5
7-2a-to-j Graphs of Small Quark Baryon Family Series Systematic Step Data	171-178, 183-4
7-2k-to-s " " Medium " " " " Data	186-193
Tables 7-4m-to-t " " " " " " Summary	186-193
Tables 7-4u-to-z Charmed " " " " "	194-198
7-2t-to-x Graphs of " " " " "	194-197
Table 7-4za Bottom " " " " " "	198
7-2y Graphs of " & All " " " " "	199-200
Tables 8-1a-e Light Unflavored Meson " " " "	203-211, 217-223
8-1a-e LU Meson 6-Quark Schematic, Graphs of " " & Pair Slide Rule	205, 213, 225-6
Table 9-1a-c Baryon Decay Neutrinos Necessities " " Summary	231-233
10-1a-b-c-d Neutrinos Expanded to Include Photons	237-241
10-2a-b Precursor Neutrinos Process	244-246
Table 10-1 " " " Data Summary	248
12-1 Working MQPP Briefing Slides	262-380
Hidden Simplicity Book CD Title Sheet	389

Foreword – – – – – Validations, Single and General

This book presents a novel paradigm that provides a coherency of concepts for all the Fundamental Particles permitting a completely new method of Classifying the Particles.

Beating the Complex Nuclear Quantum Mechanics Game with SIMPLICITY

Anticipating by 5 years the first <u>accurate</u> Quantum Mechanics lightest <u>up & down quarks' masses</u> was a major accomplishment that anyone in particle physics would think would be a center of international attention for the paper that is the prime basis of all the research in this book. But it was too simple, too obscurely published. No one but the author of the paper so far in the lead was aware of the event of having been first when it happened 5 years later with publication of the QM version. And no one else acknowledged it, even when notified with a post-print copy of the paper that was so far ahead. It was <u>too far ahead</u> in the aggressively complicated game of being first in the entire world to publish any smallest new finding in the complexly interwoven edifice of arcane probabilities of what Nature does every second in every particle of every planet or star of every gas cloud or spiral band in every galaxy, throughout a Universe. That is what the highly successful Science of the Quantum Mechanics of Nature's Nuclear particles has celebrated over the 20th Century. A simple 2005 paper's beating a 2010 Quantum Mechanics publication at QM's own game is too much, impossible; it can't be valid reading! - - But perhaps there is a sense in which all this could reasonably occur.

The initially very simple paper did become more than over-elaborated within itself in the author's attempts to comply with all the reviewer's demands for details and the consequent tendency not to leave points out. (Whoever he actually was, the reviewer appeared to have grown a habit of commenting in meticulous detail on college students' quiz papers - - all very good for teaching. He wanted every documentable bit of every small step in the logic, every possible variation of the two simple equations, no matter how useless and hard to follow, with little else. And that is all available in print and in the website for whoever wants to read it as back-up.) - - - This book condenses the resulting paper in favor of the parts that were responsible, over the next decade, for building a continuously developing new view of the sub-atomic particles (a new Micro-Quantal Particle concept, the MQP paradigm, it may become known as some day.) - - - After a typical time lapse the paper was published, with credits to the unknown reviewer and to the editor (who earned it), in the journal of the Florida Academy of Sciences, Florida Scientist, as Elementary Particle Mass Sub-Structure Power Law, Fla. Scient. (2005) Vol. 68, #3, pp 175-205; Printer's Table Erratum (2006) Vol. 69, #2, 1 page.

That FS paper worked up on a pocket calculator, in addition to anticipating the first QM accurate lightest up and down quark masses [Davies et al., Phys. Rev. Lett.. 104, 2 Apr. 2010], also was early by 3 years on a prior basis for the first comparatively accurate QM computation, by another team of physicists working on supercomputers, of the mass of the two most important larger baryon particles, the proton and the neutron. (Together these two particles are the nuclear basis of all practical earth matter.) The QM effort also included the masses of <u>most</u> of the various hadron families' baseline members. Those were <u>all</u> covered in effect, but not actually listed and claimed, in the prior paper. Significantly, the QM effort could not or did not separate the very similar proton and neutron masses, specifying them at a single mass as "the Nucleons". [Durr et al., Science (2008) Vol.322, (21 Nov.), pp 1224-7; Kronfeld, Science (2008) Vol. 322, (21 Nov.) p 1198; Wilczek, Nature (2008) Vol. 456, (27 Nov.), pp 449-50.] - - - But the simple earlier (2005) paper displayed in a published appendix the basis of separation of the two entire families of particles based on the proton and the neutron. It was earlier, and showed the system necessity for this separation not fully completed in either paper. This book sets forth to all the simple basis for the new findings and its other new results.

How Casually It All Began - - & Now How Consumingly Proven !!

When the author retired as an electronics (and later general) research engineer at the USAF Armament Research Lab, Eglin AFB, it took a while for his life-long feeling of unsatisfied curiosity about the real nature of the electron particle to come back into the focus of his daily reading. At first just keeping informed on general science and engineering progress was sufficient. Then he began noticing occasionally that even the winners of the Nobel prizes in particle physics paid scant attention to electrons in books they wrote on particles to inform their new publics and celebrate the prizes with them.

After a decade or so on more urgent interests, such as grand and great-grand children on beaches and boats, this lack of comfortable background on the electron (combined with its constant and expanding daily usage in so many heavily advertised pieces of gear for every act of modern life) began to disturb the retiree's dozing after dinner. That led on impulse to what was expected to be a quick dip into the hundreds of pages in a biennially updated report of the international Particle Data Group (PDG). (This is published again every two years with the help of the US Govt. at pdg.lbl.gov, as well as elsewhere in hard copy that might lame your wrist to pick it up.) The nine (9) sparsely abbreviated lines on the electron in the Summary Data chapters of the PDG report can hardly be understood without riffling lengthily through hundreds of other pages to unravel the limited meanings. The newcomer is left with the baffled feeling of, "What is all this?"

Of course, the answer is that the PDG report is the constantly up-dated and internationally accredited empirical data on the wide variety of characteristics of the known (and the well hypothesized) subatomic particles accumulated over a century of intense effort by literally thousands in any given day of dedicated physicists world-wide, many with full engineering, technician, and shop support. That does not make the report any easier to comprehend overall, even superficially, or page by page and condensed line by line. - - (Great effort goes into getting all the particles' characteristics recorded very exactly and completely consistently. Individual data do vary in this, rather widely. In one batch of data that is at question herein, the true geniuses of the quantum mechanics Standard Model long before worked out, and the PDG has confirmed, the exact conserved charge distributions of all the nuclear particles as added up from their component quarks. That left mainly the masses of quarks themselves widely uncertain to physicists generally until 2010, and the masses of neutrinos still debatable in early 2014. At that time of this writing, those great uncertainties are still present in the PDG accredited empirical data. Aside from its precisely known mass, its magnetic moment, charge, and exceedingly long life, everything else about the electron is very nebulous.)

So at the turn of the century it was only in the casual interest of getting a little beyond the traditional view of the electron as a "mathematical point" with "mass" and "charge" that the author was engaged in trying to grasp something of the PDG Summary Data Tables as necessary background to the electron's real nature. This again called for much further reading, often on what kinds of thoughtful research produced the best data, with the help (as an emeritus retiree) of the Lab's Technical Library, Unclassified Branch, and its fine librarians, its chief, Cheryl Mack, and Christi Rountree. A number of major government-sponsored Conference Volumes turned up, numerous papers, and several monographs with recent dates, mostly on whole-hearted efforts to understand the electron and to give it substance suitable to its importance in the science and the life of our time, which is so full of electronics. These efforts ranged over such widely varied alternatives that they were again altogether baffling. It is no wonder many gifted physicists have tried (in possible desperation) to clump all particle possibilities and their (once mysterious) distinctive forces together under such a drastically divergent notion (or paradigm) as relatively unbounded "strings"/"membranes" in high-flown theory. - - - So it was back to the PDG's primary unity in the hard observables. The hundreds of PDG participants take all this in stride as a small part of the whole subatomic particle picture, which it is necessary to see clearly in order to work with any particle. (They know equations never control Nature, just describe its actions.)

It was in trying to grasp from the outside that overview of the PDG hard empirical data, just to get a bit of a real physical handle on the electron, that the author stumbled unexpectedly into a new organizing perception of the most fundamental <u>general law</u> (or rule) of subatomic particle structure. That law led through its consequences to a couple of PowerPoint briefings of close friends and relatives, eight papers, plans for four more, and now, instead, this book. The book sums up the effort so far, with obvious plans for many more decades of work (by someone to be determined) in the advancing century.

The perception itself was very simple. Habitually sketching graphs on log vs log paper as he read, to be sure just what he had read and how it was organized, the author was looking at data on the base-line lightest members of each family of large well-known particles such as protons, neutrons, lambdas, pi mesons, and the like, in their two broad hadron classes, baryons and mesons. He plotted the two baryon and meson curves of the data's well-measured masses of those particles against his own summed totals of each one's component quark masses, as averaged over their rough estimates by the PDG at the time. Then he happened to notice in the plot the oddly different ratios of the two types of particle masses to their component sum masses. So he plotted the two curves of the ratios themselves for baryons [with three (3) component quarks] and for mesons [with two (2) component quarks] on the same graph as before. The baryon curve with its added component looked like a magnification of the meson curve. And on the same vertical line at which the meson ratio stood at 2 the baryon ratio was 3. Also odd. There each ratio was the same as the number of components; each with exponent 1. The question at once came up, did this occur because, at every equivalent sum of component masses in the two curves, they each needed the same varied exponent for the two fixed numbers of those quark components, numbers 2 and 3, as was spotted at the exponent 1? The curve of variable exponents then calculated and plotted in answer (against the same sums of each set of quark masses, as before) was, yes, a single curve for both types of large hadron particles, and surprisingly clean. (Many research writers use the reasoning of exponential curves. But this was new.) Clearly there was at work here an <u>exponential law of Nature's particle structures and their masses</u> that had not been read about in reports of prior research. Here was a new equation with new meaning within the data. It clicked on like a flashlight in a dark closet - - - (Chapter 1.)

[Much later it was found out how lucky this was; that the graphs had been plotted unknowingly with exactly the right clarifying constraint to using only the baseline lightest particle in each heavy family. (Or had that been picked up from the over-simplifying habits of the Nobel winners in public?) In any case the author was just ignorant enough at that point not to have tried to cover too much ground, that might have been included, in the famously proliferated families of the large particles. In that greater range of each particle's family of larger similar particles (with a similar name) other quirks of Nature's laws are mixed in, which would have ruined the simple curves and led nowhere.]

That exponential law led enticingly on (with encouraged labor) to considering whether the same kind of action might be happening in the same way on an extension of the same exponential curve (at its fixed asymptotic upper limit of 5 with the smallest components) to some yet smaller and unknown sub-component or two at the next lower level of particle structure within those lighter component quarks. It was also natural to include the nuclear quarks' supposedly "elementary" lepton neighbors, the electron family, and the estimated neutrinos as well [altogether a dangerous lepton-quark (LQ) extrapolation of the curve.] And, since things were getting more fundamental and hopefully simpler, it would be simplest and most convenient of all if any new very small component or two had only a single uniform mass to be added up repeatedly (instead of varying over wide ranges of mass like the various quarks.) Component charges had to be checked so that the total conserved charge in each quark, etc., would also add up properly in incremental steps of 1/3, as with the quarks in the hadrons. [For every new fan of particles those steps are 1/3 of an electron's non-varying negative charge (or of a proton's also unitary, but opposite, positive charge) on any scale of measurement of charge-force.] So it began to emerge that forceful charge and mass work together.

It was not nearly so easy to uncover next the fixed <u>power law</u> dependence of the PDG masses of all these "elementary" LQ particles on logically charged <u>numbers</u> of such simple micro-components. In fact the necessarily required micro-quantal components, such as many researchers had been trying to find and define for decades, had a mixed charge-force effect of both increasing and reducing mass that eventually did clarify the system, but only when these tiny components were grouped in conserved pairs. Further, the reduced mass effect only appeared when the members of a pair had unlike charges of positive or negative 1/6, yielding a neutral pair. This then gave three types of charges in 1/3 steps for the pairs, -1/3, 0, & +1/3. Also, the fixed exponent power law (for numbers of components to control particle mass), required by the extension of the horizontal line of the exponent's asymptotic limit of 5, continued to apply to the total number of small charged components, but not to their conserved pairs of components. Depending on their charges and numbers, the pairs were acting on mass as a separate component factor with no exponent but 1. This dual factor (times the component sum) in the resulting equation described both these micro-quantal charge-mass or mass-charge effects on LQ particle mass quite simply. - - - Though it was not as simple as the earlier exponential phase of the equation for the quarks in the much larger hadrons. There none of the quarks were neutral; all are charged either positive or negative. So the type of equation factor that described the pairs' actions inside quarks reduced to the factor 1 (and thus disappeared) for the quarks inside the heavy hadrons. In this simplification, there was essentially the same equation for both phases, with the change of phase over the four to five orders-of-magnitude extension during which the curve appeared to merge with the exponent's fixed asymptote of 5.

(Nor was the equation quite as appealing in simplicity as Einstein's famously simple equation for the mass-energy effect in all particles. - - - "Not real. Merely numerology!" said in a telcon the impatient head of a university physics department who had been sent a post-print copy of the first published paper. - - - He was evidently not the reviewer who had OKed it for publication. And also probably not the reviewer's department head. He had apparently seen too many engaging equations that finally go nowhere. - - - So acceptance was by far not unanimous. But according to the editor, the second published paper, on next consequences of the law, was reviewed yet more positively by a well-established physicist. - - - And consequences of the law continued to grow from the newly established fact that forceful charge is a major mass determinant.)

In paper one, for each quark the power law equation also gave two computed masses within the PDG uncertainties. It looked like an error. But this soon turned out to have a necessary function in causing directly (not probabilistically, as in QM) the empiricly required <u>steps in mass</u> in each of Nature's recorded PDG family series of <u>all</u> baryon particles and in at least all the PDG Light Unflavored Meson families (about 40% of the mesons. The 60% of other types still need to be checked by the demonstrated method for mesons as broken fragments of baryons.) - - - The step increases in particle mass within each family come up one by one as each light quark mass in a particle is replaced one step at a time by its heavier twin. Both the sum of quark masses and the exponent change at each step. (Smoothed into curves, these two types of step changes are not usually inverse, in opposite directions to each other, as they always were in the original exponent curve for the lightest hadrons of each family. These full family curves are strikingly highly organized in a rounded nest of curves stemming like leaves off the branch of that original simple curve. An added auxiliary law of these heavier stepped mass curves for the families is on the list of items yet to be looked for.)

Though, in its 2010 paper noted earlier, QM established only a single accurately computed mass for a quark at only the lighter value of the new paradigm's dual quark masses, it is clear, because two values for each quark inherently provide the systematic recurrent family steps in mass, that the prior wide QM uncertainties were actually more accurate (than the single definite mass) in having a wide range to include those required two separate masses for each type of quark. (QM is incomplete in having only one.)

In the up and down quarks this double mass value for each works out directly under the power law because the lower mass in each type has 2 more <u>neutral</u> pairs of quanta than in the heavier mass, where they are replaced by plus and minus <u>charged</u> pairs with the same total numbers of individual components and same net charge (or algebraic total charge.) So instead of this balance of part of the charge being struck <u>within</u> two light neutral pairs as in the lighter quark mass, it is thus struck <u>between</u> the two heavier charged pairs that increase the heavier mass for the quark, and for its resulting baryons. (In the four much heavier types of quarks from strange to top quarks, the total numbers of components, as well as the number of neutral pairs of components, also vary between the two masses for each quark.) - - (In the PDG accredited Summary Lists of hadron families there are a few vacancies in the systematicly recurring steps to be filled by these variants. A systematic search in the non-accredited PDG lists for all empiricly observed candidates for these empty steps in the accredited lists has not yet been made in this paradigm. The rule for the total number of members separated by steps required in each family is fixed by the number of types of quarks in each baryon family, whether the number is one type, two types, or even three types in the <u>strange</u> quark families and heavier.) - - In every type of quark, and in every type of baryon, the lighter or lightest baseline family member is the one that is most stably bound together and is empiricly observed most frequently. (Departures from this are very rare, but may appear with some excited "isotons" of heavier members of nuclear baryon families; or these may only be empirical confusions that are slowly being resolved in the PDG data.)

It took still longer to find in the necessarily required structures of the particles the systematic, unavoidable interferences between components that cause the observed failures of quarks to exist at all as single quarks, and also causes their ability to exist only with very short mean lives as pairs of quarks or combinations of these pairs (such as quadruplets, "tetraquarks," not triplets) in the various mesons. Only baryon triplets of quarks can be geometricly and kineticly balanced, strongly linked internally beyond bare margins, non-interfering internally, and stable as a result. (Actually, only one baryon, the proton, fully meets all of these structural requirements and does not "decay" almost indefinitely in real Nature.) Because of the numbers of ways in which interferences can occur internally and between quarks in triplets, quarks can only be very thinly scattered through the scale of apparently possible mass options with multiple micro-components. The new MQP paradigm of this book is the first adequate explanation of this effect.

As the required particle structures of the paradigm make inherently clear, quarks cannot exist alone. Consequently, they cannot gather fully preformed into baryons. To exist they require a suitably sized, brief precursor form within which their components gather in clumps and attract linkages from each other. In the clumps they forcefully sort themselves very quickly into baryon triplets of quarks (or into leptons and, as last option, also into only partially linked meson wreckages) before they separate due to excessive mass spin interactions for the strength of their net attractive forces. The precursor mass makes it totally unstable in the first cycle of swirling together. (It is not "quark soup.")

This required precursor mode of quark generation has limited linkage by pairs of quanta in only a few natural selections of types of groups. (Linkage by attracted pairs within quarks eliminated the need for the QM-hypothesized gluon particle to provide links between quarks. It also found in each link by a pair a structural limitation to three separate locked-in 60 degree phase differences in link rotation that replaces the QM need for three colors of quarks.) Structural limitations by interferences were also the factor determining why the law's fifth power of the number of components with up to 26-27 pairs of components does not generate many millions of different types of particles rather than only a proliferated hundred or two. Instant structures are the next required necessities of Nature controlling consequences of the charge-mass power law equation.

The basicly simple, but stringent, power law relation coupling charge-force to the generation of mass by the pairs became the other major influence in defining what the micro-quantum component could be in its own structure. The structure was necessarily such that all the previous relations would naturally apply. When further coupled with the spin required in nearly every particle by the PDG data, it became necessary to consider coaxial ball or conic structure for each simplest component in a pair. Uniform shapes, masses, and angular rates were required to enable exact balancing of opposite spins to zero in some cases. This provided the natural opportunity for both self spin of each micro-quantum alone and pair spin for the two of them around the center of the coaxial radius between them. (Otherwise causes would be required for eliminating one of the two modes of spinning.) Multiple pairs indicated the simplest concentric orbits in a combined sphere. Nesting synchronized orbits and axial spin sites were found for the heavier quarks. Then the directional features of the various particle forces necessarily ruled out all known component structures <u>except</u> the twin conic turbulent vortex, usually with points in (and inverted ghosts outside.) In lab tests that turbulent vortex structure directly modeled strong forces off sides and point, weak forces in a rim outwash, and electric forces only aimed outwardly from the spiral wave disk on the base of the central conic drive, with charge sign from left or right hand rotation of the cone. (This essential conserved structural charge sign is inherently not available from the symmetric sphere.)

The comparative force ranges and coupling characteristics immediately indicated that the eddy vortices in the turbulence directly model both the mass inertial coupling to the local medium and the vey weak force of gravity coupling between particles. The mass power law varied naturally with the steeply rising numbers of eddies from viscous friction between the flows of more and more gyres in a single concentric group of doubly spinning pairs. Within a balanced coaxial pair there is great reduction of friction and of mass eddies between the flows of the two gyres when they have opposite rotation and charge handedness and are neutral, since the contact flows are largely parallel. When the two gyre charges of a coaxial pair are the same with the same handedness of spin, the contact flows between them mostly rub each other the wrong way with many more frictional eddies. More eddies have greater total mass effect. Three times as much. Further, the exponential law total mass varied with quark mass eddy groups in hadrons with more separation of interactions between quark groups of gyres than within a group, and with increasing loss of full interaction of component vortices between groups as any group grows in number of vortices, as was foreseen genericly in the very first paper before vortices were involved. So the vortices relate back to the very first insight. They and their force equations fit uniquely. They were hiddenly required from the beginning.

This straight-forward vigorous vortex, with its inherent charge-mass power law of generating massively numerous small eddies of interaction energy (true mass energy) throughout groups of orbiting coaxial pairs of gyres, and with its strong and weak plus electric (& hyper-speed entangling) waves coupling forces in opposite hemispheres around each gyre - - this apparently unending, forcefully driving and centrally driven, turbulent conic vortex, is a basic unifying simplifier for all the types and families of sub-atomic particles. Now quantities of this unique kind of uniform source in a genericly specifiable space medium generate the empirical Universe of Nature in an improved majority of its subtleties and powers. [Whether the theoretical universe can be similarly recast with precision may someday be determined. This paradigm opens the door and provides a skeletal framework to make such a goal perhaps more nearly possible by humans than before. The extended linking orbits between quarks change the simplest case of fully symmetric non-electric internal forces between gyres (as in the spherical electron) to non-symmetric internal adding of electric force with extended link orbits; this requires much more, and more precise, measurement of scaled mutual gyre forces for their complete theoretical description between much more widely varied, asymmetricly angled pairs of gyres than has been done, or can readily be done, without a further well supported, formal laboratory program - - with further scaling equations for completion.]

Three, all negatively charged, pairs of such gyres, in three sphericly concentric orthogonal orbits, with full symmetry between individual gyres at all times, and outward electric force, solved at once the century-long problem of a feasible causal structure for the electron. With the equivalent three neutrally charged pairs for the mu neutrino (as then estimated by the PDG with mass upper limit just above 1/3 the electron mass), this structure also calibrated a component mass for solutions of the charge-mass power law equation for masses of the LQ particles from the smallest neutrino to the top quark - - with a few small deviations suitable for excited states of known very brief mean lives. No other combination of quantities was found to satisfy this long extended requirement of Nature. - - - What an outcome! This quick insight led to a long follow-up, necessarily including calculating a preliminary balance of internal forces in an elastic electron in two states and in the observed two sizes, both in an atom at the QM's customary less than a hundredth of the speed of light and under acceleration in colliders to near the speed of light. Finally, this insight enabled estimating instantaneous mutual forces of the largely asymmetricly angled components with only balanced departures from symmetry for an instant between two lower velocity electrons at a set separation and mutual axial angle.

As a special outcome of determining the single necessitated structure of an electron, it was also determined to be necessary that only entire particles made of pairs of gyres are limited asymptoticly to the speed of light (thus completing or re-defining Einstein's century-old restriction of natural phenomena.) Near that exterior limit, internal gyre components of particles must move very far faster than light to maintain coherent structure. So their required superphotic or superluminal force couplings to balance their orbits continuously at such speeds provides a mechanism for the observed particle entanglement that is now becoming an engineering tool. The repeated observation of actual entanglements with quasi-instantaneous couplings validates this mechanism. The EPR compromise mystique that troubled Einstein and the confusing elaborations of Bell's inequalities are no longer necessary to justify QM entanglement results. Measurement of the lower limit of that extremely high force coupling velocity can now proceed as a physical occurrence. That is a definite prediction of this paradigm.

Overlapped with establishment of the electron structure concept, there was the major intermediate step of quantitatively defining the gyre micro-quanta from necessary requirements and making lab vortex measurements with scaling equations to electron scale. It was confirmed that these vortices must necessarily be centrally driven gyres like the tornadoes and hurricanes with their central heat engines, which scaled directly from the lab equations in observed detail. This extended even to the locations of tornado formations within hurricanes that are vertically compressed in scale due to the relative shallowness of the atmosphere. (Weather experts reject this empirical approach as simply too turbulent to compete with a laminar vortex model in tub drains that has been a standby for centuries. The rejection will probably hold until there is a complete fluid mechanics set of equations for all the separate simultaneous turbulences and their forces with full theoretical accounting for the varying density of the atmosphere with altitude. The classic scarcity of full solutions for the Navier-Stokes equation will doubtless play a hand there.) - - - In the electron under constant conditions all six gyres in three coaxial pairs are always fully symmetric with each other in equal concentric circular orbits in a sphere, as are the gyres in the Usual Leptons. Full geometric symmetry that satisfied those essential cases, with limited asymmetry cases, permitted holding the lab mutual force data to an achievable introductory quantity. (Angular asymmetry opens up a quasi-infinity of data needs for full coverage of non-symmetric linking gyres in quarks and between quarks in hadrons, as well as between particles generally. More precise and complete lab gyre force data, including filtered frequency bands for the six known types of turbulent force coupling and general reduction to equations for the asymmetric mutual forces, may take some man-decades of research.)

Even before that, many mysteries of the particles, such as the stepped mass series of the families of the everyday heavy particles (noted earlier) and quarks and the electron family, began to fall completely into place with surprisingly little effort and clear-cut numerical calculations for each mass. In the case of the hadron families this came from substituting the heavier mass of each quark for the lighter mass in a perfectly regular sequence. Such a process accounted for four members of a baryon family when all three quarks are the same, six members when two quarks make up the three, or eight members when all three quarks are different. Some members have closely valued "isotons" from excited states in a single member group, and some members are entirely missing to date, though predictable for eventual observation to complete the regularity of the family. From this it was clear that some PDG families needed to be reclassified from the historical assignments to fill out the regularities of all families. (See website appendices for the full story and samples from papers in the chapters.) As fragments of broken baryons, meson families were similar and never stable, but much more complicated because some breaks must necessarily have occurred in the middle of quarks rather than always between quarks to have the empirical masses actually observed in experiments. (The full story on mesons will continue to develop. New fragments keep coming, and some are unlikely to be expected.) For four and six quark mesons, pairs of quarks or quarklets must join first, and then the couples join only two items at an instant in sequence (in precursors) to calculate the PDG measured masses.

Overall, the simple equation in its two exponential law and charge-mass power law phase forms continued to describe mass generation processes in particles. At first the particle mass data at hand were from 1984 to -86, and deviations of calculated from observed masses were in some cases large. With the Particle Data Group data of the late 90s, deviations shrank to <10%, and their number fell by half. With the PDG data of 2002, they moved closer, < 2% with only 4 really beyond uncertainties. With the 2003 interim PDG data, the deviants decreased to 3. When 2004 PDG data were available, the LQ data trend did not continue; six deviations were noted, one at 3.3%, the rest <2%. Deviations from the empirical PDG LQ data have continued to be up and down by small amounts. This is a broader validation of the basic law publication and its later consequences than the specific anticipations at the beginning of the Foreword. Still, the most general validation of the mass-charge law and of the various papers with online appendices that follow from it, as condensed in this book and proposed in a new particle paradigm auxiliary to QM, comes from the numerous close correlations between this body of newly initiated, turn of the 21st Century research and the 20th Century's much more elaborate, widely developed, & intensively used, QM of the Standard Model (SM).

The Many Striking Correspondences with Quantum Mechanics' Standard Model

Several points of direct correlation with QM were noted earlier in telling how this book and its new insight into the subatomic particles originated very casually before becoming an intense and sustained research. Obviously almost every topic in the book deals with the same particles as have been pointed out by a century of particle QM in the Standard Model. It is the matches between them that make new contributions to understanding the mysteries of QM or to filling in its blanks that will be of most interest. These correlations could not have been found without the bank of PDG empirical data.

First and foremost, the many unsuccessful efforts of the practitioners of QM, from Nobel Prize Winners to numerous other writers, to find a widely useful "preon or parton" component for the still "elementary" SM particles has a very strong correspondence to the new finding herein that spin-charged, centrally driven, turbulent conic vortices of uniform mass (when alone) fill this need as pointed out earlier (especially since the QM feature of "color" comes from the way the quantum acts in links between quarks.) This is especially notable since the gyres occur in coaxial pairs that can spin with an added rotation mode in orbits in sphere-like structures with multiple types of interactions in inherently explanatory directional patterns that generally fit the requirements of Nature's measurements of forces and momentums. (This is probably the longest sought concept consummation of many efforts at finding a mechanism for spin since the early attempt to model the entire atom as a single vortex. It is predictable that refinements of the initial assumption, that mass is evenly distributed throughout a particle sphere, will be made.)

Second in correlations with QM and Standard Model particles is the new finding of a first structure that provides a single source for all the known forces of particles and does so in the correct orders of their strengths as well as with the observed directions of the forces. This even goes so far as to provide a source of multiple tuned selectivities within bandwidths and of force for effectively unique entanglements at a comparative velocity of propagation that matches its extreme requirement in QM.

Third in correlation by filling a gap between the QM & SM paradigm of Nature is the new finding of a completely <u>causal</u> structure paradigm that utilizes probabilities only in limited cases when factors of interest are unknown or uncontrollable. In causality, the symmetric gyre forces, dimensions, and velocities at lab scale have been measured in the lab and their calculated forces, dimensions, and velocities at scale match the SM and empirical PDG data. Mass energies are generated exponentially with the number of components as the energies of eddies from the same viscous couplings that generate strong force, charge, etc., to provide data matches across several types of data.

Fourth in major correlation by difference is the great simplicity of many of the new equations such that a significant amount of particle computation can be transferred to pocket calculators versus supercomputers. (First-time physics readers may still want a dictionary at times.) This ties in with another marked difference, that the new paradigm has had only about a man-decade of effort and is presently quite incomplete compared to the QM/SM, but can rapidly expand given a small fraction of the continued SM effort.

A general important correspondence is that the QM axis of summation of effects at arc-cosine $1/\sqrt{3}$ from any orthogonal body coordinate is explained and confirmed by the paradigm's usual vector sum through the centroid of the spheric octant between the 3 primary orthogonal axes of the sphere of orbiting components in all <u>U</u>sual particles.

Many Matched Paradigm/SM Features are Especially Related to NUCLEAR <u>QUARKS</u>.

There are probably more matching features involved with quarks than with any other type of particle. The SM identifies a need for explaining the non-existence of single quarks, at least in ordinary earthly conditions, but is less definite in the explanation itself. To provide for all the observed variations in particles, the paradigm must usually have one more component pair orbit out of six in each quark sphere than can without risk synchronize its pair of gyres with those in all the other orbits unless that critical orbit is forcibly attracted outward toward two other quarks while also attracting their linking orbits to avoid internal interferences in all three quarks. This makes quarks link each other mutually in baryons with these expanded orbits (or briefly in mesons) as quarks' only existence. The linking is by a combination of the electric and strong forces of components. This unavoidable paradigm action between three potential quarks in a loose precursor clump due to particle density inherently accounts for the SM/PDG lack of single quarks and also replaces the SM necessity of calling on additional specialized gluon particles for linking and binding quarks in particles. This too is a correlation.

Since such geometric linking orbits in the paradigm are always occupied by charged pairs of components, the inherently stable proton and stabilizable neutron are aided in linking together by the electric influence of the pairs in such orbits in the quarks. The paradigm's proton and neutron bonds also result in stable two and four (or six) baryon sets that can further interact to build the more massive atomic nuclei. In the SM the same exact equivalent result comes from much more elaborate interaction of "gluons", both electric and "color" charges, and freely imaginary virtual particle effects.

In the paradigm the unique baryonic stability of a proton arises from the perfect orbital balance and minimum mass of its two up quarks with their high ratios of quantal positive charge-to-mass (and to internal momentum) held closely by their orbital links and charge forces to the one balanced negative down quark. These electric forces support the point-thrust and reduced "strong" side forces from the proton's gyres within the quarks. At this whole baryon's two-step greater level of organization than in the gyres themselves, such side/point forces are not sufficient (even with electric support) to stabilize the slightly heavier neutron with its imbalance of one orbit in each of its necessarily modified down quarks, lower electric charge-to-mass ratio, but yet stronger total side cohesive forces than in the proton due to more gyres in each of the neutron's two down quarks than in an up quark. The proton's net stability margin of cohesive forces is great enough to stabilize the neutron's baryon triplet when the two link into a synchronized deuteron (^{2}H) with positive charges attracting negatives all around the meeting plane between the baryons. SM puts excessive reliance on the strong force.

In brief, the paradigm effort has also experimented with singles and symmetric pairs of centrally driven, turbulent conic vortices under a wide range of conditions in the lab and has reduced the data on forces and velocities to scaled equations for routine calculation on vortex components of subatomic particles. Since the gyres in the elliptic expanded linking orbits of the quarks are no longer in symmetry with the greater number of sphericly orbiting gyres in the quarks, the equations do not now yield complete force balance solutions for the quarks, and parts of the solution must still be estimated until many more non-symmetric measurements are made. (As there are no unsymmetricly expanded orbits in the electron, its internal force balances between orbiting gyres were completed numericly.) SM relies on large numbers of supercomputer probability runs.

In the baryon families of the paradigm with their systematic steps of masses from substitutions of the dual masses of quarks, there must be 4 basic step groups (if no vacancies) in a family series with only 1 kind of quark in its three quarks, 6 step groups with 2 kinds of quarks and their dual masses, or 8 step groups with 3 different kinds of quarks. Additional family members may occur within any group by additionally excited levels due to a pair of gyres occupying an orbit site that is not the most balanced and deeply bonded orbit. SM is considered not so simply and clearly defined in classifying the many series or families of particles that have these steps of mass increases.

There are also three Series Prototype Plans of quark micro-quantal charge sets in the neutral 0, +, and −/++ plans of baryon series. This is due to the net −1/3 charge (with some neutral or oppositely charged pairs) of three kinds of quarks, and net +2/3 charge of the other three kinds of quarks. Only three baryon charge plans can arise from quark combinations. SM is not as clear in such charge distribution prototypes.

Accordingly, some PDG/SM baryons have been reclassified to family series with the same charges and quarks, at times as isomers. (See Vol. 2 or on-line appendices.)

The three charged gyre pairs in sum equator orbits that link the quarks of any baryon necessarily fall into three orbit phase angles 60° apart with reference to each quark's internal plane of orbit symmetry. In becoming a baryon during a precursor stage, within one orbit cycle these pre-quark planes order themselves by mutual forces between gyres, being mutually driven into orbits such that pairs in those phase angles synchronize all the pair orbits of the entire baryon. This results from the repeated phase angle sequence of 1 to 3 beginning at 1 after the odd pre-quark (if one is inherently odd) in whatever rotation sense the predominant charge rotation sense of the precursor group of gyres may be. Any excess or smaller cluster may form leptons or be driven away. If such an ordering does not occur, with three sets of quark links being attracted into place, from mutual gyre forces within one orbit cycle, the entire precursor of individual pairs flies apart, and no quark is formed. Other clusters of pairs may interact as precursors at any time, if the local density of pairs is sufficient under gravitational compression that large enough clusters occur. Otherwise, only smaller leptons may occur in smaller clusters at lower densities of pairs.

Such 60° phase angles in equator orbits of quarks cannot be exchanged in an ordinary reaction of a baryon since any significant change of phase brings about an interference during the reaction that destroys baryon triplet structure. Consequently the three bound phase angles of quark linking orbits in the paradigm are the basis for the mysterious three quark "colors" required in Quantum Chromodynamics (QCD) to track which quarks can exchange in which ways in a QCD particle reaction. This can only be overcome by full disruption of a baryon and break-down of its assembly in a new precursor phase. Such reaction intermediates are called recursors in the paradigm.

Other Matched Paradigm-QM Features are Especially Related to <u>ELECTRONS.</u>

The paradigm's mutual force equations for symmetric gyres were run numericly and instantaneously with the three orthogonally orbital gyre pairs for simultaneous convergences in internal force balance, gyre flow, orbital velocities, and exterior electron dimensions matching the two supposedly inconsistent QM atomic orbital and collision conditions in Mac Gregor's 1992 The Enigmatic Electron. In atomic orbit and other cases of sub-photic electron velocity the structure has about the Compton radius then required. Forces of acceleration to near photon velocity compress the structure and gyres by orders of magnitude toward the ultimate gyre cores, meeting PDG collision radius data. The electron's dimensions were found to be elastic with dual convergences at computing cut-off to within 2% deviation in dimensions and internal force balance.

Dual dimension drivers model each other's effects interchangeably with different rotating velocities at nested drive core contours so that the external effects of different outer dimensions are interchangeable. Likewise static QM matches were summed (using unsymmetric force estimates as noted earlier) to 12% deviation in force at cut-off between the gyres of 2 separated static electrons, each with 3 pairs of gyres at velocity.

The paradigm's turbulent spiral wave subgyres across gyre cone bases generate charge forces directed only hemisphericly outward from each coaxial pair's cone bases, smoothly creating the triply overlapping sphere of constant force. Electrons have none of last century's QM difficulty of internal, quasi-infinite charge force of self-repellance that defeated prior electron designs by requiring an impossible point charge, an infinitely strong material sphere rotating at precisely the equatorial speed of light, etc.

This perfectly balanced structure inherently causes the electron's QM/PDG stability and extremely high lower limit for the empirical mean lifetime. Yet this electron particle is only an assembly of driven circulations and waves in six interacting vortices.

This orbital structure derived from the quantal charge-mass power law and lab scaling data equations generates the required QM spinning summations at equatorial light velocity analyzed in last century's summary electron by Mac Gregor (above.)

This constrained electron structure of viscous flow velocities internally necessitates existence at the gyre center of superphotic drive force sub-components that are also exhibited at distance by presently observed quantities of empirical entanglements for which the paradigm thus supplies a radiation mechanism and some relevant equation additions to resolve this 100 year QM mystery.

The synchronized orthogonal orbits of this electron necessarily exhibit around the summation axis a QM "Zitterbewegung" pseudo-vibration when all gyres simultaneously transit the equator at 60° apart and orbit to 35.3° from its poles twice in each orbit cycle.

The muon and tauon members of the electron's Lepton family are excited, heavy, multiple pair spheres with QM-like `very short mean lives due to low cohesive force-to-mass ratios from linearly added forces and exponentially increased masses, both with numbers of gyres in gyre pairs, under the charge-mass power law of particle mass.

There Are also Matched Features Especially Related to <u>NEUTRINOS</u>.

For baryon decays, the PDG data tables and other QM/SM sources commonly show empirical equations that add up (like acceptable equations) only in net energy and net charge shortcuts, not in conserved charges, nor usually in quarks themselves. To balance such shortcut equations for the charges of the established SM quarks, etc., it is necessary to add an input impact neutrino of some kind (or neutral baryon or meson) to all but one PDG (2006) case (out of about 20) of an accredited 30% to 100% baryon decay channel with adequate definition of the observed channel reaction output. A similar number of these cases also require additional unseen output, in other neutrinos, for real charge conservation. Such shortcomings include a now classic SM beta decay of a neutron to a proton plus e⁻ ν_e. (& Similar 2012 PDG Summary Data.)

Using the paradigm's strict quantal charge accounting under the charge-mass power law, half of these few most visible PDG QM/SM decay equations can only be balanced by a neutrino input with much greater mass than the prior PDG upper limit for the tau neutrino. These well documented baryon decays prove by their weight of data evidence that such neutrinos necessarily exist. (Proof valid as that for many baryons.)

Further, this result of the empiricly based mass-charge power law, in conjunction with the long established baryon mean lifetimes, demonstrates that the decays are not spontaneous, but are caused by impacts of neutral particles (largely neutrinos), and that the abundance of the specifically noted neutrinos alone must necessarily be far in excess of prior estimates, with neutrino mass totals that necessarily must be a major rather than very small contributor to dark matter. This implies that all decays of particles which are found to exist, even very briefly, including "oscillating" neutrinos, should be examined by the paradigm's conserved accounting method for even greater presence of neutrinos.

Yet further, the frequently "observed" W and Z particles and the recent "Higgs" demonstrate that neutrinos, and/or a precursor/recursor (a true Higgs), can exist up to and including a "potential" mass, if ever completely formed to one structure, somewhat similar to the PDG stated mass of the top quark, but only for extremely brief periods.

The paradigm structures and power law also demonstrate a quantized pair mechanism for neutrino masses that leads directly to stable astrophysical micro-neutrinos. The pairs are evidently conserved, except in pulsars, nova stars, and black holes that appear to emit in polar jets separated gyres moving with continuous acceleration from their constant point-thrust to observed velocities, of which one series of papers identifies some speeds as superluminal. These jets are observed to excite impacted particles to powerful emissions of radio, gamma ray, and optical energy before dying out at great cosmic distances (many parsecs.) The die-outs are evidently by impact-decelerated re-combinations into normal pairs of gyres in ordinary new neutrinos capable of further re-combination. By this paradigm mechanism black holes could slowly evaporate and bring long-term regeneration to stabilized portions of a cosmos.

In impact on an electron in an excited orbit around a nucleus, such a micro-neutrino, formed of two coaxial base to base conic gyres with almost total suppression of the generation of massive eddies, can absorb the excitation energy of the electron by being slid apart far enough to come into a yet more stable side-by-side photon configuration that is flipped by the decelerating electron into any of several rotary polarizing motions with respect to the continuing motion of the particle. The exposure of the base charge spirals of the particle raises its mass generation by several orders of micro-magnitude, and the rotary motion of the charges in any of several constant or varying planes sets up an accompanying field train along the gross motion vector somewhat like that of a long-line antenna with its net linear motion of charges. This rotary motion and its field is a by-product that very slowly decays over cosmic time periods, simulating, but only simulating, a Doppler red shift as the particle travels on at the speed of light until its rotary energy is absorbed resonantly by another electron that is initially in the lower energy state of the first emitting electron, or a red-shifted resonant state, or a free electron state. At that point the photon is flipped back into the coaxial state and continues as an ordinary micro-neutrino with reduced mass and obscured electric spiral wave systems on its two conic bases. The paradigm thereby closely fits the observed data and constrains acceleration of undue expansion of the QM universe.

In these senses of the paradigm there is little difficulty in observing vast numbers of neutrinos. The bulk of the PDG Data Sheets on baryons, etc., are heavily loaded with observations in the name of decay channels. Exploitation of that established bank/reservoir of data, through this paradigm or otherwise, will completely change the status of the Neutrinos as a class and their significance in Nature. The full significance is most readily available by this new Micro-Quantal Particle (MQP) paradigm of all the particles, including a Planckian ethereal material within which their gyres must necessarily churn.

For neutrinos with simple structures, antineutrino mirror images are only oppositely viewed along the sum axis; neutrinos can be their own antineutrinos, without QM incompatibility. The observed QM/SM antiparticles are only re-assortments of conserved coaxial pairs in collisions and nominal PDG decays. Link orbits in quarks have one 6° clearance; refined lab data may show asymmetric matter forces there.

Comparing electron mass, 0.511.. MeV, to the PDG upper mass limit <0.19MeV for the muon neutrino, yielded its 0.17.. MeV with 3 neutral pairs in the paradigm, the original critical ratio in the mass/charge power law of 1/3 for the generation of interaction mass by neutral quantal pairs vs mass generated by charged pairs (-- the real "Higgs.")

In summary, this book shows many fruitful correspondences and correlations between well established Quantum Mechanics and the Standard Model, with their PDG data tables, and a new paradigm of construction of the subatomic charged leptons, quarks, and neutrinos, from micro-quanta of charged mass in usually coaxial pairs of spinning conic micro-vortices. These structures scale by velocity/dimension/force equations from lab measurements of turbulently driven conic gyres, combined with a general mass-charge power law of particle interaction mass-energy derived from the PDG data tables. These numericly defined quanta orbit in spheric structures that can be synchronously linked in quarks under an exponential extension of the power law to form the nuclear particles, other hadrons, organized basic nuclei, and atoms with causal stability. The dozen summarized papers of the paradigm and condensed chapters in this book resolve many serious SM uncertainties, supply consistent causal structures for many QM mysteries and arbitrary factors, systematicly reclassify the subnuclear true particles, greatly simplify particle math, and yield testable numerical consequences. This provides a physical foundation for potential development of a true unified field theory of all the forces and actions of the sub-atomic and nuclear particles, and other advances of new physics and cosmology, over coming decades of the 21st Century AD.
Fred E. Howard, Jr.　　　　Ft. Walton Beach, FL USA　　　　21 April 2014

A Hidden Simplicity
of NUCLEAR PARTICLES

SECTION I

The Charge-Mass Power Tangle

(A Simple Twist that Multiplies Nature's Cross-Links)

Chapter 1.

A Sub-Structure Power Law of the Masses & Charges of "Elementary" Particles

SUMMARY: A simple numerical basis for making real and humanly understandable the micro-quantum bits of uniform mass and plus or minus 1/6 electric charge in the sub-structures of the smallest known (and supposedly elementary) particles is observed in the international Particle Data Group's tables of experimental (empirical) Standard Model (SM) particle data. Necessarily, new dependences of particle mass on both charge and number of components are required. A very simple math equation for these physical relations computes whole sub-nuclear particle masses at the known conserved electric charges consistent with the Quantum Mechanics SM for the hazily known quarks, for their anti-quarks, and for the leptons such as the even less known neutrinos and the everyday electron in electric lights, TVs, and pocket cell phones. This computing capability in the basic equation comes from its fixed fifth power of the number of the necessary micro-components in any such particle, times the simple sum of the micro-masses of all those components taken one-by-one, multiplied by another simple sum of fractions 1st of the connected pairs of the components that are charged, and 2nd of the pairs that are neutral with a modifying fraction for the type of particle built up this way in Nature. A variation of the equation, for a variable power exponent of the known number of quarks in a larger nuclear particle times the sum of the quarks' masses, was found first & applies to the proton and neutron that make up the large nuclei of the material atoms of Nature (and to the other observed, but unstable, baryon & meson hadrons.) This empirical (practical) observation of a new type of power law basis for systematic regularity in particle structures parallels to a degree many recent theory attempts that failed to give a general capability. This successful new finding provides a basis for determining later the more complete useful characteristics of these particles at the basic level in our Universe, and in that way for filling in many missing parts and mysteries of the Quantum Mechanics theory that has produced the nuclear weapons, nuclear reactors for electric power, and many industrial and medical advances of the last century, like TVs, cell phones, and MRIs of damaged brains.

In the field of sub-atomic particles, the international Particle Data Group (PDG, www.pdg.lbl.gov) listing of charges of 1/3 the charge of an electron in the Standard Model (SM) quarks is an essential practical part of Quantum Mechanics (QM) theory about how the material world and universe of Nature is put together at the lowest, sub-microscopic level (Eidelman, 2004.) This observed and theoretical certainty has led many distinguished physicists, including some Nobel Prize winners, to theorize about yet smaller, fractionally charged, preon/parton sub-particles of which all atomic particles would be made, especially the sub-nuclear quarks and anti-quarks, or at times even the leptons, like the electron. In this repeated effort physicists built up complex theories with elaborate special classifications of types of particles and their components to meet all conditions (for instance, e.g.: Haisch et al., 1994; Treiman, 1999; Salam. 2000; Bandos et al., 2001; Dugne et al., 2002; Luty & Mohapatra, 1997; Kim, 1998; Gsponer & Hurni, 1996; Bergshoeff et al., 1988; Pati et al., 1981; Pati & Salam, 1983.) None of such QM-SM researches has yet succeeded in winning general acceptance by other physicists as determining (defining) the <u>new natural laws</u> of the various parts of atoms.

This chapter is an easier reading condensation and up-date in emphasis of the essential parts of the introductory foundational paper of a new paradigm (structural plan) of the sub-atomic particles. That paper was originally published in the peer reviewed journal of the Florida Academy of Sciences, which holds the copyright to that original edition, including most of the original figures that are used here with new titles and inserts. The original paper, complicated under peer review, appears in *Florida Scientist*, (2005) Vol. 68, Issue #3, pages 175-205, with 2006 follow-up, and in **www.electron-particlephysics.org** and at www.particlephysics.info, the author's websites. The chapter's scientific references are listed by the alphabetical names of authors in a final book appendix (& in a Vol. 2 of Tables.) The plan of the book is to have a chapter-per-prior-paper or two or per-lecture. It follows the author's progress in developing through a series of papers & briefings a new overview of the finest level of particle structure that makes up Nature as we know it, a useful auxiliary to Quantum Mechanics' Standard Model that a junior-college-level public also can appreciate.

Also, there have been many attempts, including some within the PDG (such as Manohar and Sachrajda, 2004), to determine masses or, at least, mass ratios for the leptons and quarks (LQ). (e.g.: Treiman, 1999; Salam, 2000; Bahcall et al., 1998; Rodejohann, 2002; Fukuda et al., 2000.) Among the LQ particles in the PDG reports, only the masses of the electron and its family member, the muon, have been very well measured. These are known to at least 8 significant figures, or to about 1 part in 100 million. All the other LQ particles are very poorly known, typically to a few percent of uncertainty in mass, or even less accurately, often far less. In other particles, only the masses of the large and important proton and neutron, among about a hundred others of their general type, are known to accuracy like that of the electron and muon. Clearly, comprehensive efforts to make accurate mass determinations for all particles have been unsuccessful too—as at pdg.lbl.gov. (The impact of the abstract theories of "strings" or "branes," and their equivalents, on such widely realistic physics objectives must also be generally considered inconclusive until wide-spread acceptance and combination with the QM and the SM are achieved throughout the world physics community.)

Those types of difficulties in gaining more and more thorough knowledge of Nature appear to lie in finding its most critically essential parts and their real actions. These essentials must be concentrated at size levels well below the long known atom and also further below the next lower level in sub-atomic particle structure of the well-known proton and neutron, the main baryons, as well as the less constructive mesons.

In overview, this introductory basic chapter of this book shows how that zone of difficulties in understanding particle masses and charges is observed in a fundamentally new and very simple way that is the underlying key to all the structures of Nature, and to this entire book. Even in the first chapter this key leads to a number of other solutions of particle problem areas. That major change comes about through a fresh quantitative view of the conventionally "elementary" (non-dividable) LQ particles at the second sub-atomic level of particle structure and their necessary internal parts and actions at a third lowest level below the atoms. In this approach, real solutions to such deeper problems unfold naturally and quite simply from beginning well above them where a new basic generating of mass was observed unexpectedly between fairly well known particles at the upper sub-atomic levels. (Howard, 2005, from which this chapter is condensed.)

This exactly calculated, quantitative observation involves the electrically charged protons and electrically neutral neutrons and the other SM hadron particles. These are made up of the SM quarks, all of which are known to be electrically charged, either positive or negative. (Charges are always conserved and add up linearly, like pennies.) The prior empirical data on these particles are all recorded in the hundreds of pages of the biennially up-dated PDG Summary Data Tables (pdg.lbl.gov) accumulated over the 20th century and the beginning of the present century. There the hadrons include the three-quark baryons such as the proton and neutron, which primarily make up the plus-charged atomic nuclei, the main massive (and usually stable) bodies of the atoms. The hadrons also include the dual-quark mesons that are never stable and do not usually appear in an atom's nucleus (but do appear in various PDG/QM/SM particle "decays.")

With the SM meson and baryon particles this chapter first notices in their PDG-measured masses a strange numerical variation with an odd systematic decrease of exponentially increased masses built up from the increasing sums of the PDG-listed masses of their quark components (Eidelman et al., 2004). At the smallest sums for the smallest members of each family of these hadrons the exponent of their number of parts approaches a high upper limit. At the larger sums the exponent becomes small. This mass relation is both plotted in curves on (logarithmic) graph paper and stated in a very simple equation. (While PDG values for these quark masses have varied since this first paper was written, and will again vary biennially, their very broad uncertainty at that time was found to indicate more accurately than did any single value of small uncertainty the widely separated set of dual values of mass for each quark found to be required in this chapter and further proved in later chapters. The lower one of those dual values is the one that applies in building the named leading member of each family of hadrons that is involved initially in this chapter. Adjusting the displayed logarithmic curves for the most accurate later PDG values will shift the curves to the left slightly, but will not change the conclusions from studying them. The original curves are used to emphasize this point.)

Following tentatively that newly derived number and mass relation of the well-known first-level sub-atomic hadron particles with their second-level quark and anti-quark component particles, this chapter next explores mass and number relations between the PDG SM second-level particles and a new third level sub-structure with the necessary general number values. At this level these relations are forced to include a specific dependence of particle mass on charge. This firm empirical dependence is new in particle physics. It joins with the earlier new exponential relation of the number of components with particle mass to develop jointly into the controlling basis of all particle masses and, through that, of many other particle features (properties or characteristics.)

The equation for the relation between hadrons and their quark components thus extends (at the upper limit or asymptote of its highest power at the lowest component sums) to the relations between SM leptons and quarks/anti-quarks (LQ) themselves and their new components defined in the chapter. Trials in those LQ types of conventionally elementary QM particles yield estimates of component sub-structures with new micro-quanta of mass and conserved 1/6 fractional charge within particle masses matching the SM values. In combination with the new term for dependence of particle mass on charges, the new equation becomes broadly useful. It expresses the typical exponential mass increase in the LQ classes of particles at a fixed power of the number of the components. [This empirical power law is only distantly similar to some fairly recent theory (e.g.: Haisch et al., 1994; Kim, 1998; Bandos et al., 200I; Luty and Mohapatra, 1997.)] As a power law it provides a new basis for systematic regularity of the leptons and quarks/anti-quarks in particle masses and quantal sub-structures with conserved charge wherein the neutrino leptons extend into the Extremely (E) small micro-mass ranges found necessary in particles from stars. Substitution of defined terms leads to a slightly simpler equation format that is particularly suitable for the Usual (U) range of the bulk of the LQ particles that have primary emphasis in the first two chapters. These equations provide a new basis for reorganizing and retesting particles such as uncertain quark masses (Manohar and Sachrajda, 2004), neutrino mass oscillations (Bahcall et al., 1998; Rodejohann, 2002; Fukuda et al., 2000), etc., in theories and experiments of particle physics. There are further predictive implications, including (in all chapters) a simpler regularity in proliferations of the hadrons and in neutrino "mass oscillations."

These determinations effectively resolve the lower level particle mass and charge difficulties first noted above, that were inconclusively dealt with in efforts prior to the peer-reviewed and published source document for this chapter (Howard, 2005, which was originally derived during a review of the PDG listings.) Examples from that source and its appendices in all three levels of particles clarify mass and charge interactions generally, and point out a resulting cause-and-effect understanding of other important particle properties. (Later chapters and appendices show the resulting detailed determinations for some more complete characteristics and capabilities of the real particles at all sub-atomic levels. Particle spins, or particle angular mass momentums, become necessary in those chapters. Some features are not yet studied in this way.)

(Since fundamentally new factors are included here, this overview is extensive so that the first-time reader in sub-atomic subjects, in particular, or those familiar with other approaches, will have an organized sense of where the discussion is headed as it uncovers the essential mathematic equations that give quantitative or numerical validity and completeness to any hard knowledge or science.

A Sub-Structural Equation for <u>MASSES</u> of Composite Nuclear Baryons and Mesons

Any quick survey of all the sub-atomic particles in the PDG tables of data (www.pdg.lbl.gov) shows that each family of particles may have a very wide range of apparent sizes in terms of masses (which would have weight in the earth's field of gravitational attraction near its surface.) Mass is often measured most conveniently in millions of electron-Volts (MeV) of Einsteinian mass energy (a very small amount to us) or in billions (giga-eV, GeV.) Typically, the electron family of Leptons ranges from the electron's just over a half MeV of mass energy to the tauon's well over 1700 MeV. The various other family ranges overlap with this wide range. Distinguishing particle types by mass across whole family ranges has been confusing. It needs a simpler view. (Many discussions never get beyond showing only the smallest member in each family.)

Looking at the family types by the well-recognized PDG numbers of members in each family appears simpler initially. All of the 3 LQ families (and the QM boson family) have far fewer than 10 members each in most 21st century editions of the PDG tables. The baryon kind of particles has well over 100 members that are broken into 7 families in various ways, and the meson class numbers about 200 members that are broken into 7 or 8 families with a large number of what appear to be sub-families. However, all of these 300 plus baryons and mesons are very closely related as hadrons by being composed of quark components (including reverse anti-quarks in mesons.) The hadrons are distinctive in this scale and are evidently important in numbers of types in families.

The bulk of the PDG listings pages also are devoted to the well-related baryons and mesons. Likewise, the great bulk of the mass of visible, readily observable, and testable Nature is in the baryons in the nuclei of the atoms. Any attempt to understand

Standard Model/PDG Particle Data (2000±)

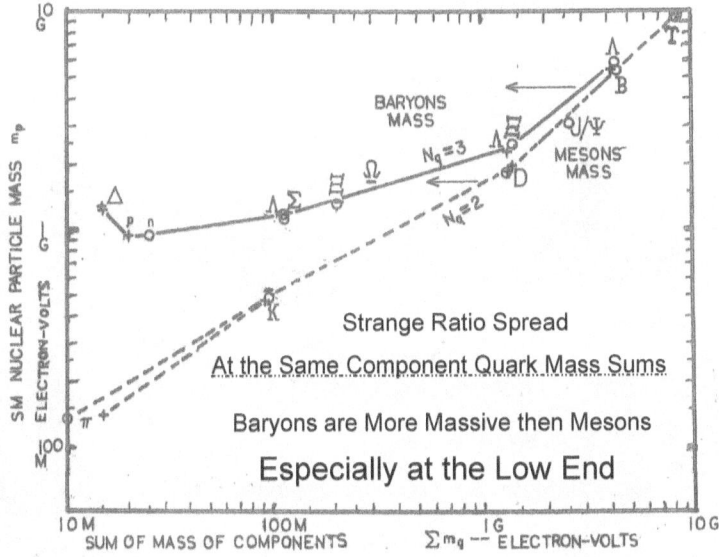

Fig. 1-1 The masses in electron-Volts of the lightest members of each Nuclear Hadron family m_p plotted vs the sums of masses of their Quark/Anti-quark components Σm_q. 2 quarks in Mesons, 3 in Baryons. + & - are charged hadrons. Circles are neutral. Letters are for PDG particle names in Greek or English.

the particles, or any one particle (or any few), cannot expect to do so without some comprehension of these hadrons. And, as the overview indicated, the hadron masses of the smallest member of each family have a distinctive family relation to the masses of their component quarks. (This eventually gives a newly found meaning to mass.)

To follow this line of thought, a graph plot of hadron masses gives a picture that can be understood readily. The masses of the smallest hadron of each family type and its quark components are near the top of each PDG list of SM particle facts, and have a very direct meaning. The smallest members of each family are the most frequently observed and usually have the longest data record. Accordingly, Figure 1-1 displays SM data from the PDG (Eidelman et al., 2004) on hadron sub-atomic particles which are well known from typical overviews in the literature (e.g.: Close et al., 1987; Treiman, 1999) to be composites of quarks/anti-quarks. The two curves show whole particle masses in electron-Volts for the baryons (3 quark components) and mesons (in QM 2 components, quark and anti-quark.) These particle masses are plotted against the sums of the most widely used average values in recent decades of the very uncertainly known masses of the well-recognized component quarks/anti-quarks. Something is different

Systematic Variation of Exponent y

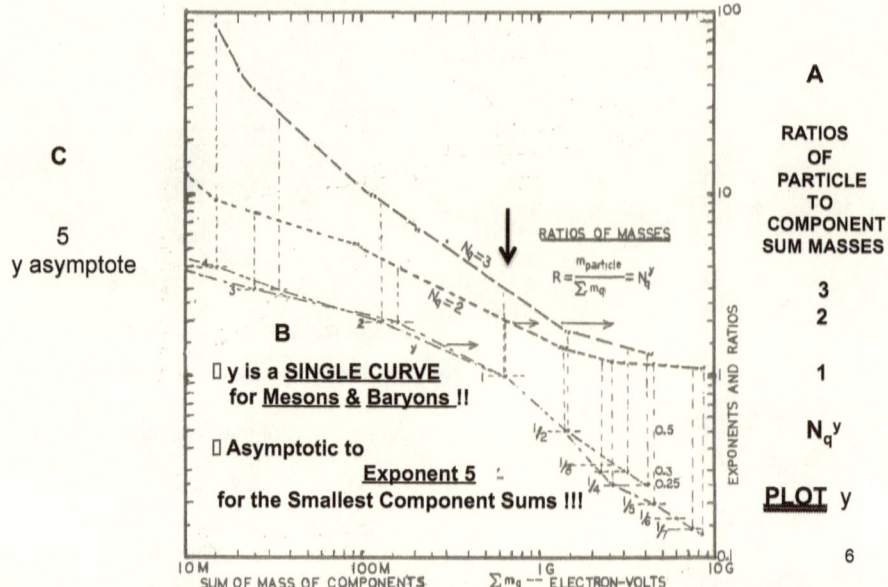

Fig. 1-2 The Ratios R of nuclear particle mass m_p to the sums of quark component masses Σm_q in the thought process that the ratios raise, going from Section A to Section B (the plot), and finally to C (See text). It was noted first that the N_q =3 value in baryons is at 3 under the big arrow where the meson N_q =2 value is at 2. The Ratio for each particle class at a component sum of about 700 MeV is the same as $N_q{}^y$ where y=1. Plot y at all component sums; which is even more surprising. Y makes a fairly smooth single curve!! (Except for an evidently excited & anomalous Δ this also applies to a curve of y vs m_p, not shown.) The curve approximates 0.1 at about 10 GeV sum and approaches, but does not reach, a limit or asymptote of 5 at about 10 MeV sum, approximately the lowest component sum for baryons and mesons. between baryon and meson masses toward the lower end of total component mass, something much larger than the 3 quark and 2 quark difference. It is the ratios of particle mass to summed component masses on the plot axes that appear significant.

In Figure 1-2 (Fig. 2 of Chap. 1) Section A highlights the values on the two curves of the ratios under the arrow in Section B. These two curves show the ratios R of those SM whole particle masses m_p to those sums Σm_q of component quark/anti-quark masses for smallest family-name particles of the two types of hadrons. (Here *q* refers to the quark/anti-quark class of components without distinction between them at this point since those of the same type would normally have the same mass, though opposite in charge, where conserved component charges total up to particle charge.)

Graphing the PDG nuclear baryon and meson mass data in this way reveals that the varying numerical relations between the composite particle masses and the SM quark/anti-quark sub-structure masses in both these types of hadrons fit a single derived simple function curve, and its algebraic equivalent:

$$R = \frac{m_p}{\sum m_c} = Nc^y$$

$$m_p = Nc^y \sum m_c$$

(Equations 1-1a & 1-1b) [Chapt.1]

This relation equates the ratios to the **N** number of component quarks/anti-quarks (fixed at 2 or 3 for each curve) raised to a positive **y** power, which then varies with the sums in an exponential law. That law identifies the SM mass increase ratios for hadron particles over the summed masses of the quark/anti-quark components within the particles. These ratios fall toward +1 in Figure 1-2 for both types of the SM heavier hadrons. In the two-part curve **y** is plotted for whole number and fractional values derived from the two ratio curves, with small discrepancies from a single regular curve that approximates 0.1 at high masses. With the lighter component sums, in Figures 1-2 and 1-3 the value of the exponent **y** crosses the value 4 and asymptoticly approaches 5.

This simple exponential law equation means that summed component masses in hadrons Interactively generate more particle mass by the number **N** of interacting component quarks with the N exponent approaching an asymptote of 5 for the lowest sum masses. Thus the greater part of the PDG empirical mass energy of the lighter particles actually arises by a sub-particle process of interactions between components!

But on the right side of the graph, where quarks are 10^3 times more massive, **y** is the opposite by $10^{-1.7}$ times, far lower in amplifying the quark sum. In this action quarks can only be globular clumps of smaller reactors that progressively mask, interfere with, or dilute each other's mass-building interactions from separated quark-to-quark between their larger clusters (but not within the 2,000:1 growing quark masses themselves!) In other words the mass-generating interaction exponent **y** for 2 or 3 quarks progressively declines to 0.1 with growing quark sum mass in hadrons; which is a contradiction in interactions over this range. Globular quarks as components cannot be the real hadron reactors. There are necessarily interactive micro-components within quark globules. This firm numerical relationship, between quark components of a particle and the particle mass, locks the components into being a sub-structure for the particles all along the curve. This applies to the quarks of all the types used in structuring the particles graphed in the original basic plots of Figure 1-1. It does not yet structure individual quarks themselves, but does raise the question of similar reactive relations inside them.

Search for Micro-Components in Extending the MASS Equation to LQ Particles.

Same plots extended to LQ Range.

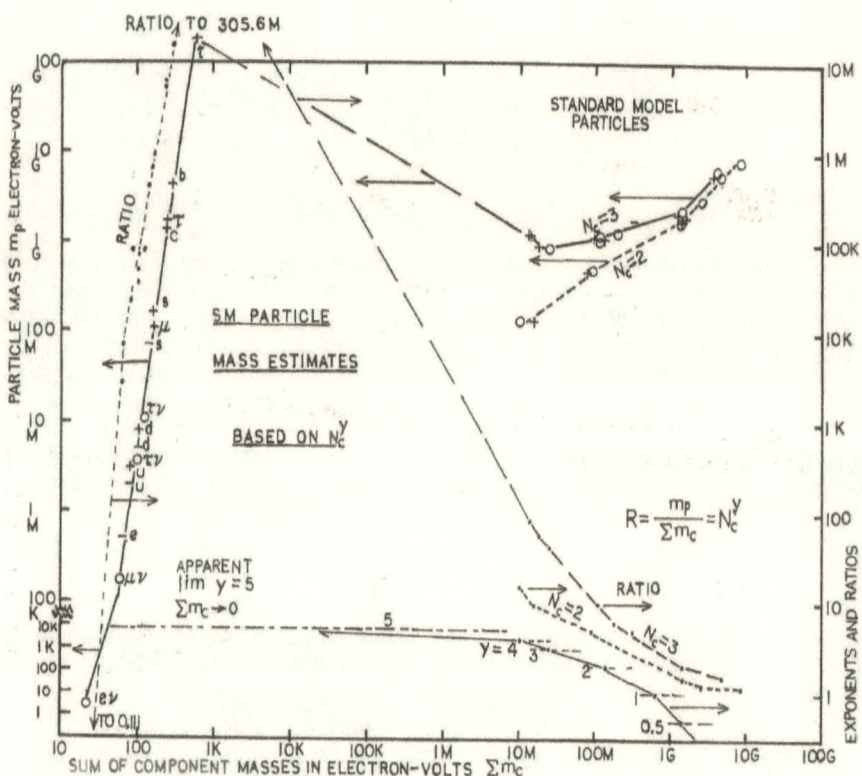

Fig. 1-3 Extension to LQ Particles of the Exponent y at the Asymptotic Limit 5. Exponents and Ratios scale on the right as before, and Particle Masses scale again on the left. The LQ particles are far to the left due to the very small size of the Sums of their Micro-Quanta on the bottom scale. The mass and mass ratio of the top quark are off the scale. (This is inconsistent with the uncertain PDG status of a top hadron that is necessary for the quark to exist.) Dual values are shown as antiquarks for the 3 smallest quarks; for the 3 large quarks dual values are too close to plot separately. NOTE the break in particle scale at the lower left to contain the Electron Neutrino, e nu, on the one page at 5 orders-of-magnitude (OM) smaller mass than the electron. The almost straight line of the LQ masses takes a small apparent angle at the scale change. -- At lower right the end of the y curve is cut half an OM for convenience.

By observation on a graph the apparent limiting value of **y = 5** in the generalized composite sub-structure mass equation (1-1b) is readily extended toward very low mass sums of components (as shown in Figure 1-3) to apply to more nearly elementary PDG LQ particles which (unlike the hadrons) are not composites of quarks/anti-quarks, but of micro-sized components. The equation output becomes just the quarks/anti-quarks and the leptons (LQ), including the charged electron family and the uncertain neutrinos.

For more than a decade the PDG accredited lists showed the neutrinos in three primary types (electron neutrinos or nu_e, muon or Greek letter mu neutrinos, and tauon or tau neutrinos) with upper limits in mass, in which the two higher limits overlapped the Usual LQ particle mass ranges noted earlier. (Many new QM particle papers still use that classification. So does this chapter's source cited earlier. The names for these neutrino mass ranges are still very useful.) Recently, the PDG tended to drop the types, partly because the QM astrophysicists studying supernova exploding stars insist that micro-mass neutrinos (Extreme mass ranges) are required in their field, and also because other QM physicists are looking for much more massive neutrino-like particles (which are covered in later chapters as neutrinos beyond nu tau, or nu super-tau.)

In reviewing most previously well-established PDG particle family types (except bosons), the lightest, the electron neutrino, has long had a PDG upper limit that is five orders-of-magnitude (OM, factors of 10) below the masses of the electron and the mu neutrino at the bottom of the U LQ range. The electron neutrino is, therefore, definitely in the E range, and does not need to be generated by the Mass Exponential Equation initially in looking for a micro-component for the Usual particles. It is only needed to see whether a single law can also extend to the Extreme particles (which it does in the final generally applicable law equation.) The QM and PDG definition of the nu_e is that it appears in particle reactions associated with the electron and was long estimated to have a mass <3 eV. A sum of a few masses of this size range, if multiplied times their number (around 5 to 7) taken to the 5th power, could generate a mass value in the same general order as the electron mass of 0.511 (rounded) MeV. This sizes an LQ problem.

There are other estimate factors involving neutrinos. The muon neutrino (nu_μ, or nu_{mu}) with a generally accepted PDG upper mass limit of <0.19 MeV has always been associated with reactions involving the muon particle. That mass is very indefinite, too much so to be the principal influence on mass of a micro-component. However, the facts that it is neutral and is limited just above 0.17 MeV, 1/3 the mass of the electron, make it a possibility, or first quick estimate, that U neutral masses might have that ratio of 1/3 to charged masses, whatever the mass of one micro-component might be.

More broadly, since neutral, positive, and negative charges, as well as 1/3 steps of charge, are required for various SM particles, the simplest micro-components would be positive and negative 1/6 charge quanta, of a single uniform mass, appearing normally in conserved (or permanently organized) pairs. In such a structural plan any particle could be neutral or of any multiple of 1/3 charge with only two basic kinds of components in the required number of types of these pairs of components. This conforms to the SM conservation of charge in all classes of particles.

In such a plan the simplest electron with its minus 1 charge would have 3 pairs made up of **$N = 6$** negative 1/6 quantal components of conserved charge with mass at a rounded 10.9525 electron-Volts of mass each. This yields a 6 component **$\Sigma m_c = 65.715$** electron-Volts mass. That multiplied by the 5th power of 6 yields a rounded 0.511 MeV particle mass for the electron, with a quantum at each end of a balanced 3 dimensional set of orthogonal axes with plus and minus locations. The positron (electron anti-particle) would be fully symmetrical with this electron. If coupled with the observation about the muon neutrino and neutral pairs at 1/3 the mass of charged pairs, this also derives a mass of 0.17 MeV for that neutrino consistent with the prior PDG mass limit. Most simply, this observation depends on the electron's having no neutral pairs of quanta, and the mu neutrino's having no charged pairs. In repeated calculations with all the various value options in the electron and various exponent powers, etc., no other trial value of such a calibrated universal component mass and charge was found as suitable for deriving under a power law SM/PDG-consistent masses of LQ particles in the Usual range over 6 OM from the muon neutrino to the top quark (as in Fig. 1-3.)

Condensing that search: LQ particle masses m_p that fit the empirical data in the Usual range arise in a strict general particle paradigm of structure that fits Nature as observed and recorded by the international PDG. The LQ particles must necessarily be composite and are constructed of micro-quanta of uniform mass that are charged either a conserved -1/6 of the electron charge or +1/6. These quanta interact exponentially to generate more m_p mass by the constant 5th power y of the number N of components interacting within an LQ particle times their sum of masses. Micro-quanta exist in the bulk of Nature only in conserved pairs that are charged +1/3, -1/3, or neutral 0, in which neutral pairs generate only 1/3 as much mass as charged pairs in the Usual LQ range. (However, there is an Extremely small range of m_p mass in which the interaction mass energy of neutral pairs varies inversely as positive integral powers of **3**.) The baseline electron structure has 3 negative interacting pairs. The baseline mass micro-quantum has 10.9525…(rounded) eV of mass energy calibrated on the PDG electron. Charge is conserved, adds in linear algebra, and is the true mass interactor. In math, therefore:

Exponential Mass Law Extends on Asymptote
to "Elementary" LQ Particles as
Mass–Charge Interaction Power Law

$$m_p = N^5 \sum m_c \left[\frac{n_\pm}{n} + \frac{n_0}{an} \right]$$

- N = Number of microquanta components c, $N = 2n$
- m = mass of particle p or c in eV mass energy
- n = number of microquantal pairs (in regular LQ particles)
- $n_\pm$ = number of charged pairs of microquanta (")
- n_0 = number of neutral pairs (")
- m_c = 10.9525… eV, $\Sigma m_c = m_c N$ in all LQ particles
- $a = 3^x$, integral *exponent* $x \geq 1 = 1$ in the Usual range
- **Forceful Charge is a Microquantal Mass Interactor** [14]

(Equation 1-2)

This is the baseline general form of the Mass-Charge or Charge-Mass Power Law equation. Table 1-1a lists the computed LQ particle masses in E and U ranges

Table 1-1a: Equation 1-2 Computed LQ Particle Masses versus 2004 PDG SM Values

Particle & Charge	Parts	Rng-Pairs	Calc. Mass eV	SM Mass	Deviation	%Dev
Electron n'trino	0	2 E 1+ -	2.8846 eV	<3.0 eV	Zero	0%
Mu neutrino	0	6 U 3+ -	0.1703 MeV	<0.19 MeV	"	"
Electron	-1	6 U 3- -	0.511 MeV	0.511 MeV	"	"
Positron	+1	6 U 3++	"	"	"	"
Up quark min.	+2/3	8 U 2++, 2+ -	1.914 MeV	1.5-4.0 MeV	"	"
" " heavy		8 U 3++, 1- -	2.871 MeV	"	"	"
Down qu'rk min	-1/3	10 U 1- -, 4+ -	5.11 MeV	4.0-8.0 MeV	"	"
" " heavy		10 U 1++, 2- -, 2+ -	8.032 MeV	"	0.032 MeV	0.4%
Tau neutrino	0	12 U 1++, 1- -, 4+ -	18.169 MeV	<18.2 MeV	Zero	0%
" " alternate		12 U 6+ -	10.900 MeV	"	"	"
" " "		10 U 1++, 1- -, 3+ -	6.5715 MeV	"	"	"
" " "		8 U 1++, 1- -, 2+ -	1.914 MeV	"	"	"
" " "		6 U 1++, 1- -, 1+-	0.3974 MeV	"	"	"
Strange q'rk min	-1/3	14 U 4- -, 3++	82.467 MeV	80-130 MeV	"	"
" " heavy		16 U 2- -, 1++, 5+ -	107.19 MeV	"	"	"
Muon	-1	16 U 3- -, 5+ -	107.19 MeV	105.7 MeV	1.5 MeV	1.4%
Charm q'rk min	+2/3	22 U 6++, 4- -, 1+ -	1.1665 GeV	1.15-1.35 GeV	Zero	0%
" " heavy		24 U 4++, 2 - -, 6+ -	1.395 GeV	"	0.045 GeV	3.3%
Tauon	-1	24 U 3++, 6- -, 3+ -	1.744 GeV	1.777 GeV	0.033 GeV	1.9%
Bottom q'rk min	-1/3	28 U 4++, 5- -, 5+ -	4.0218 GeV	4.1-4.4 GeV	0.0782 GeV	1.9%
" " heavy		32 U 1- -, 15+ -	4.4106 GeV	"	0.0106 GeV	0.24%
Top quark min	+2/3	54 U 7++, 5- -, 15+ -	170.986 GeV	174.3± 5.1 GeV	Zero	0%
" " heavy		52 U 10++, 8- -, 8+ -	172.1 GeV	"	"	"

- - - - - -

Table 1-1b: Equ'n 1-1b Computed Primary Baryon Masses vs 2004 PDG SM Values

Particle	Charge	Parts Sum eV	Exponent	Calc. Mass eV	SM Mass	Deviat'n	%Dev
Proton	+1	uud 8.939 MeV	4.2359	938.2679 MeV	938.27203	0.0051	0.0005
Neutron	0	udd 12.136 MeV	3.9588	939.5139 MeV	939.56536	0.0515	0.0055

(It can be useful to photocopy this table and keep it in the book as a place-marker. Add Eqs. 1-1 & 1-2.)

for reference and compared to the original 2004 SM listings by the PDG, which change slightly every two years. Due to PDG changes the deviations between the computed and current observed mass values also change biennially, but these deviations remain small at most PDG re-evaluations, as shown next in Figure 1-4 (from a later paper.) -- Table 1-1b covers the two primary baryons used most often versus constant PDG data. --In more detail the last column of Table 1-1a shows that all but six of the listed masses originally computed from the power law equation for SM particles were within the PDG accepted range of the empirical data (Eidelman et al., 2004). Two of the six are shown as heavy alternative masses for family baseline particles that have no deviations. The deviation from the SM data of one of these two alternates is less than 0.5%; the other has the largest deviation listed at 3.3%. Three others of the six deviant cases are within

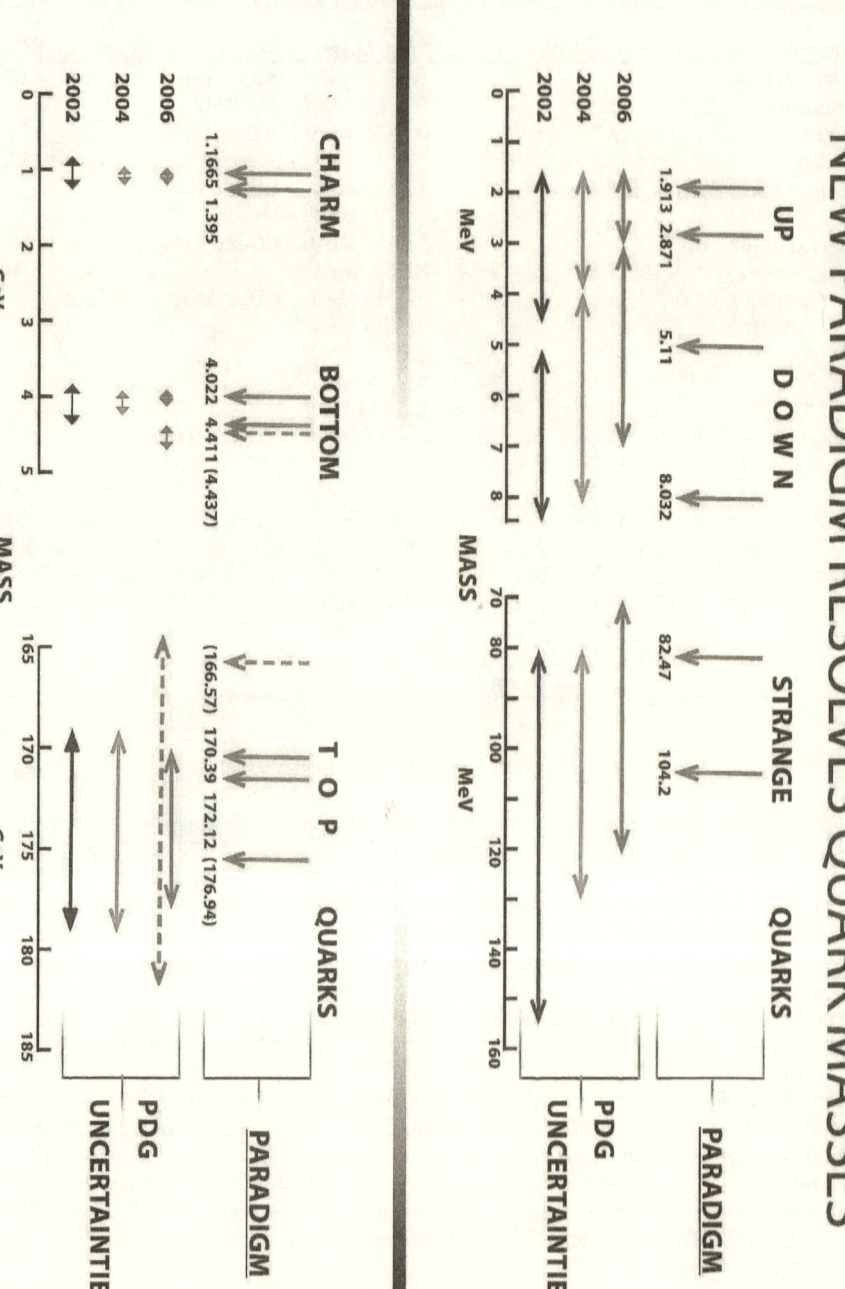

37

Figures 1-4 and 1-5 (From original Paper 3) Recapitulate Effects of the Law Equations

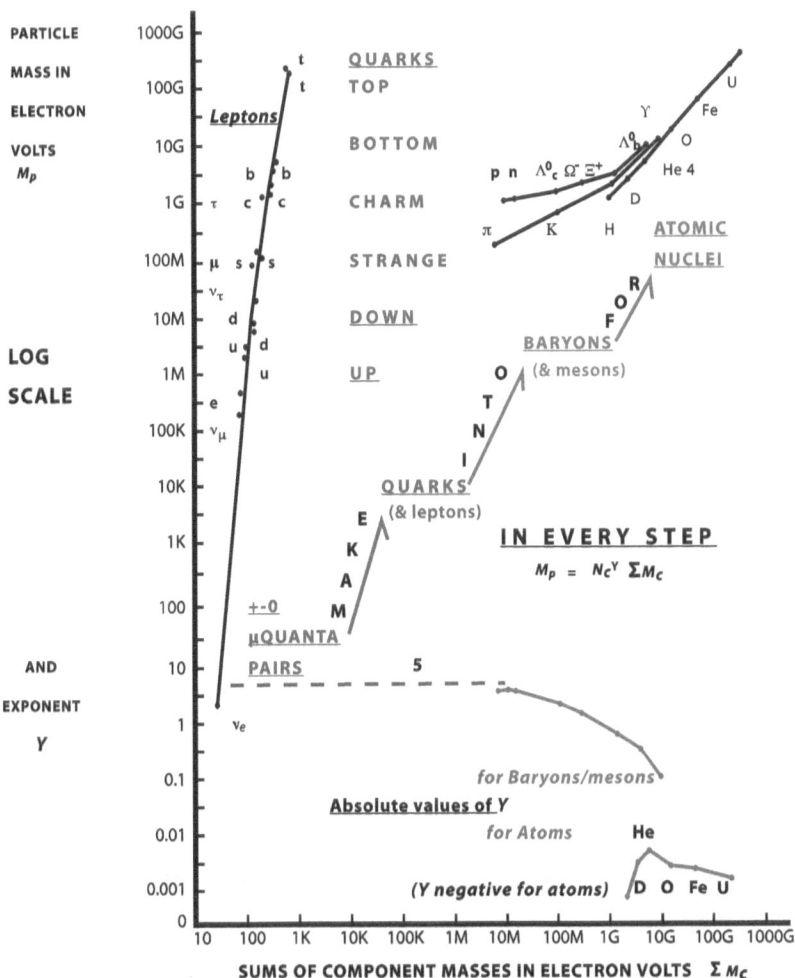

New Empirical Paradigm Scale to Atomic Nature

FIGURE 8 OF PAPER 3

2% of the SM/PDG empirical data. One is within 0.25%. The three with the largest deviations are among the larger particle/anti-particle sets above 1000 MeV (1.0 GeV) in mass, where slightly divergent structures might tend to occur. (Briefly excited/irregular PDG structures cannot be excluded.) Again, no other value of exponent *y* than 5 was found to be suitable for particle mass calculation over the LQ range of SM particles.

There are several adaptabilities in basic Equation 1-2 for particular areas. It is immediately obvious first that, since all quarks are charged, the bracketed fractions for components would reduce to (1 + 0) if applied to quarks in hadrons, and that the only real difference from the Equation 1-1b is that *y* in Eq.1-2 is at its asymptotic limit 5. (A general equation for *y* in both areas has not yet been sought. A curve-fitting equation approximating *y* in Equ.1-1b for the baseline lightest family members of every family in the hadron range is given later.)- - -For the electron neutrino 2.88 eV in E range $a = 3^5$. For the even smaller micro-neutrinos m_p in the milli-eV to micro-eV E range listed from cosmic references in Appendix A of the original published paper (Howard, 2005, see Vol.2) the exponent of 3 must go much higher. This effect of *1/a* in the equation could be shown in Fig. 1-5 (left) by a short constant line at 0.333 below the Usual LQ particle masses down to below nu_μ, with an arc spur for E from there to 0.0041 below nu_e, and a continuation arrow straight down off the scale to 0.00000002 or lower for low micro-eV neutrinos with $a = 3^{16}$ or more (which is all consistent with particle structure in later chapters herein.)- - -For masses of multi-baryon nuclei in all of the atoms beyond hydrogen a phase of Equation 1-1b with negative fractional exponents is required in a Macro-Atomic Nucleus range, as shown added at the right side of color Figure 1-5.

A form of Equation 1-2 that is useful for some revealing calculations in only the LQ Usual range (where $a = 3^1$) is obtained by using the listed substitutions to yield:

$$m_p = N^6 \frac{2}{3} m_c \left(0.5 + \frac{n_\pm}{n} \right) \quad (1\text{-}3)$$

where the bracketed term (or F_0 in Howard, 2006), varies cyclicly between 0.5 and 1.5 as mass increases, first by slowly increasing the total number of pairs, and second by increasing, at each new total, the number of pairs that are charged (one at a time from zero pairs to that total.) (The form of "*a*" here avoids the original papers' special rules.)

BROADLY CLARIFYING CONSEQUENCES of the CHARGE-MASS Power Law

It is clear from the equations that the broadly generalized structures of particles in the PDG hadron and LQ ranges must generate stored mass energies (and therefore mass gravity forces and inertias) that depend primarily, not on accumulated QM "strong" or "weak" forces with large quarks in the hadrons, nor on "weak" or "strong" forces between QM components in the LQ particles, but instead: 1. On the bare numbers of interacting components at each of these sequential stages of structure. 2. On the closeness of globularly gathered interaction groups of pairs of component quanta in the LQ and the partial shielding or separation of these close-knit interaction groups in the hadrons. And, 3. On the necessary distribution of the mass-generating interacting force from conserved "electric charge" between charged or self-neutralized conserved pairs of micro-quanta at the smallest and simplest of the general ranges of hadron, LQ, and pair structures. (These three numbered factors must also apply directly at the Extreme micro-masses of the cosmic and electron neutrinos that can only be made of one adaptive pair of quanta, as well as at the heaviest Macro-ranges of multi-baryon atomic nuclei beyond hydrogen's single proton.)- - -It is this importance of the interaction of charge that outweighs everything else on this point, whatever charge really is (as the equation itself must also determine)- -It is this forceful charge-mass action of the micro-quanta, cross-coupled as that is with all other characteristics the quanta have that do not generate mass, which is the unexpected twist of actions into the generation of gravitational and inertial mass, multiplied with all the quantas' cross-linkages to other varied particle properties. This action gives our surrounding Universe the essential capability of pulling together massive stars and galaxies, planets, comets, asteroids, moons, other satellites, planetary atmospheres and seas, and the moving or static clumps of living matter attached to earth, including ourselves. Charge and its resultant mass product are major initial interactive (**unified**) force elements in natural structures.

Accumulations of all the known particle forces must be acting secondarily to hold various accumulations of PDG mass energy together in particles with various stabilities. Three (non-boson) SM particle types, the baryonic proton, the leptonic electron, and (due to their dimly observed arrival from cosmicly distant supernova star explosions) some almost undetectable assortment of leptonic neutrino pairs (and their antiparticles), plus the baryonic neutron (if, and only if, it is stabilized by being tightly bound to a proton), all of these are each observed to be held together in stable balance (at or below typical heat levels in planets and small stars) for cosmicly very long mean lifetimes. (Here balance may be either a highly kinetic balance of all forces and mass momentums in intermeshed spinning flywheels, as the light electron structure later shows, or an almost static balance of shapes, masses, and forces, as between the mainly internally rotating heavy quarks of a relatively slow-moving proton.) Since masses build up exponentially with number interacting, and forces only accumulate by linear addition of numbers, the long-lived particles tend to be the lightest members of their families or types. They are the ones with the higher force-to-mass ratios. The remainder of the heavier subatomic parts of ordinary (non-bosonic) massive Nature listed by the PDG (at pdg.lbl.gov) are marginally to fully unstable structures and are likely to be broken apart with various very short mean lifetimes, typically millionths of a second or much less. This all has a strong influence on what particle structures can be.

Structural balance clearly becomes more critical and less often possible as mass increases with numbers of parts, depending on the particular parts. The break-over from balanced stability to an unbalanced instability for individual baryons occurs with the neutron (only a small 0.138% heavier than the stable proton.) For individual charged leptons, the break-over mass would be well above the electron, in the long gap to the unstable muon, which is over 2 OM heavier. For individual neutrinos, the lifetime evidence is still scant except for the very small ones, some of which appear to travel to the earth from distant parts of the detectable Universe. That requires a very long lifetime (in some form.) For grouped baryons in nuclei, the rebalancing of neutrons by balanced bonding with protons in deuterons is sufficiently stable that an entirely separate structural system of balancing multi-baryon nuclei, initially based on those stable couples, enables the natural systematic assembly of cosmicly stable nuclei up to element #82, lead, with a very few isolated heavier ones beyond it.

Here generic, necessary, structural charges are generating only observed (and thus necessary) mass in general terms. [The necessary specific forms of complete structure to generate all the well observed forces, masses, and lifetimes of individual particles are resolved after Chapter 2 in the continuing development of this Micro-Quantal Particle (MQP) paradigm (structural plan) of sub-atomic Nature. To this point, only exact numbers of quanta with generalized charges, masses, structures, and groupings of the required components of the particles have been directly involved with the power law equation. There are many other consequences of that far-reaching law.]

Some More Immediate Consequences of the CHARGE-MASS Power Law.

The PDG uncertainties of the neutrino masses in stating only their upper limits (Table 1-1a) are not at all uncertainties of whether neutrinos exist. Many neutrinos are definitely required in QM to account for the disappearances of readily observable mass energy or kinetic energy in particle reactions that can be described in no other way than to assume an output neutrino that is not observed leaving the reaction. Since QM does not have a way to account for the components of all particles, including especially the components of quarks as well as of the electrons, many such uncertainties in QM and its SM are unavoidable. The situation is even more uncertain in QM when unrecorded neutrinos must necessarily have made inputs of mass energy and kinetic energy into a reaction with conserved energy totals. The existence herein of the fundamental Mass-Charge or Charge-Mass Power Law Equation for necessary micro-quanta within LQ particles and its related Exponential Law Equation for quarks within hadrons, both derived directly from the QM PDG bank of accredited empirical particle data, now make it possible, and eventually obligatory, to account strictly for the necessary conserved charged or neutral pairs of micro-quanta in neutrinos both in the inputs and outputs of particle reactions in order to make the reaction equations balance in the most basic sense, like other math. The process for this accounting is condensed here in later chapters from a website paper on detailed neutrino structures. (Many of these required neutrinos are much more massive than the prior PDG upper limits of tau neutrinos, with numbers of them still smaller than the other known neutrally charged particles that might have the same kind of reaction. There is no available alternative to using these newly required super-tau neutrinos found herein for reaction accounting as may be needed.)

This kind of accounting can only be done thoroughly when reaction data are complete and specific on the full set of observable input and output particles, which is not always available. In addition, input neutrinos may be larger than the minimum requirement for a particular reaction, yielding an additional unobserved neutrino output. This will result in findings of minimum limits of neutrino mass inputs rather than the prior upper limits. (A later chapter will cover some outstanding hadron "decays" that are necessarily the result of neutrino impacts, beginning with the classic outstanding case of the severely unbalanced equation for the "decay" of neutrons to protons with outputs of electrons and neutrinos. The currently QM-accepted equation is balanced only in net charge, not in conserved charge and not in the conserved micro-quanta that carry the conserved charge, especially in quarks, as the Power Law will demonstrate. The quanta and their mass accounting take precedence over the accounting for the division of total energy which must be done second in order, with final balance of both.) The full result is that input neutrinos are not generally as difficult to observe as has been assumed. Their evidence can now be shown to be present in the impact causes of many reactions that have been thought to be "decays" of baryons. Full use of this new ability to observe neutrinos may require man-years of re-oriented review of old research records to bring them out where sufficient data were recorded.

A similar but deeper obscurity hides neutrinos that are not taking part in particle reactions, but are passing through volumes of space that may be otherwise fairly empty or may be in solid or liquid bodies such as inside the earth. Much reported "dark matter" establishes that some such form of total fragmentarily observed mass is present in every studied galaxy. (Drees, M., and Gerbier, G., 2006, in the biennial PDG report Yao, W.-M., et al., 2006 and subsequent issues) Calculation of the neutrino masses required to yield the available SM/PDG "decay" observations should at length make an appreciable contribution to those mass totals in the local volume of this galaxy. (A further correction should also arise from a revision of the accelerated expansion of the space between galaxies determined from red shift of light photons as proposed in the last chapter here. This may in addition return Einstein's cosmology constant to his initial preferred value of 1.0. Such consequences of the Power Law Equation 1-2 would reach back about a century, in an architecture of old complexities.)

Relevant to the determinations of neutrino masses are the PDG Note (Olive, 2004) giving the cosmicly determined upper limit of total (or additively combined) mass for all types of stable light neutrinos as <24 electron-Volts, the strong possibility of finding eventually a non-SM fourth kind (or more) of neutrino of uncertain flavor discussed in a PDG Note (Kayser, 2004), and especially the PDG uncertainties about "oscillations" of the three SM/PDG neutrinos between different apparent mass levels discussed in a PDG Note (Groom, 2004). (These notes have been continued in later year PDG reports available at pdg.lbl.gov.) Without the generic micro-quanta structure these PDG observations are very difficult to organize into a coherent system. The "oscillations" are particularly troublesome [as was noted in Example 4 of original Paper 1 Appendix C (in Vol. 2 here.) There, with the aid of tables such as Table 1-1a above, it could be seen readily that the impact of an energetic electron neutrino hitting a quark of an ordinary neutron (Case 1) might break the baryon down yielding either 3 tau range neutrinos, or 2 heavier tau neutrinos and a mu range neutrino, or 1 tau neutrino of 2 alternative sizes plus a mixture of 4 or 5 electron and mu neutrinos (or other much larger super-tau neutrino possibilities required in a later paper and a later chapter herein.) Of these listed alternatives, a heavier tau neutrino in any event would be the one most likely to leave a recognizable second impact in a neutrino-sensitive detector. This would indicate in a PDG Note that an expected, very light, electron neutrino in a stream of them had unaccountably "oscillated" into a much heavier tau neutrino at some point in its path. With expected mu or tau neutrino streams (from a known reactor source) and other resonant detectors, other similar PDG "oscillations" could indicate a reduction of neutrino mass at some probability. A re-classification of these QM/PDG mysteries is indicated wherever they are identified, as in the later chapter here.]

A second interesting kind of early consequence of the LQ Charge-Mass or Mass-Charge Power Law, and its source from an exponential extension of it over the general structure of baryons, was the initial clear-cut and decisive basis by the combination of these laws for separation between the very similar PDG masses of the proton at 938.272... MeV and the neutron at 939.565...MeV (Howard, 2005). This was accompanied in the original published paper by a demonstration of the microquantal basis for the re-classification of the scrambled PDG Nucleon family series of baryons into two very regular series, one charged like the proton, and one neutral like the neutron (though this break-out was not formally fully completed and claimed.) The Table C2 for Example 3 of Appendix C in the original published paper demonstrated that under the law equation the ratio of the *Σmc* sum of each of the newly determined quark masses to the sum of the proton's quarks for its series family increased by a perfectly regular 0.11 per family member by stepwise substitution of the heavier mass values for each quark from Table 1-1a above. Likewise, that ratio of the neutron's quarks for its family increased by a regular 0.08 for each step to the next family member. This controls the regularity of these two singly charged and neutral series, which are prototypes for all the various related baryon series. (A similar ratio control holds for the one other doubly charged prototype pointed out in Table C1 of the original published paper, informally completing coverage of the regularity of baryon series generally, as further shown in a later chapter here.)

The third interesting early consequence of using the Charge-Mass Power Law Equation 1-2 came up through its adaptation to Equation 1-3 above [as employed in the second paper condensed herein (Howard, 2006) published in Florida Scientist, 2006, Vol. 69, No.3, pp 192-215, Sub-Structure Laws of Particle Masses and Charges; A New Systematic Classification of Subatomic Particles (See Vol. 2 here.)] In the law the cyclic variation in charges of the quantal pairs causes the cycles of mass increase of the LQ particles in Figure 1-6 next. The logarithmic decrease in spacing of the cyclic lines on the logarithmic scale indicates a very systematic regularity of the spacing of the masses at which particles might appear in Nature. This is further shown at left by the crossing curved lines of dots for each such opportunity (though the dots are omitted on the right for clarity since there would be too many of the crossing lines too close together toward the upper right.) There are 435 such numerical sites, yet only 20 to 30 sites are shown as occupied depending on the allowance for uncertain neutrinos. This is another major factor restricting particle structures. A very large percentage of these site opportunities must have built-in limitations of stability, or interferences, that stop them from working.

That leaves the question of whether such a high level of undetermined blockages of particle function prevents all the apparent regularity in the cycles from appearing in the particles' rate of distribution across the parallel spread of mass-charge sites. The

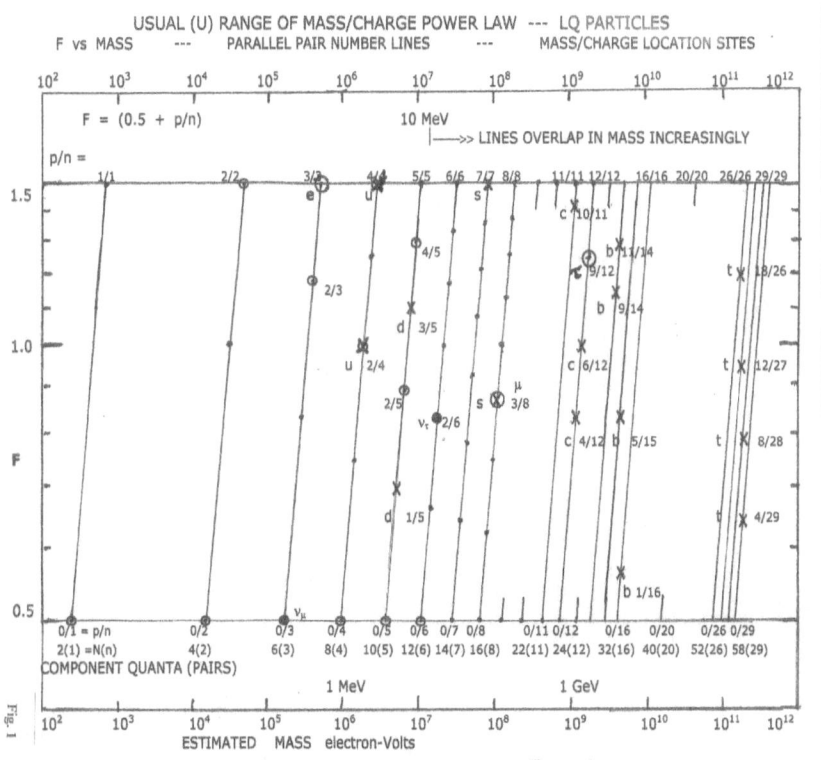

Figure 1-6. Logarithmic Graph of Factor F in the Usual Range of Equation 1-3 Versus LQ Particle Mass. Cyclic Factor F varies between 0.5 and 1.5 as component quanta pairs in the particles are increased from 0 charged pairs p out of n =1 pair in a particle to 29 charged pairs out of 29. After the first eight cycles at left these lines are omitted for clarity wherever there are no particles on a line in Table 1-1a as increased in Fig. 1-4. Xs indicate Quarks, larger circles indicated Charged Leptons, and small circles indicate some Neutrinos within the PDG upper limits. Dots on the first 8 cycle lines indicate sites that could have been particles if something else (to be determined) had not interfered. Note that the dots also indicate sweeping curves of even and odd possible charges of particles across the cycle lines. Dots are omitted for clarity above the cycle lines for 8 pairs. (This was Figure 1 in original published Paper 2.)

quarks have the widest spread of particle masses to test such a search for some kind of regularity that might remain across the graph of rising mass because of the increasing density of packing of sites on the doubly logarithmic scale, and to check whether the regularity has meaning for structures, if it does appear. Figure 1-7 highlights locations of the quarks including the various triple possibilities for the higher mass types of them that appear over a series of the biennial PDG reports to be acceptable under the PDG

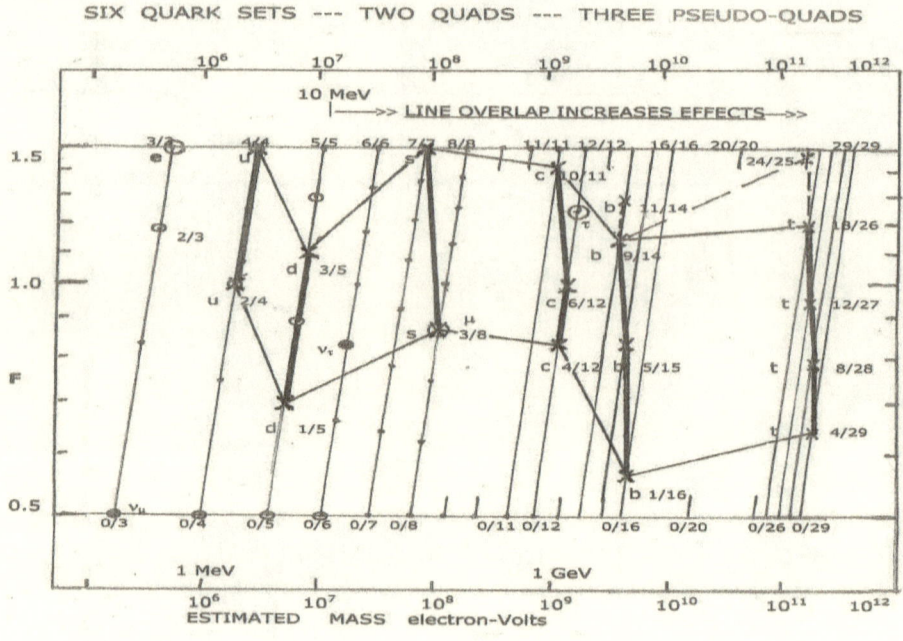

Figure 1-7 Cyclic Regularity of Two Masses per Quark for Six Quarks. Reversal and break-up of the quark mass line slopes begins with the pair-line overlap and increases. Two families of quarks with three in each family are indicated by u-u, d-d, s-s similarity to c-c, b-b, t-t in shape.

uncertainties on quark mass like those values shown in Fig. 1-4. There is an apparent kind of rough regularity that is clearly visible. But its meaning for structures is difficult to interpret beyond the point that the higher quark masses become very far separated numericly. Another meaning that is immediately readable has to do with the PDG, QM, and SM system of dealing with quarks as if they arise naturally in three families of two. The quarks were discovered over the last two-thirds of the 20th Century in these three QM groups of two "flavors" with an initial uncertainty of the relation between "strange" quarks (#3) and "charm" quarks (#4). It is definitely true that the first two, the "up" and the "down" quarks, constitute the great body of stable baryons that make up the cosmicly balanced and stable particles of the Universe we have always known. The other four, including the "bottom" and "top" quarks, which were established late in the century, make up most unstable hadrons that also appear in cosmic ray showers and similar very brief occurrences. Still, the graph of rough regularity of quark masses indicates that #3, the strange quark, should be more closely related to the up and down than to the last three. Also, the charm appears part of a cycle including the last two.

Whether the new systematic classification of these quark particles into two families of three, as shown in the second published paper (Howard, 2006) and followed herein, is predominantly more fruitful in accounting for the comparatively low frequency of observations of the heaviest baryons will depend in later chapters and continuing efforts on the difficulties of establishing balanced stability with low cohesive force-to-mass ratios for such large numbers of quanta in a coherent structure system for the three heaviest quarks. (See Vol. 2 reference tables of baryon structures for later chapters.)

The continuing large PDG uncertainties on Quark masses, as indicated in Figure 1-4, are clearly confirmed here as more accurate than QM distinctive single masses by tending to include the dual masses for each quark required in the original two published papers for the structures of the prototype families of increasingly high mass of baryons under the Charge-Mass Power Law (and condensed in this chapter, Howard, 2005 and 2006). It is forecast that most of those empirical PDG uncertainties will be necessarily expanded until PDG adopts the two required certain masses for each quark.

[Since the original published papers which this chapter condenses were written (Howard, 2005 and 2006), ten years of further application of the new law equations has revealed which options in their logic have proved to be the most essential and necessary contributions to a full paradigm of the sub-nuclear particles for inclusion here. Other options that may be employed and back-up tables are included with the complete original texts in a second volume of appendices along with the complete reference lists beyond the primary references listed herein. Note that gluons are no longer considered necessarily includable, as in original Table 1, but that table's added small neutrino options in a Medium range between the E and U ranges, with varying a, are expected tp be required eventually in more energetic pseudo-"decays" and neutrino "oscillations" observed in the QM/SM. Structures for these options are included later here, as well as structures suitable for symmetric self-antiparticle neutrinos generally.]

In <u>General Summary</u> of Chapter 1, condensing the two introductory published papers that informally initiated this book's Micro-Quantal Particle paradigm of the sub-atomic particles a decade ago: Overall close adherence to the SM/PDG data of the particle masses computed from a very simple equation (on actions of the particle charges) accounts systematicly for the seemingly irregular progression of the lepton and quark/anti-quark masses, and also for an extended progression of the baseline lightest members of the families of the proliferated hadron particles, as well as providing a basis and three prototypes for the regular internal progression of masses of the various baryon families individually. In addition, the implied general structures here are sufficient in numbers of still incompletely specified components for correlation with many as yet unspecified variations of special relative properties such as those labeled by QM as "charm", "strangeness", "flavor", etc., for lack of more descriptive names, and also those whose names have more definite meanings, such as spin, symmetry, anti-particles, stability, lifetimes, the nuclear forces, electric field directions, etc. These factors are not all suitable to work out separately until the most rigorously necessary or definitive structural requirements have first been met. As a general LQ correlation, in the heaviest quarks/anti-quarks most of the component interaction capability must necessarily be either engaged well within a bounded separate assembly of components, or shielded, with little of that capability interacting for mass gain externally between complete quark/anti-quark configurations in the heavier hadrons (where the ratio of mass gain beyond the sum of component masses approaches 1 in Figure 1-2.) Certainly, the light quarks/anti-quarks, with a necessarily higher ratio of external exposure of any sub-structure components from within the smaller component assembly, do make up the light baryons most capable of both the more energetic force interactions per unit mass within an entire baryon and greater stability (as in the proton.) The lighter electron/positron also, with even less total internal interaction mass and mass gain, could likewise have its six charged quanta on the implied group surface for a greater proportion of exterior interactions. Such external interactions of particles may be largely geometrically quantized attraction, repulsion, twisting, or combinations of these, between configurations of sub-structural components with more or less surface exposure of many variably polarized, neutral pairs, with very short range net exterior forces, and/or with numbers near the particle surface of un-neutralized charged pairs with longer range force fields. These general properties of necessary particles and their controlling law equations establish broad guides on particle structures for the remainder of this book and for its MQP paradigm of the sub-atomic particles.

More broadly, charges of quanta are fully conserved because the (presently undefined) active structures of quanta are, as far as can be observed, permanent and indestructible. Mass is an interaction product both within and between quanta, an interaction of structures. Mass is evidenced as mass interaction energy, which appears conserved as long as the structure of the numbers of quanta that are interacting within a particle is unchanged. If more quanta are brought into a particle, the interaction and the mass interaction energy increase. If quanta are separated out of a particle, the constant creation of mass interaction energy decreases to the levels in the two (nominal, or more) separated structures individually, and the masses and mass total decrease accordingly. The mass energy level created in the single particle before separation is not destroyed, but goes out in a weak mass interaction energy radiation wave still capable of interacting with the original force (along with the continuing full charge force wave) when it arrives with the speed of light at other particles made of quanta with mass interaction energy and acts with both the charge electric and the decreasing mass gravity forces.

In that frame, a simple, independently observed, equation systematicly correlates the composite masses of the PDG-listed SM hadron particles with the masses of their quark and anti-quark sub-structures. The equation also extends with a minimal (self-consistent) phase change to more elementary SM quarks, anti-quarks, and leptons, including neutrinos and anti-leptons, to compute their masses and charges as composite structures of micro-quanta of 1/6 fractionally charged sub-particles with informative accuracy. This is the simplest basis for the generalized structure of all these sub-nuclear particles. Tying in the top quark provides a regular systematics of mass-generating charge throughout the QM proliferation of particles. This simple numerical link to the data of empiricly observed SM particles for computation of sub-structural elements, and for a regular power law of masses, provides a departure point for re-examination of the broader uncertainties of Quantum Mechanics and the Standard Model. These include not only the outer structure of particles reorganized generally in this chapter, but also the necessarily required inner detail structures and originations of quarks, electrons, oscillating neutrinos, photons and other bosons, and the decays and reclassifications of the families of famously "proliferating" hadrons, as well as many, if not all, related phenomena, such as a basis for full unification of the natural forces as the forces of structured micro-quanta within particles.

REFERENCES OF SPECIAL INTEREST
[Complete Reference lists for this Chapter and for the original published first two papers condensed herein are included in Volume 2, Appendices & Reference Tables.]

FOREWORD Main References

DAVIES, C., and LEPAGE, G. P., et al., 2010, Determination of the masses of the common up and down quarks, Phys. Rev. Lett. **104** (2 Apr. issue)

DURR, S., et al., 2008, Ab initio determination of light hadron masses, Science **322**, 1224

HOWARD, F. E., JR. 2005. Elementary Particle Mass Sub-Structure Power Law, Florida Scientist, **65**, #3, 175-205 (See Vol. 2, A Hidden Simplicity, Appendices.)

KRONFELD, A. S., 2008, The weight of the world is Quantum Chromodynamics, Science **322**, 1198

WILCZEK, F., 2008, Mass by numbers, Nature **456**, 449

CHAPTER ONE Main References

EIDELMAN, S., et al., 2004, Particle Data Group biennial report, Phys. Lett. B 592, 1, pdg.lbl.gov, including:
    AMSLER, C., and O. G. WOHL, 2004, Note on the quark model.
    GROOM, D. E., 2004, Note on understanding two-flavor oscillation parameters
    HOEHLER, G., and R. L. WORKMAN, 2004, Note on N and Delta resonances.
    KAYSER, B., 2004, Note on neutrino mass, mixing, and flavor change
    MANOHAR, A. V., AND C. T. SACHRAJDA, 2004, Note on quark masses.
    NAKAMURA, K., 2004, Note on solar neutrinos
    OLIVE, K. A., 2004, Note on low mass neutrinos.
    VOGEL, P., AND A. PIEPKE, 2004, Note on electron, muon, and tau neutrino listings.

HESTENES, D., AND A. WEINGARTSHOFER, (eds.), 1991. The electron, new theory and experiment, Kluwer Academic Publishers, Boston, MA. 405 pp.

HOWARD, F. E., JR. 2005. Elementary Particle Mass Sub-Structure Power Law, Florida Scientist, 65, #3, 175-205 (See Vol. 2, A Hidden Simplicity, Appendices.)

MAC GREGOR, M. H., 1992. The enigmatic electron, Kluwer Academic Publishers, Boston, MA. 165 pp.

PARTICLE DATA GROUP (PDG) BIENNIAL REPORTS
    See Eidelman, also Yao, and pdg.lbl.gov

YAO, W.-M., et al., 2006, Biennial PDG Data Report, Jour. Phys. G, **33**, 1, pdg.lbl.gov, including up-dates of the above and:
    DREES, M., and G. GERBIER, 2006, 22. Dark Matter

SECTION II

The <u>Necessary</u> Particle Sub-Structure & Quantum Formations

Chapter 2.

Inherently <u>Required</u> Sub-Structures for General Particle Functions & Micro-Quanta

SUMMARY: The very generalized sub-nuclear particles of Chapter 1 are found to require very specific and detailed structure to be capable of carrying out their general particle functions with systematic regularity. Regular cycles occur naturally as synchronized orbital movements in a basic spherical form that is the simplest structure for globular gatherings of required numbers of sub-particles, starting with the simpler and lighter LQ particles that occur in great quantity with much background data. The observed functions and forms of complete particles also require that sub-particle components interact cyclicly as constant micro-quanta to generate mass both internally and between larger particles. These smaller micro-quanta are resolved by examining also their other most pervasive, or wide-spread, overall general functions for their necessarily jointly required forms with new criteria of balanced simplicity of large numbers in cyclic spinning combinations. Then a single kind of natural fluid vortex, with its symmetrical opposite, is found suitable for lab experimental models. These spinning gyres, with the capabilities of two basic symmetric quanta in balanced coaxial pairs, also spinning on a perpendicular axis through the pair axis center, demonstrate <u>a single unified source for all the natural forces and actions of the sub-nuclear particles, including closely interrelated forces of electric charge and mass gravity under the Charge-Mass Power Law</u>. In all well-known Usual particles, assemblies of pairs orbit in synchronous interaction in balanced spheres with inherent linkages. The further the quantal pairs are tested, the more exact the fit to Nature. That is explored widely in this chapter (and later.) Each quantum becomes both a coherent kinematic particle at a building block level and an interactive fluid structure of currents and interpenetrant waves linked within the larger sub-atomic structures. The forms of the structures are shown in this chapter. The quantized force, size, density, viscosity, and current velocity measurements with mass-based scaling equations appear in the next chapters. A further lab program is indicated in an advanced tank facility.

    In Chapter 1 the fundamental Charge-Mass Power Law of Numbers of general structural interactions was clarified finally as being the primary law at the most fundamental level of sub-nuclear particles down through their being constructed of generalized, electrically charged micro-quanta. These micro-quanta are like the uniform bricks used as building materials for a wide variety of solid brick houses (built as in the past.) There is a transition level between the livable homes in the houses and the bricks out of which they are made. Still, the nature of the houses is dependent on the specified hardness and form of the bricks, whether they are simple dried mud brick or are made from a finer clay that has been fused together in a high-temperature furnace. There is another transition level between the bricks and the clay. If the house is very large, the construction blocks must have been actually melted together and re-solidified over geological time, as with harder granite stone from a quarry. The properties of one level of those structures must necessarily be determined exactly from the properties of the other. This chapter must find from the Power and Exponential Laws, the accredited PDG data tables, and other confirmed QM functional knowledge, the necessarily required structures both within the subatomic particles that the PDG has identified and below them in the next transition level of the micro-quanta and their principal structure. [In a later chapter or appendix it will also be necessary to specify the gross nature of a both fluid and frozen-impeller sub-structure out of which gyres form (at the level of the fired clay in bricks or a melted geologic magma hardened into granite stone.)]

The close relation of the Power Law of LQ Charges and Masses with the Exponential Law of Hadron Masses emphasizes continuity of structural relations across all the sub-nuclear particles. This related progression of structures is also emphasized by the almost universal finding of PDG listed "spins" in particles generally, though the exterior structure of entire composite baryons, especially bonded baryons in nuclei of atoms (Bass, 2007), may not spin in the same sense as their components (as when multiple houses with people moving in and between them transition into forming a town or city.) Another common particle property noted in Chapter 1, and listed by the PDG when it has been measured, is that of mean lifetime with its dependence herein on structural balance and balance of forces and inertias, especially in any spinning structure. Such broad features place severe limitations on specifiable structures of particles and components and are guides to all the necessities. Simplest structures that meet the combined necessities are the only ones found in Nature unless a particle is considered particularly "excited." Natural structures that appear repeatedly in relatively large numbers must, therefore, have forms that inherently combine forces and functional actions. The necessities are stringent. (Howard, 2010)

Due to the earlier critical part the Electron played in establishing the mass quantum and the modifying effect of charge of the paired quanta in the Charge-Mass Power Law's general definition of the LQ particles (Chapter 1), the electron must also be a particle of prime necessity for determining the forms of particles and establishing structures that are inherently consistent with that law. The six electron quanta with like charges in pairs on orthogonal 3D axes around a common center provide perfect initial balance that is maintained in synchronized interaction orbits around a perfect sphere. That matches the extremely long mean electron life of greater than 4.6×10^{26} years shown in the typical international Particle Data Group (PDG) accredited data listings (Yao et al., 2006, pdg.lbl.gov, continued biennially for decades.) So does the small number of 6 components and the PDG mass listing of 0.5109989...MeV (carried as 0.511... rounded herein for convenience in Power Law computations, Table 1-1.) This mass is near the low end of the Usual LQ particle masses. It indicates low inertia to force ratios in internal dynamic balancing to match long life (in later chapters.) It is the lightest well measured PDG mass. The electron's other PDG properties are very nearly ideal for a sphere. The electron is also one of the most frequently observed particles. (Only the proton, a much heavier and more complex baryon particle, but very nearly the lightest baryon, is competitive with the electron in these selection criteria.) All of this information qualifies the electron at low velocity (or in the ground state) as a potential prototype for the well-known particles. There is the exception that the electron structure is not considered known at all, and it is usually treated as a mathematical point with no form. That does not, however, contradict or disprove that the known evenness of its all-directional electric field strongly indicates that the electron is a sphere, and that most of the numerous prior attempts to establish an electron structure have been consistent with the shape of a sphere. In the presence of the spherical electric field of an electron at rest with a long-continued, very small PDG electric dipole moment of $(0.07 \pm 0.07) \times 10^{-26}$ ecm (electron-centimeters), recently restated as less than 10.5×10^{-28} ecm, the requirement from the directional evenness of six distributed microquanta under the Charge-Mass Power Law is that the simplest electron must necessarily be a sphere.

Necessary 3 Orthogonal Spheric Orbits
with ONLY **1 primary** octant
shown for <u>majority</u> RH spin quanta in coaxial pairs

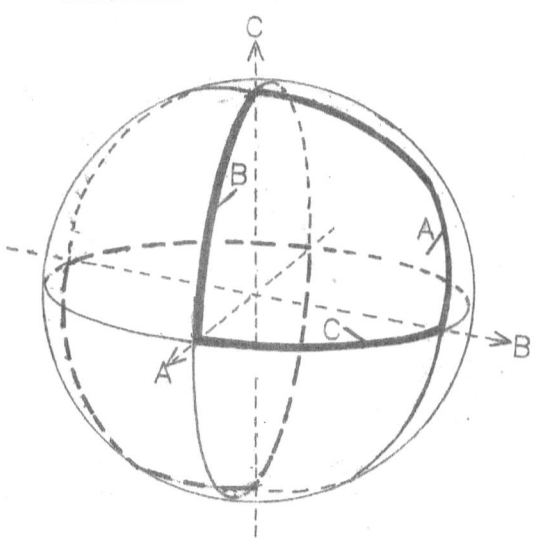

Fig. 2-1 A Basic Prototypical Sphere Structure for Usual Particles Based on Electron Needs. (See text.) There are three concentric, orthogonal orbits with only 1 primary octant of the sphere, around which all orbits move in the same direction, no matter what the individual quantum spin directions or handedness may be. Each orbit around its own axis is in the spherical direction of the handedness (with respect to the outside of the sphere) of the predominant number of quanta of the whole sphere (if there is a predominance such that the sphere is not neutral in spin rather than being neutrally balanced in spin of quanta. Quanta are not shown for clarity of orbits.) For regularity, balance, simplicity, and spheric perfection, the quanta are assumed to be orbiting in coaxial pairs with synchronized starting positions of all their co-axes at 45° on the middle of the leg of the primary octant. (In this case, shown for all quanta in right-handed spin outside the sphere, whatever one octant they all are simultaneously entering, or leaving, or at the middle 45° for, automaticly becomes the primary octant.) Orbits are at a uniform rate.

Accordingly, Figure 2-1 shows that particles necessarily have a basic framework form of three available, concentric, orthogonal (perpendicular, 90°, right-angled) spheric orbits as required for the electron or positron balance properties. This also matches the required orbits for the mu neutrino at the PDG mass limit with the same number of components and pairs in Table 1-1, but neutral pairs where the electron pairs are all charged. Both types of these smallest Usual particles can each have this structural form in the simplest balanced symmetry. The form provides for particle and quanta spins in either right or left hand rotation for two kinds of micro-quantal spin. Each quantum in a pair can spin at a uniform rate separately right or left on its internal end of the co-axis, and at ninety degrees to that axis both may also spin together on their orbital axis through the center of the sphere, also at a uniform rate for synchronization.

Ref. Summation Axis
through centroid of primary octant

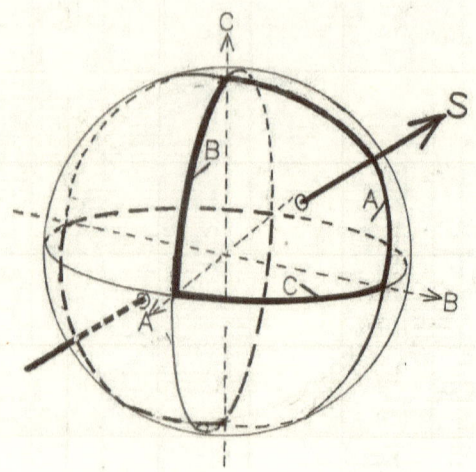

Fig. 2-2 The Sphere Summation Axis S Through the Primary Octant's Centroid. All the quantal spinning of individual quanta and the synchronized orbital spinning of pairs of quanta, each at its proper uniform rate, are summed on the Summation Axis S (or S0) of the sphere through its primary octant's centroid. That octant is determined by the orbital relations of Fig. 2-1. (The rule there is very simple, but, without good 3D perception, a small plastic ball, preferably transparent, and colored markers may be useful.) Continuing the right-hand orbit assumption, the predominant orbits around the primary octant indicate the handedness of the sum itself. The pairs are each separately alone on orbit A or B or C.

In Figure 2-2 the Summation Axis S for the combined orbital spins, in this simple case, clearly is through the centroid of the primary octant, as it always is. For larger particles with more component quanta there must be growth of the synchronized pair orbit possibilities. But the symmetric balance around the S axis must be preserved for such a particle to exist. The first increase of orbital sites for pairs is at A', B', and C', which each lags its ABC pair in orbit by 45 degrees. All six pair sites can possibly be occupied. The next site for a single pair is on the Sum Equator, which is at the angle 54.7° from the C orbit here in Figure 2-3. With ABC starting at 45° on the primary octant, the SEq pair site must be near the C axis and must lead in the predominant direction by 7.5° ahead of the vertical (here) plane of the C and S axes, or at 60° intervals around the SEq orbit in order to reduce orbit interferences between pair sites to a minimum. The next two usable orbits are called the – – and ++ orbits, as shown next in Fig. 2-4

Sum Equator Orbital Plane
Where any coaxial pair autobalances, like any pair spinning on the S axis

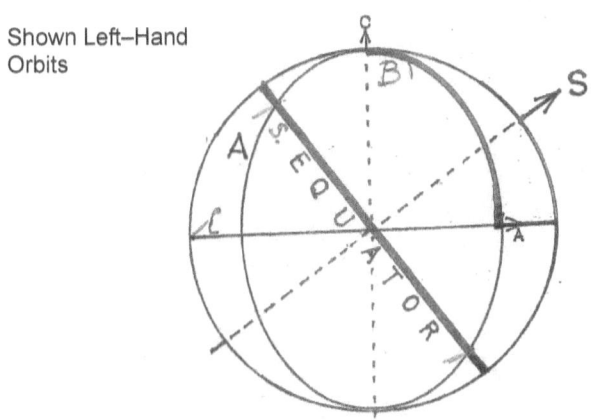

Shown Left–Hand Orbits

Fig. 2-3 The Sum Equator (SEq) Orbital Plane. This plane is seen on edge at 54.7...° rounded from the C orbit plane, or the angle whose cosine is $1/\sqrt{3}$ (arccos $1/\sqrt{3}$), just as S is from the C axis. The remaining angle between SEq and C axis is 35.3...° rounded, which will be used to set the next orbit.

and subsequent figures. The – – orbit plane is symmetricly opposite the SEq plane around C pole at 35.3° from C pole in the C-pole-S plane, which leaves – – orbit 19.4° from S in the figure. The ++ orbit plane is symmetric to the – – orbit at 19.4° on the other side of S in the figure. Both these orbits go around the S pole in the same hand direction as ABC orbits. Since ++ is near C orbit, its pair site lags C site in passing through the vertical CS plane by 22.5°, or half way between C and C' sites as projected on the C orbit plane. Then at the same synchronizing start instant, the – – site near to the primary octant (the front face for sync) is 45° behind the ++ position (as seen from that position), but going in the opposite direction with reference to the C pole and axis, or to the ABC primary octant, or to the – – and ++ crossing points with the C plane and SEq plane (See Figure 2.5.) They will be 90° apart 22.5° later than shown In their start points (Fig. 2.6 main and right side panel) when – – hits CS plane and ++ hits C plane.

The next figure, Fig. 2-5, summarizes in a perspective view these orbit sites for coaxial quantal pairs and adds the new feature that coax pairs may also spin statically on axes through the centroids of all the octants, as at S0, S1, S2, and S3, with four pairs of quanta. This gives a total of 13 pair sites. These 26 quantal sites appear to provide for all the LQ particles in Table 1-1a (Chapter 1) except heavy bottom and top quarks. But

TWO OTHER ESSENTIAL ORBITS

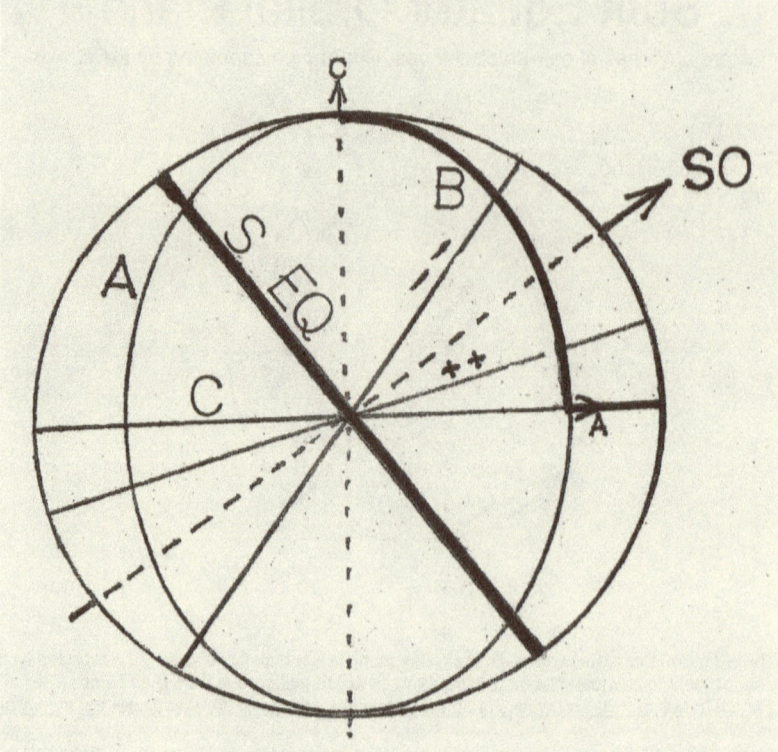

FIG. 2-4 Edge View of the Minus-Minus and Plus-Plus (− − and ++) Orbits. The − − orbit plane is symmetricly opposite to the SEq (sum equator) orbit plane around the C pole at 35.3...° (rounded) from C pole in the C-pole-to-S0 plane of view. That symmetry puts the − − orbit plane 19.4...° from S0 in the figure. The ++ orbit plane is symmetric to the − − orbit at 19.4...° on the opposite side of S0. The orbit edges are seen as straight lines. Note that the C and SEq orbits are seen in the same way, and there are four orbits passing through the congested traffic point on the sphere in the center of the figure, as well as on the opposite side of the sphere. - The outline of the sphere is not an orbit, but the arcs of A and B orbits with the line edge of C orbit outline upper and lower octants. S axis is often labeled S0 (zero) to distinguish its octant from the 7 others. B axis is beyond the sphere directly behind A axis. (The − − and ++ orbits also balance around a possible orthogonal pole where SEq plane crosses S0-C plane of view.)

there is already a conflict of two orbits at the four-orbit crossing. Since both are needed, this requires new self-sorting Precursor assemblies in which pairs can be attracted into expanded linking orbits, as in Fig. 2-6. (This is the first adaptation needed, even with smaller quarks and leptons, in building sites for the observed particles from the basicly simple, orbital interaction structures. Adaptations grow for further necessities in more Extreme requirements, such as cosmic ray showers, etc., as well as the Usual particles in ordinary stable star and planet systems arising from the structures' natural capability.)

++ & – – Orbital Planes Total 6

One Symmetric with S Eq around C axis, then both balance around S0 Ref in same direction as majority of charged pairs –**High Traffic** at C Crossings is **CRITICAL FEATURE for CONFLICTS, BALANCE, and SYNC in Quarks**

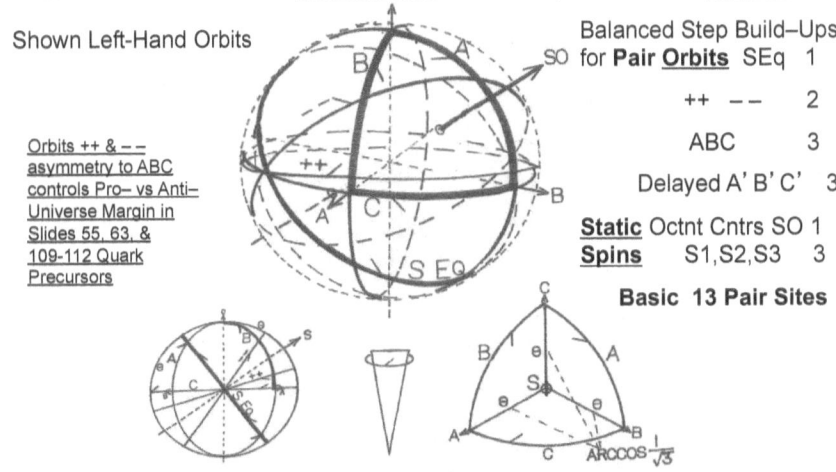

Shown Left-Hand Orbits

Orbits ++ & – –
asymmetry to ABC
controls Pro– vs Anti–
Universe Margin in
Slides 55, 63, &
109-112 Quark
Precursors

Balanced Step Build–Ups
for **Pair Orbits** SEq 1
++ – –   2
ABC   3
Delayed A' B' C'   3
Static Octnt Cntrs SO 1
Spins S1,S2,S3   3
Basic 13 Pair Sites

TRAFFIC JAM NECESSITATES QUARKS   52

Fig. 2-5 Interactive Orbits for Pairs of Micro-Quanta & Initial Added Sites. (See text.) Slide numbers link later, including a Briefing in Vol. 1's last chapter, where an Anti-Universe effect (left) of this structure is interesting. Only the Primary Octant around S (or S0) is drawn (lower right), but its internal angles apply similarly to octants around S1, S2, and S3 and the opposite octants with reversed spins around the axial arrow tails. The numbered arrow-heads fit the octants with bases on the Primary Octant & 2 out of 3 arcs in the predominant direction. The 15° cone is the impenetrable angle on the sphere surface for each spinning quantum, and gives the minimum arc separation of quanta centers for no orbit conflict. The Table at right shows the numbers of balanced pair sites in groups, which total 13 basic pair options. Four pair sites included are the non-conflicting static spin sites for pairs in the eight centroids of octants. Orbits of at least two quantal pairs conflict if confined to the surface of the sphere, but quark SEqs do expand.

At each point in this series of figures the individual sphere of orbits looks like a particle, t is still only a skeleton of structural forms that true components of particles can take in us cases. That is particularly true of Fig. 2-6, since one assembly of components cannot to this expanded SEq form by itself, and the pair of quanta in the SEq orbit has a risk of sion with another pair within a small fraction of an orbit if the SEq pair stays on the surface e sphere of interacting orbits at which there is a balance of all the mutual forces and ias for each quantum. To go to any of these circular or quasi-elliptical forms, and stay in orm, the pair of co-axial quanta in the SEq orbit must be attracted out of the position of the on the surface of the ball of orbits by two other similar spheres and SEq orbiting quanta, on each side to the front of the primary octant. In fact all three of these distinctly similar, ot necessarily exactly similar, configurations must come through a somewhat less similar

Sum Eq. Quasi–elliptical or Circular Orbit

<u>If & only If</u>, <u>*TWO*</u> other prequarks pull pair into radius 2x orbit, or <u>quark fails–</u>

Quark "**<u>COLOR" & "GLUON"</u>** Function Capability by Mutual Sorting

S Eq & Other Orbit Sync Points Inc. A' ,B' ,C' **+ SO,S1,S2,S3 Static in Octant Centers**

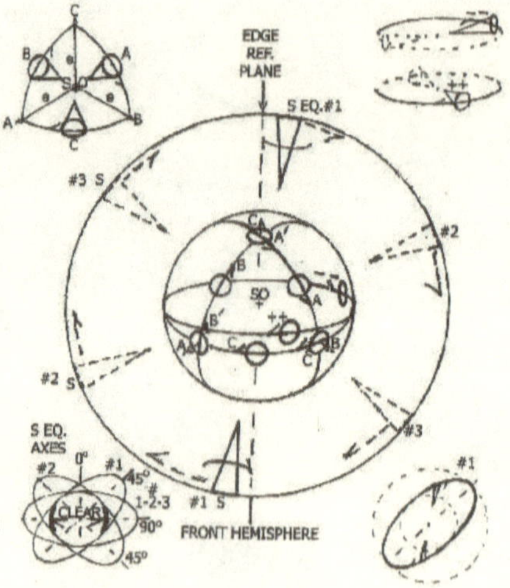
S Eq #1–2–3
single alternates
necessarilly
indicate "Color"
(Ellipse–like orbits
shown tangent for
clarity need minor
radius near 1.5 in
magnetic moment
estimates.)
45° Axis Alternates
Circular on sphere
Front Views
Only
Non–symmetric SEq
Quanta Forces
Estimated
Symmetric Forces
Scale from Lab Data
EDGE
REF.
PLANE
S EQ.#1
#3 S
#2
#2 S
#3
S EQ.
AXES
#1
0°
45°
1-2-3
90°
#1 S
FRONT HEMISPHERE
CLEAR
SO
54

Fig. 2-6 Sum Equator Quasi-elliptical/Circular Linking Orbit & Orbit Closest Approaches in SEq & Axis Octant. In a formative Precursory gathering of pairs of co-axial quanta that interact to force each other into any of a series of relative positions, if there are enough of the suitable charged and neutral pairs, they can, in less than an orbit cycle, sort themselves into three associated prequarks in close triplet formation. In those relative positions, they can eject all the excess pairs that do not clear the beginning orbits. But as an almost ejection, their SEq orbits (which are mostly initially conflicted at the C crossing of four orbits) are attracted into expanded ellipse-like links or circular links, which-ever fits the sphere ball as an instant finished quark into the baryon triplet linked form. Failures to complete a formation simply scatter, possibly repeating the process in a fraction of an orbit cycle to re-assort as three smaller leptons, including single-pair electron neutrinos or other multi-pair standard or non-standard neutrinos such as small mu neutrinos, or small tau neutrinos. Electrons, positrons, or mesons also may complete the assortment if there are too many pairs. The conserved pairs do go, as they must, to some finished set of particle states. In the process of forming what-ever quarks result, all of the ellipse-like orbits finish in 45° major axis angle <u>options leaving the conflict points on the sphere clear, with constant angular velocity orbits, not elliptical.</u>

near-pre-quark approximation of one of these forms within a Precursor, larger globular assembly of quantal pairs within a very brief fraction of the time period of one orbit. - - If that does occur, the three may continue to link up as structural forms from Fig. 2-6 in a true baryon triplet, as in Fig 2-7. - - In that triplet, and in later additional reactions, it will be important that the two expanded SEq orbit links (of Fig. 2-6) in any corner of the triplet eliminates the need for a QM "gluon" pseudo-particle to be imagined to provide the corner linkage between each two quarks at a triplet corner. It will also be very

Quark Triplet Orthogonal S Eq. Corners
Proton Prototype Structure

(S0 Sums Spins at 54.7° tilt on corner axis – Quark Spheres Tangent)

(#quarks/quanta)
Odd down is #3.   **(1/5)**
2 ups are #1 & #2 . **(2/4)**

ALL ORBITS SYNCHRONIZE IN A <u>TRIPLET RING</u>

ALL BALANCE & MIN. MASS SO 10^{28} YEAR LIFE

S Eq Quasi-Elliptic Orbits Shown Tangent at R= 1 for Clarity

But Magnetic Data Require Min R Near 1.5

NET SPIN AXIS DOWN HERE WITH S Eq CORNER

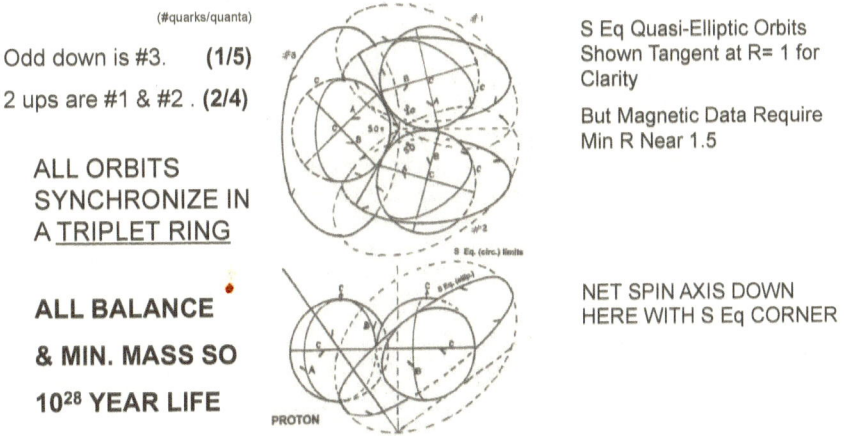

Fig. 2-7 The Basic Baryon Prototype Structural Form in a Proton. The overall structure of the proton in very nearly the smallest possible mass of a baryon quark triplet ring in a 90° corner of SEq orbits.

important for reactions that in a triplet no two quarks can have the same Fig. 2-6 60° separated location of the single SEq coaxial pair of quanta for each quark, and further that, once set, this location cannot be changed by sliding from one 60° position to another without a conflict that completely breaks up the quark. This eliminates the QM need for the set of three imagined "colors" carried by the quarks to control where each quark can fit in the baryon or meson product of a reaction. Numbers by the 60° sites replace "colors." (In atomic nuclei 22.2 ±1.0 % of all quanta are in these linking orbits.)

    If the set of three pre-quarks does not quite come together in a fairly symmetric, no-conflict, pre-baryon triplet of pre-quarks just before they all reach a form of Fig. 2-6, then none of them will quite reach one of these forms, and the whole group may fall apart into other smaller configurations. Or possibly two members of the initial triplet of pre-quarks may continue to develop into laterally bent near-misses of the links of a Fig. 2-6 form and keep the two together as a meson particle fragment of a baryon, while only one pre-quark falls away and is repelled into breaking up. Or that lost single pre-quark may even join up in another meson with another single failed pre-quark, if another is by chance nearby and close to the suitable mutual 3D angular symmetry without being off-set too far. - - It is also possible that two such bent meson di-quarks may quickly come

Sphere 6–Orbit Sync Ref Angles

(See Interorbit Clearance Angle Table 6-5)

- All 6 Orbits Clear 1 Static Spin Pair in Octant Centers
- A, B, & C at 45° in Primary Octant (& Pair–Mate Opposites Implied always)
- A', B', C' Trail A, B, C by 45 ° (Sync at Poles) in Left Hand rotation
- ++ Orbit Trails C & Leads C' by 22.5° LH (in offset toward C pole)
- ++ & – – are 90° Apart in Arrival at Crossing
- – – Leads at SO–C Vertical Plane <u>front</u> by 67.5 ° in Left Hand rotation
- SEq #1 "   "   "   "   " <u>rear</u> by 7.5 ° LH or

  (Alts. for Single Pair) #2 leads by 67.5 ° or #3 by 127.5 ° (RH No Change)

  <u>If SEq Still on Sphere SEq #1 & #2 conflict with A' , #3 with C' & ++</u>

  (6.5 to 13° cone overlaps) <small>(Conflicts Block 1 of any 2, Not Both).</small>

  (Color limits: EXPANDED SEq 60° options do not change for normal reactions due to blocking conflicts in transition, but major Precursor-type destructive reactions can re-organize SEq)

FIRST STRUCTURAL SITE CONFLICTS
INCREASE RARITY OF SYNCHRONIZABLE PAIR SETS

Fig. 2-8 Sphere Orbit Internal Sync Start Sites and Conflicts If SEq on Sphere. Internal orbit conflicts add to the baryon conflicts between tangent spheres (Quark chapter. See Interorbit clearance angles in Vol. 2 Appendices.) Precursive quarks and quanta must have the right charges to pull themselves together in baryons. These factors compound within increasing numbers of quanta pairs in heavier quarks to cause <u>the very wide spread between the few possible combinations of quanta in quarks (Figure 1-7, Chapter 1.)</u>

together with a single quark of each di-quark aligned well enough to join together in an angularly twisted line of a tetra-quark meson with bent quarks only on the open ends of the four quarks. It has been noted in PDG appendices on quarks that there are strong indications that four quarks may come together (Amsler & Wohl, 2004, etc.) The quark Appendices of this paradigm (Vol. 2) show that both four and six-quark meson forms, as in Fig. 2-9, are required to generate some of the PDG reported masses of mesons.

However, all these required structural forms for particles of various kinds are also dependent on there being a compatible and, therefore, required structural form for a standard quantum that fits both all, repeat, all the particle structures and the PDG observed actions of particles, such as spin, and also provides a form of charge which can generate mass in accordance with the Charge-Mass Power Law. Quantal structure must also generate forces of at least the other PDG listed types [if not at least one more force which QM needs to perform all its observed (and currently increasingly reported) functions, but which QM has long avoided designating as a force (through a mysterious

Meson, 6 Quark Schematic

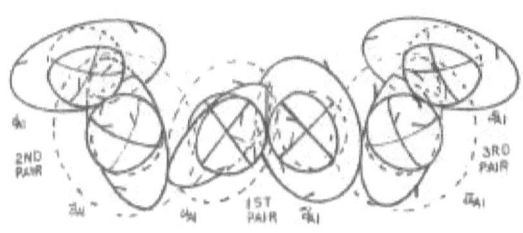

RANDOM MESON A-B-C AND S EQ. ORBITS

ILLUSTRATES OPEN ENDS OF ANY MESON CHAIN

– ENDS FAIL SEQUENTIALLY – <u>STRONG FORCE INEFFECTIVE</u>

Fig. 2-9 Structural Form of a Six-Quark Meson. A tetraquark meson is similar to this chain of six (shown in extended form for clarity) with omission of a diquark from either end of the chain. There is also the possibility that one or both of the two links between diquarks will join in a form of link that turns them toward each other rather than extended. Still, the 90° linkage angles are such that the two end quarks <u>cannot close a ring by linking with each other and satisfying the end forms (except in baryon triplets.)</u>

century-old compromise and a limiting lack that have between them caused continuing controversy, to be fully resolved in necessary structure over the next six chapters.)]

All of these needs must be met by the structure of one kind of form for micro-quanta in any particle. For the whole basic particle form required by fundamental Mass-Charge Power Law and its Exponential Mass Law extension, six orbits and four static spins in octant centroids are sufficient for most and the most frequently seen particles. A cylindric doubling of the quanta pairs in the static spinning sites at the spheric octant centroids and a short-lived second outer shell of spheric orbits are added when needed for the heaviest PDG-observed bottom and top quarks and their rare particle series.

The many properties of particles have two common factors that appear repeatedly, spinning and having two alternate or dual aspects of some characteristic or property that is important. In the necessities of the Charge-Mass Power Law equation, newly discovered in the PDG Summary Tables of accredited (approved) particle data as the fundamental basis for the Micro-Quantal Particle (MQP) paradigm initiated in

Chapter 1, there were dual plus or minus charge force interactions in quantal pair duos that had two controlled levels of generation of mass. In the necessities of 20th Century Quantum Mechanics that established the particle phyics of nuclear fission and fusion, a world-changing science, there is great emphasis on dual plus or minus spins for almost all sub-nuclear particles. There are also dual North/South magnetic spin interactions and dual strong/weak force interactions. Gravity force interactions are necessarily between two or more masses. Thus, micro-quantal dual right/left spinning interactions must have some structural basis for these empirical interaction necessities.

More particularly, there must necessarily be an inherently <u>unified</u> source for the two most disparate forces, which the Charge-Mass Power Law demonstrates to be completely interdependent. Both charge force and mass gravity force are interacted through <u>dual spin structure</u> above. A simplest criterion needs a specific single source of that. Right/left spinning strongly implies either a spheroidal structural form for the basic quantum, or a conic form (which might include a rounded pear shape or a cone with uncertain or imprecise edges.) There is also the earlier necessity for definitely handed spinning of quanta in coaxial pairs. Spinning spherical forms have symmetric right and left hand spins on opposite ends of the spin axis and are indeterminate in direction of observed rotation. This requires the general conic shape. Multiple coaxial conic pairs orbiting synchronously in a concentric sphere of orbits of limited size are naturally adaptable to interactive forces and less critical in radius with <u>conic points inward</u> as the simplest and therefore necessary <u>U</u>sual form. That also requires uniform quantal mass, spin rate, orbital angular rate, and quantal dimensions. PDG data often requires exact balancing of opposites to zero. Pairs can balance perfectly for consistent stability and long life only if there is also a quantal fly-wheel balance of forces and inertias in exactly symmetric regularity of internal interactions. Low mass-to-force ratios with low masses are necessarily most stable, since under the Charge-Mass Power Law total masses (and inertias) of LQ particles increase as an integral positive power of the number of components but PDG conserved forces such as charge force increase in simple linear addition with number. All of these influences naturally select for and require conserved uniform small pairs of symmetric conic quanta with coaxial points Usually inward (PI).

Multiple dispersed interactions of various numbers of quanta both inside a spherical group and between groups also require, either an impossibly complex system of quantal "gears," or herein basicly one kind of quantal vortices in a viscous fluid with interpenetrant waves and currents. A system of several separately observed types of interactions directly requires turbulent vortical quanta with a number of consistent sub-gyres spinning at constant fairly uniform rates within resonantly coupled, distinctive frequency bands. Directionality of interactive gyre couplings requires separate coupling modes off the sides and conic bases. Experimentation with symmetric cone pairs in the most common consistent fluid, water, demonstrates (Fig. 2-10) an exact quanta form fit.

Here each type of current mutually attracts forcefully any co-directional currents and repels with frictional pressure the contra-directional current by the Bernoulli effect (Bernoulli, 1738). In addition to the spiral waves of about six bent and tapered vortices, each spiral in the turbulent Gyre Diameter (GD) area of this spheroid is jacketed by yet

QUANTA REQUIRE A TURBULENT CONIC VORTEX

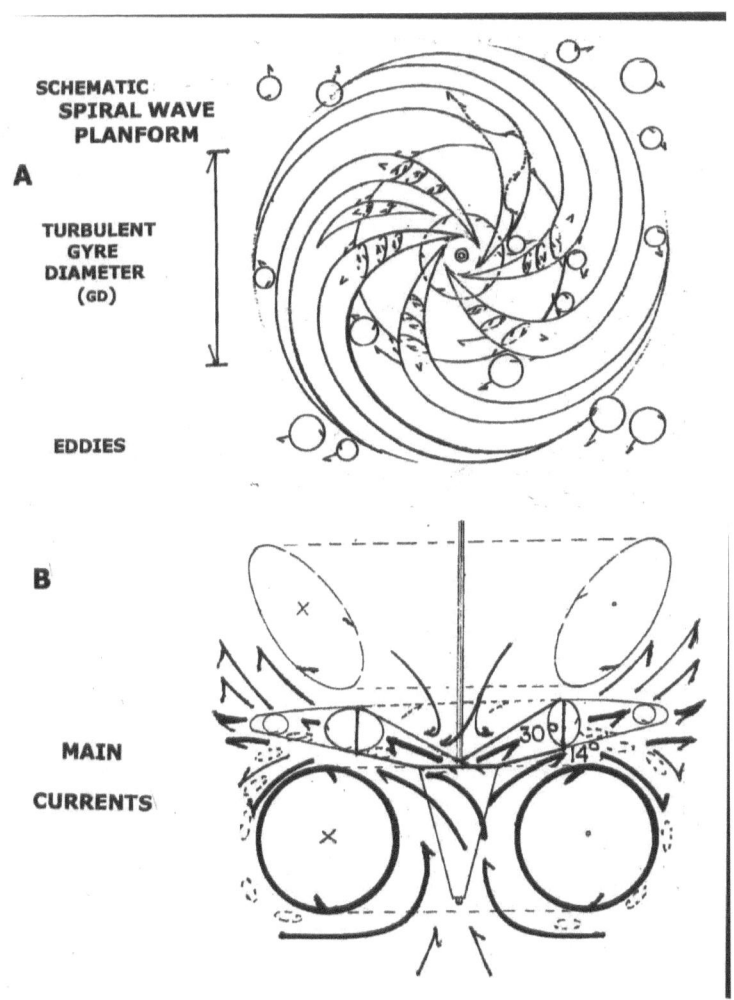

Fig. 2-10 A Centrally Driven Turbulent Conic Vortex with Toroid Rings. A. View of Spiral Wave Planform in Lab Water. B. Axial Cross Section. In B the boldness of flow arrows indicates measured current flow velocity. In A spiral waves fade into out-going smooth circular waves that die out with travel. (See text.)

another (almost fractal) coat of smaller out-going sub-spirals wrapped halfway around each large main spiral, but not on its bottom where the large spiral is in direct viscous frictional drive contact with the very thin boundary layer of water on the base of the unpolished wooden cone and then with the side outflow. There are also other broken wavelets between these sub-spirals around each large spiral wave. The result is that out to the GD it is not possible to see clearly through the turbulence with the eye. (Instead, ordinary walnut meat fragments with the distinctive broken shapes of their dark outer coating are traceable for continuously changing velocity vectors in each type of wave in short-exposure motion picture frames with constant frame rates between 30 and 100 frames per second. Walnut bits have about the same density as tap water.)

At the GD near the 45° trail angle of the main spirals the disk of strong turbulence abruptly subsides into smooth outer spirals which further out fade into very smooth circular outgoing waves that fill an outer disk volume of water. Much energy of the turbulence goes at the GD into the viscous frictional generation of large numbers of small, fairly similar, right-and-left-handed eddy gyres, which drift away and mix into each larger circulation as an additional energy storage until they decay and are continuously replaced. Reliable spiral wave plan data are obtained with the top of the spirals about a base diameter underwater. This spacing will exclude the weak upper toroidal rings. If the vertical drive axis is raised until the GD disk comes out of the water, it has sufficient internal momentum to raise the turbulent volume about a centimeter or a half-inch or more above the surrounding water with its smooth waves, and the abrupt edge of the GD turbulence is very clear. Inside its currents Bernoulli mutual forces are very strong.

Fully submerged right angle drives with long, calibrated pendulum mountings enable measuring flow velocity and mutual force between two gyres in the nominal plane of their axes. In that experimental plan the flow patterns below and above the spiral wave disk are very clear for good data, including the swarms of eddy gyres. These are seen to drift axially along the cone sides in the accelerated laminar flow which shoots out at 14° beyond the cone base in a lateral spiral under the spiral wave disk to drive a strong lower toroidal ring, a cone point-thrust jet drive, and a weak outwash. Later chapters give data, equations, and continuous scaling (from modeling recorded weather data on observed tornadoes and a strong hurricane driven centrally by heat of water condensation, with gravity and Coriolis effects) down to predictedly observable, if not already effectively well observed, gyre pairs within sub-nuclear particles of many types. (In this the MQP paradigm is possibly the lineal, though initially uninformed, descendent of a distinguished 19th Century physicist's forgotten attempt to model the atom with a single vortex. Bernoulli mutual current forces were well known.)

The outer flow patterns of these turbulent conic gyres are quite flexible, with increasing complexity of change with approach within the spiral diameter until, within the GD spacing of axes, paired drive cones are usually pulled into contact and stay forced together in the circulating water. A wide variety of cones, including metal and plastic cones, and fluids show the same effects and provide data that reduce to the scaling equations. The 30° right circular wooden cones of about 15 cm or 6 inches length (made of unpolished, but smooth, maple wood in a school wood shop) are the most widely informative with the readily obtainable electric drives, calibrated test pendulums, water tanks, and fluid viscosities and densities (where dissolved sugar and temperature are quite effective. A good physics and chemistry handbook is useful.)

Before any measured data are taken, the structures of some of the sub-gyres in Figure 2-10 indicate directly the kinds of mutual force interactions for which they are responsible. The strong and accelerating flow along the sides of the cone, aided by the inflow from the strong toroidal ring around the sides, is definitely involved in producing the QM strong force. In fact the toroid itself contributes to that force between pairs. A further component of it may not be immediately obvious until testing raises awareness of the presence of another influence; there is a distinct intake of fluid from the surrounding medium around the point of the cone into the side accelerations, then out from the vicinity of the cone base at the 14° angle beyond the flat base. That is an inefficient, but definite, form of the classic jet propulsion process yielding a point-thrust force on the cone [which must be subtracted from many other measurements to clarify their sources of force, and then totaled in net strong attractive force with cone points Usually in (PI).] The opposed outwashes of that laterally ejected fluid fraction between two adjacent gyres of any rotation handedness are sources for the QM weak force (that is difficult to separate from a net sum-of-forces measurement unless relative gyre positions are carefully adjusted.) Since orbiting pair cone bases are Usually faced outward on a sphere (See electron-positron chapter), the hemispheric directivity of the spiral wave disk off the cone base is clearly the source of electric charge force interactions. And the shedding of eddy gyres under the GD margin on the 45° slope of maximum bend of the spiral wave sub-gyres is the highly appropriate source of initial very weak force of gravity coupling between multiple frictionally generated eddies in the lowest band of rotation frequencies. These couple well to the local medium for inertial proportionate drag (also a true stability force.) When two conic gyres are interacting, if they are of opposite rotation handedness at their bases, then their flow currents in the sectors facing each other are essentially parallel and will generate few additional eddies between them compared to the case when they have the same hand rotations and all their interacting flows are opposed with high viscous friction to generate large quantities of eddies. The effect is heightened at an exponential power of the number of vortices interacting within a particle due to the severe turbulence around the accelerating base portions of the electric-charged spiral waves within the GD. All these waves are a continuous radiation of the naturally formed and conserved gyre quantum with unknown source. There are further factors of very close match of structural forms between these vigorously driven turbulent conic vortices and the necessary requirements of the PDG and QM observed activities of the particles. The data and equation chapters will emphasize most of them. In structural form alone it is clear from the number of equivalences that these are not mere modeling check points of particles, but are detailed and informative structural congruences over extremely large ranges of scale of forceful natural gyres (from the quanta of particles to galaxies) to be studied closely throughout the hidden simplicity of the Micro-Quantal Particle Paradigm as an instructive working auxiliary to particle Quantum Mechanics.

REFERENCES OF SPECIAL INTEREST
(Full references, tables, and appendices are in Volume 2.)

BERNOULLI, D., (1738) Hydrodynamics

HOWARD, F. E., JR., (2010) Striking Correspondences of the Micro-Quantal Particle Paradigm & Quantum Mechanics, www.particlephysics.info

SECTION III

Quantal Vortex Formation Data & Structural Equations For A Particle

Chapter 3

Turbulent Conic Vortices -
Lab Structural Gyre Data, Flow Ratios, Macro-Scaling Equations

SUMMARY: This is the first of two chapters primarily on newly measured lab-scale data of centrally driven turbulent conic vortices as the century-long-sought and now very necessitous quanta within the SM and PDG empirical sub-nuclear particles. Except for being at empirical laboratory scale in water as the simplest and most nearly ideal lab fluid, these gyres are otherwise fully congruent with the micro-vortical smallest quanta of the Micro-Quantal Particle Paradigm of the sub-nuclear particles herein. As set forth in prior Chapters 1 and 2, these micro-quanta, the basis of this particle paradigm, are derived by inherent necessities of the data of the accredited, empirical Summary Data Tables of the international Particle Data Group Report (published in full biennially by Lawrence Berkeley National Laboratory at pdg.lbl.gov and by a major physics journal.) These newly measured lab data on vortices and empirical equations derived from the data are briefly validated here by macro-scaling up to correlate numericly with the similar, centrally driven, violently turbulent, and forceful vortices most often observed by human beings (especially in the United States) in the fluid of the intensively studied atmospheric air from the earth's surface to the outer mesosphere. The first of these two empirical lab chapters is limited to measured flow velocity structures of these lab (and atmospheric) vortices. In the next chapter measured mutual forces of these highly turbulent lab vortices are reduced to empirical equations that are also briefly validated numericly by some direct correlations with violent weather vortices from surface to mesosphere. For instance, in its landfall Hurricane Ivan brought 23 recorded tornadoes overland in one most distant outer rain-band. The scaled turbulent gyre diameter equation developed in these two chapters defines Ivan's outer limit of tornado violence within five percent accuracy. Thus the lab data equations in these two chapters are ready as natural phenomena for generalizing empiricly to scale down next to the micro-quantal vortices required as natural component parts by the PDG particles in a necessary space medium.

    The previous chapter ended with a necessity for exploring uniform conic vortices that couple between any adjacent pair of them with vigorous multiple forces, even in surrounding conditions that are otherwise relatively quiet. That requirement eliminates the common drain vortex that is driven conicly near the fluid surface in typically laminar quiet flow by static external fluid pressure. Having numerous vortices near each other in many particle cases further emphasizes that the drain vortex is not the gyre type of interest. Neither does the fairly common vortex that is externally driven by adjacent differences of momentum along the edges of one or two currents that tend to set up gyres along a line appear to be suitable. Only gyres that are centrally and strongly driven to multifaceted turbulence by some form of smooth rotary conic impulsion and are completely surrounded on their exteriors by much less disturbed fluid are appropriate for the previous chapter's needs.

    Considering natural vortices for a suitable centrally driven conic vortex model at once suggests the tornado. It has strong forces that could interact well. It is centrally driven vertically by a heat engine up-flow combined with the rotary twist of geo-gravitational Coriolis effects (sometimes called a rotary force with respect to local coordinates.) Numbers of tornadoes may appear distributed along the irregular line of a front, but they may also appear in groups inside a single thunderstorm, or in a large group in a number of large storm cells inside a hot and humid high over the Central US, or in the rainbands of a hurricane at landfall. In fact, a hurricane itself is not unrepresentative of this type of vortex, though compressed vertically to a frustum of a conic drive by the relatively shallow depth of the lower atmosphere. Or one of the giant storm supercells is highly representative of the type, though its strong inner rotation is hidden within a huge outer cloak of only marginally developed cloud. But rotational turbulence and spiral wave outflow around the upper conic base of the central drive, with inflow into tip-to-base feed currents, all three of these gyres have in plenty. And multiple tornadoes have been observed interacting around a single thunderstorm. (Forbes and Bluestein, 2001) All of these great vortices have the same causal chain of quasi-fractal multiple natural actions that automaticly adapt somewhat differently only because of differences in drive scale and in the local fluid medium (gaseous fluid in this case, with many small airborne droplets and dust particles.)

Good correlation with these varied storm structures at large scale would effectively validate the natural applicability of any lab-scale vortex demonstrations and detail measurements made under controlled and repeatable lab conditions. Such validation over wide scales would confirm the general natural chain of multiple vortical actions at all scales, subject only to adaptation to the general factors of the local medium. In the particles the medium is called space. Fortunately, for interstellar space at cosmic scale the PDG has already found it necessary (Amsler, 2008, etc.) to accredit some characteristics of space. It is sufficient for scaling between components of particles to make reasonable assumptions of how the cosmic medium must extend into the fairly similar ratios of great emptiness within the subnuclear zone. Fortunately, again, it has already developed implicitly in Chapter 1 and explicitly in Chapter 2 that the important interactions in this zone are few and are far below the emergent complexities of the electric and (its subordinate) magnetic fields. The important interactions occur in the causes of those fields of force from the empirical fluid mechanics that is inherent in the Charge-Mass Power Law and its continuing consequences in the necessities of centrally driven turbulent conic vortices.

This is the first chapter on numericly measured lab vortices, including briefly correlations with very energetic cyclonic weather formations from the surface to the mesosphere. This introductory chapter is limited to vortex currents. The lab experiments are in water with submerged and surface conic vortices driven by central cones (like weather heat engines) to a high level of rotational energy and turbulence for measurement of both the anatomy of water parcel velocities (in this chapter) and also the forces developed by single and dual gyres (in Chapter 4.) Initially intended as preliminary screening trials for later research elsewhere, the experiments had unexpectedly informative results and were taken as far as was practicable (without significant funding.)

An initial literature search found much complex analysis of vortices for other cases. Evidence was incomplete on secondary and tertiary flows in turbulent gyres and on gyre forces. The rarity of such data contrasts with the large body of literature on vortices that are not as directly relatable to gyre forces, such as laminar surface gyres in liquids, 2D gyres without secondary flows, and isolated cylindric or toroidal gyres. The exploratory data herein show significant direct correlations with the strongest types of weather vortices, such as hurricanes, supercells, and tornadoes, and add to the scant empirical data on forces and secondary to quaternary flow structures in turbulent conic gyres at all prior data scales.

Many natural vortices are axially short, conicly tapered, and centrally driven by a heat engine (plus Coriolis momentum and pressure effects.) The gyres often naturally develop spiral waves, in a disk of centrifugal flow away from a broad upper end (or cirrus cloud cap) and in-flow into a smaller lower end (or eyewall.) This may be only the lower hemisphere of a fully immersed gyre with a disk-coupled hemisphere above it. There may be toroidal rings around hemispheres completing a full sphere of conicly driven activity in one combined vortex.

Such lab gyres were viscously driven to high spiral-wave turbulence in city tap water at 20° C by a spinning, unpolished, right circular maplewood cone of 30° whole cone angle and 7.94 cm base diameter, on a 6 mm shaft with a cap nut (acorn nut) replacing the tip of the cone. For flow velocity measurements (herein) the cone was directly base-driven by a 10mm (3/8 inch) electric hand drill, in CW (clockwise) or left-hand base-referenced rotation in the plan view. (For a penetrating screw or a nut going onto a screw that would be known commonly as right-hand rotation because the opposite side from the viewer's body is turning to the right.) Angular rate was optically measured under load for peripheral (tangential) velocity V_P of the cone base, with less than 4% maximum deviation from the average 872.9 cm s^{-1}. This provided known velocities on a simple boundary layer surface of viscous rotational flow in the main body of the vortex. (In force tests changes of motor drive gears 16:1, viscosities 4:1, cone areas 5:1, etc., enabled general scaling equations.)

Flow trajectory and velocity vector measurements were obtained on fully immersed and near-surface gyres by stop-motion video (at 30 frames per second, from 2 meters distance) of neutrally buoyant and individually distinctive, irregular particles in the 0.25 cm size range (preferably broken walnut meats) against a 5 cm white grid on black, 20 to 25 cm behind the center of action. This yielded definite primary, secondary, tertiary, and some quaternary gyre flow velocities in turbulence.

Water surface distortions were minimized by a calibrating grid on the tank bottom. Without corrections for out-of-plane and curved paths, measurements of lower limits of apparent velocities in cm sec^{-1} and velocity tracks in the scaled plane of view (Figs. 3-1, 2, 3) are shown at median locations for sample bins, with estimated one sigma errors of about 10% and 5°. However, instantaneous velocities are continuously changing in such turbulence, so that in one frame interval fast particles are often decelerating in 50% slower regions due to loss of shear and pressure wave momentums to surrounding water. It is estimated that true peak velocities must be at least 50% higher than measured averages over frame intervals, but comparative speed ratios are less affected. The sizes of flow structures are not affected by the velocity uncertainties; repeatedly measured dimensions are at ±5% error.

Figures 3-1, 2, and 3 summarize in scaled plots the interacting flow structures in turbulent lab vortices. Figure 3-1's six subfigures a through f introduce flow current velocity data in centimeters per second by steps for clarity, starting with viscous coupling from the drive cone to water at the drive cone's surface in Fig. 3-1a. In Fig. 3-1 each subfigure panel a through f has the fully submerged vertical view directly above its plan view below it on the page. Below each plan there is a partial vertical view of a similar vortex in which the base of the drive cone is submerged beneath the surface to the diameter of the cone base so that the surface ripples may be seen above the disk-shaped volume of full turbulence out to the Gyre Diameter (GD.) At the GD the turbulence abruptly subsides to smooth continuations of the spiral waves, and the continuously fed energy of turbulence converts into the energy of the viscous eddy subgyres drifting through the currents of both the upper and lower hemispheres of the full vortex circulation as shown in Fig. 3-1e. Subfigures 3-1a, c, and f are full views of the vortex at its stage of development, and the other three subfigures are half views. Figure 3-2 combines the flow velocity data for planforms of spiral wave cores in the highly turbulent disks of either near-surface or fully immersed vortices, and for the cone-base portion of near-surface elevations. Figure 3-3 shows the elevation of a fully immersed, sphericly complete, turbulent conic vortex with its flow velocity measurements in centimeters per second (cm/s or cm s^{-1}.)

Right-half elevations are tank side view (with internal structures in dashes.) Left-halves are partial cross-section through the cone axis, with the cone surface and flow trajectories near it. Structure contours outline significant changes of velocity. Particle tracks on structure lines or dotted with arrows show velocities projected in the view plane in cm s^{-1}. Near-surface gyres driven at cone base diameter below the surface approximate fully immersed vortices. Angles use degree notation since it is compact.

The Figure 3-1a column of three views displays the primary stage of viscous coupling of momentum from drive cone surfaces to adjacent fluid (correlating to concentration of buoyant heat release, Coriolis effect, and altitude ceiling for Hurricane Ivan's driving eyewall which is a frustum of a cone.) Fluid acceleration on the cone sides is smooth in a nearly laminar layer; on the base there is lateral jerk (rate of change of acceleration) in a thinner turbulent layer. There is a conic boundary of shallow mixing between V_S and V_B flow vectors as they spread centrifugally from the base tangential velocity of V_P = 872.9 cm s^{-1} (like scaled cirrus tops of supercells):

$$\frac{|\mathbf{V}_B|/|\mathbf{V}_P|}{|\mathbf{V}_S|/|\mathbf{V}_P|} = \frac{0.349}{0.262} = 1.33 \tag{3-1a}$$

where also $\qquad |\mathbf{V}_S| = 1.47 |\mathbf{V}_{SU}| \tag{3-1b}$

and V_{SU} is the mean upward velocity along the side at 156 cm s^{-1} (as in eyewall convection.)

The near-surface gyre (Fig. 3-1a bottom) at standard base immersion repeats the immersed vortex. The plan view of schematic particle trajectories in the base primary flow neglects its fine turbulence. Turbulences (Fig. 3-1a-c) die abruptly at a definite GD (shown with tangential velocity though the mean flow is at 45° to the radius.) Figure 3-1b (left view only) plots a core rotary component of a single spiral wave and a partial

69

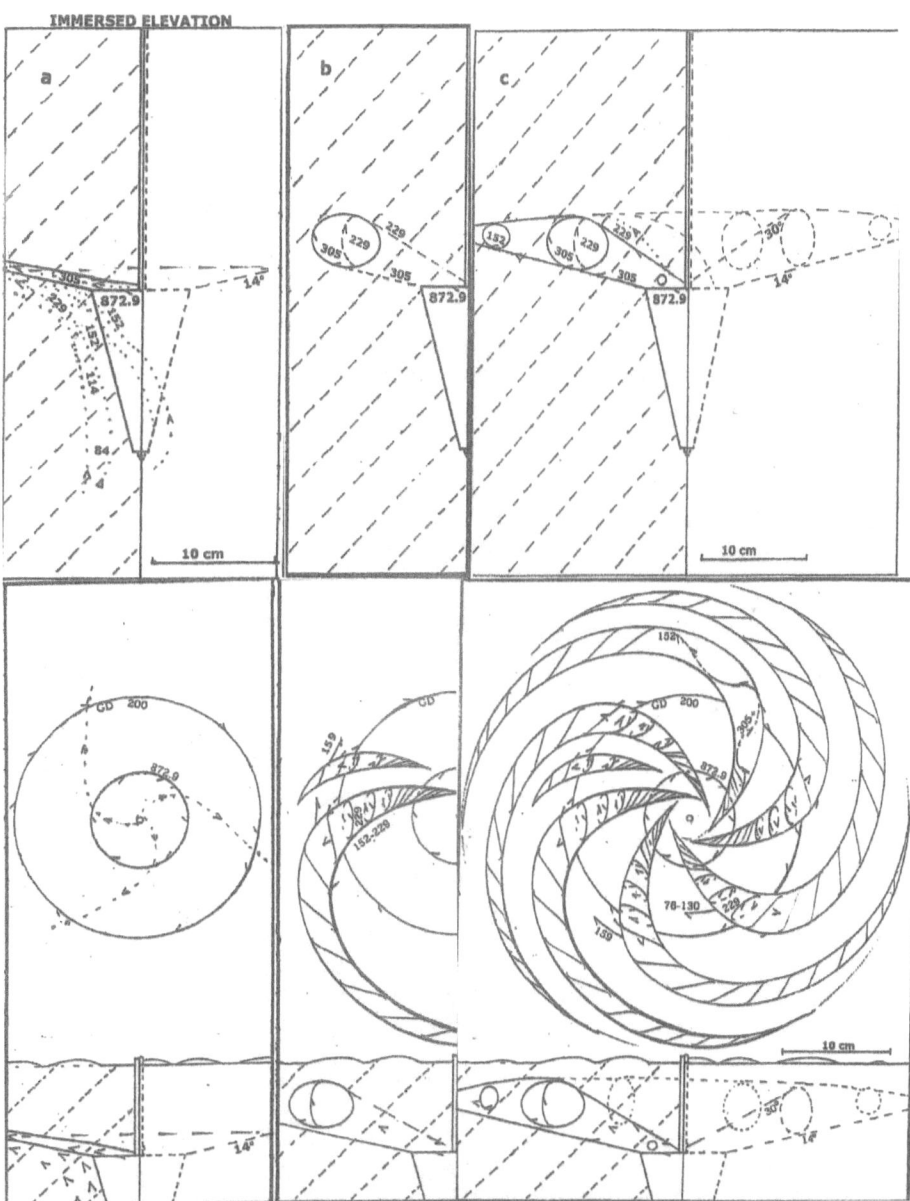

Fig 3-1a to c Step-by-step viscous coupling of lab vortex drive from a 30° wood cone to scaled flow structures of immersed and near surface turbulent conic vortices with parcel velocity tracks in centimeters per second. These centrally driven lab flow structures scale up four to six orders of magnitude to storm supercells, tornadoes, and hurricanes with tornado bands out to the GD turbulence limit. (See text.)

interstitial wave as tertiary parts of the secondary spiral-wave disk of gross centrifugal turbulence (Fig. 3-1c) coupled viscously from the base primary flow sheet (Fig. 3-1a). The vertical section shows the wave core thickening by $\pi/11.25$ (16°) above the $\pi/12.82$ (14°) upslope of Fig. 3-1a, and also the centrifugal velocities of the bottom and top of the spiral core. In the core diameter peak at the GD, a cut-away displays the vertically rotary core component V_R (model-related to convective surges in Ivan's rain bands in an expanded version of this chapter in Volume 2):

$$|V_R| / |V_P| = 0.262 \tag{3-1c}$$

The tertiary spiral wave core is a conic vortex itself [re tornado forces (Howard 2009)], embedded in the base flow cone and wrapped to the GD limit by vibratory outer shells (not shown) of quaternary subspiral-gyre fragments that conflict viscoelasticly in mixed pressure and shear wavelets. At the GD the turbulent momentum couples to the surroundings more rapidly than it is added from the base flow; the peak turbulence abruptly subsides; the decelerating core bends more than $\pi/4$ (45°) away from the main radius; and to 2 GD the spiral wave changes into a circularly spreading, smooth wave system. At cone $V_P = 872.9$ cm s^{-1} in 20°C lab water, GD (Fig. 3-1a-c) is 2.56 d_c (cone diameter) or 20.3 cm average, with ±1 to 1.25 cm impulsive fluctuation at each spiral. [Scaling GD Eq. (3-2), defines the outer limit of Ivan's violent tornado band.]

In Figure 3-1c about six spiral-wave and interstitial cores (with subspiral wavelets not shown) create the strongly visible turbulence of the subplanar spiral-wave disk within the GD. The conic inner tips of the waves and the smaller wavelets constantly vibrate vigorously about 0.5 cm. One wave plots the typical corkscrew core trajectory of water particle velocities to a peak radius of a spiral at about 1.5 cm at GD. The entire disk plan rotates at 2π-4π s^{-1} (radians per second) or 1-2 revolutions per second for tangential velocity V_D at GD. [These upper spiral wave cores, driven by the relatively moving surface of the cone base, also model details for rotating fluid over a relatively fixed surface.]

The flow off the base and sides of the cone slopes at 1 in 4, or $\pi/12.82$ (14.04°), above the base plane (Figs. 3-1a & c, & 3-3), giving this cone-shaped centrifugal flow its upward flow component of momentum (Eq. 3-1d). The wave disk's turbulent volume within the GD fills the space marked 14-30°, average $\pi/8.2$ (22°), above the plane of the drive cone base. At an average centrifugal wave velocity V_W of 267 cm s^{-1} this mass of fluid increases an upward component of added momentum (and reaction force):

$$\mathbf{P}_{DU} \propto M_W \mathbf{V_W} \tan(\pi/8.2) + M_S \mathbf{V_S} \tan(\pi/12.8) + M_B \mathbf{V_B} \tan(\pi/12.8), \tag{3-1d}$$

where the fluid masses are those in the scaled centrifugal flows above the volume centroid of the drive cone and between the perimeter of the cone and the GD [re tornado forces (Howard 2009) and Chapter 4.]

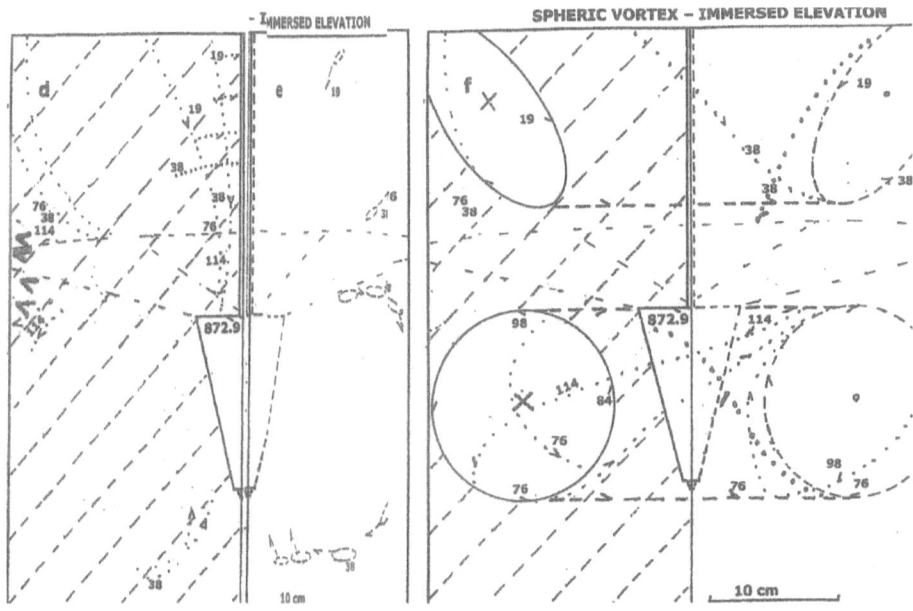

Fig. 3-1d to f continue the step-by-step viscous coupling development of the prior Fig. 3-1a to c in the elevation views above and the matching plan views below (on the next page.) In this sequence the energy of the flowing current spreads throughout the single vortex involved in its various current forms. The current velocity numbers shown, however, are the fully developed current values in centimeters per second (cm s^{-1}) as before. (The lower half of this figure continues on the next page.)

    In these wave structures the most outstanding feature is the spiral wave planform in the thick disk of turbulence attached viscously to the base of the drive cone. It is the force couplings of these spiral waves through the long known Bernoulli effect, even though mixed with other flow wave influences out to very large distances, that generates the force of charge between vortices. This coupling is aided by the resonant nature of the waves in their comparatively high frequency from the turbulence in which they are involved. Whether the charge effect is positive or negative depends entirely on whether the vortex spins to the left hand or to the right hand direction on its base with the thumb of each hand pointing away from the cone. With two adjacent cones rotating in opposite senses, the spiral waves from the two cones in their strongest flow away from the drive cones will be flowing in parallel in most cases, particularly if the cones and the associated gyres are arranged symmetricly to each other. This is almost always the type of arrangement in the bulk of the cases with any two gyres inside one particle sphere, especially with the coaxial pairs of gyres that are always forced into remaining coaxial. As Bernoulli discovered between two and three hundred years ago, two parallel fluid flows in the same direction at about the same energy level will experience

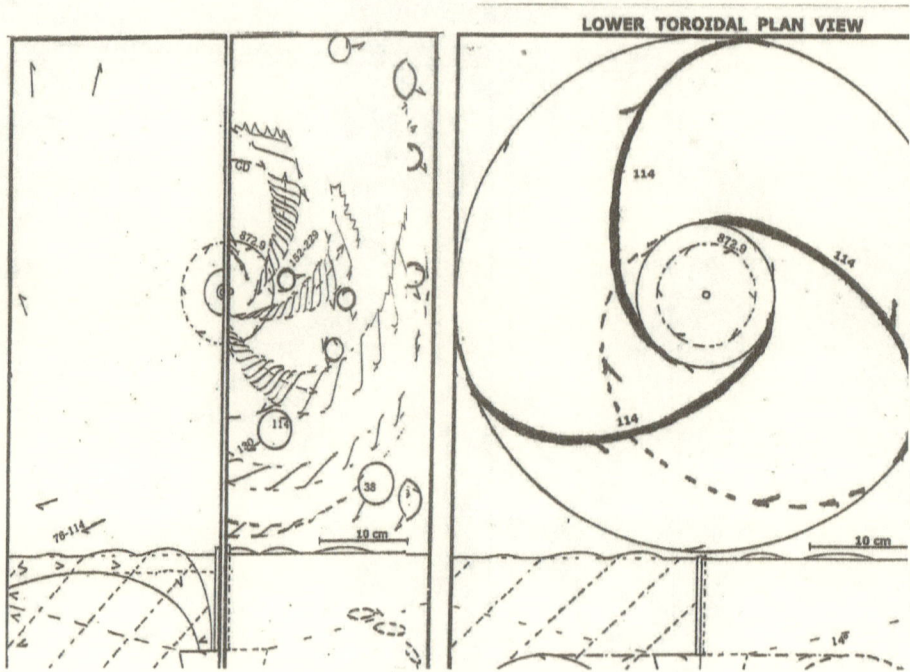

Fig. 3-1d to f Lower Half Continued from Above. This shows above the individual plan velocities of the outer feed into the lower toroid, the spiral wave feed of energy into the small eddies that escape into the toroid, and the velocities in the upper surface of the lower toroid itself, all of which match the cross sections in the upper halves of the figure in the page before. At the bottom of this half of the figure is shown a partial cross section of a near surface gyre with the cone base at one base diameter depth.

a drop in their internal pressure effectively in the directions toward each other at right angles to the flow direction. That yields a force pushing the flows together cohesively, and also pushing the two gyres together in the case that they are rotating in opposite senses. But if the two symmetric gyres are rotating in the same sense, then their flows in contact will be in opposite directions with a viscous build up of pressure pushing them apart. With numbers of on-average symmetricly arranged gyres the pressures in either predominant sense can be impressive, or if the charge is neutral, no pressure at all.

It is also the spiral waves that generate most viscous eddies, which appear as mass. This mass of eddy energy is generated initially in each vortex at the points at which the spiral waves bend through 45° from the radius and cross the GD circle. The GD limit is at that radius from the cone center because the spiral waves lose much of their turbulent energy (into the eddies) and become smooth spirals from there outward. When many gyres are present as quanta in a particle, more eddies are generated between their flows by viscous friction in accordance with the Charge-Mass Power Law.

If cone drive stops, initial numbers of spiral waves (Fig. 3-1c) settle to six, but the spirals briefly continue advancing as before. Then they reverse, circling against the established primary vortex with a wheel-like counter-drive from the prior internal rotating momentum of each spiral until collapse. In reversals the waves change from full spirals to diametric bars with spiral waves trailing from each tip [correlating to upper layers of a filled storm center vortex {or barred spiral galaxy forms {Longair 1996}, implying dissipation of galactic momentum drive or a past fringing collision with another galaxy.}] Here wave reversals confirm core storage of high tangential momentum per cc P_{IW} near the GD at $V_{IW} \geq V_R$ (Eq. 3-1c), a lower limit [as with surging convective momentum from Ivan's surface spiral waves (See expanded version in Vol. 2)]:

$$\mathbf{P_{IW}} = M_{IW}\mathbf{V_{IW}} \geq 229 \text{ gm cm s}^{-1}\text{cc}^{-1}. \tag{3-1e}$$

Figure 3-1d (in a left-half section) shows inflows to and outflows from the primary drive cone flows (Fig. 3-1a) with curvature into large amounts of recirculation (Fig. 3-1f.)

Figure 3-1e (in a window view) shows typical rotational velocities, core sizes, and tracks of small quaternary eddies. These subgyres are conic in shape, but decay slowly. Their steady lab population throughout the main vortex is constantly shaken by turbulent disk reactions, disrupted, and renewed by centrifugal interactions of tertiary spiral-wave cores near the GD, with rotating senses depending on origin at a core's leading or trailing edge. Eddies are ejected into the outer shells of toroids (Fig. 3-1f), largely in lower hemisphere recirculations [where presence of eddies would contribute pre-tornadic rotations to Ivan's spiral-wave surges and to Fujita's multiple vortices in storm-cell tornadoes (Forbes and Bluestein, 2001).]

In a top elevation (Fig. 3-1f), separated by space for the spiral disk and primary side outflow driving them, core outlines of the wave-pulsed upper and steadily strong lower toroids summarize internal velocities of spiral corkscrew particle tracks around parallel circular axes. The middle view tracks typical particles on the top of the lower toroid. This toroid core and surrounding flow actually occur as multiple, flexible, spiraling shells of particle and eddy tracks which both merge with the core on the top outside and peel away carrying eddies in recirculations from the toroid core toward the cone to spiral into the primary drive flows (Fig. 3-1a) [like Ivan's spiral rain bands; or just within the GD, like Ivan's right-front tornado band; or in the upper toroid, like mesospheric downdrafts above storm supercells (Howard 2009).] The toroid also peels away into fluid outside it [transferring forceful circulation power outward (Howard 2009).] Observed velocity tracks V_T in the outer sides of the lower toroid are vector sums of the vertical V_{TV} and horizontal V_{TH} components, with a speed ratio to the cone V_P (that is scalable to Ivan):

$$[|(\mathbf{V_{TV}} + \mathbf{V_{TH}})|/|\mathbf{V_P}|] \approx (|\mathbf{V_T}|/|\mathbf{V_P}|) = (114/872.9) = 0.13. \tag{3-1f}$$

Figures 3-2 and 3-3 show the scaled locations of the combined flow velocity data from the different flow structures (Figs. 3-1a-f) in the planform (Fig. 3-2 upper panel) of the fully immersed vortex (Fig. 3-3) [or the hemispheric surface vortex (Fig. 3-2 below.)]

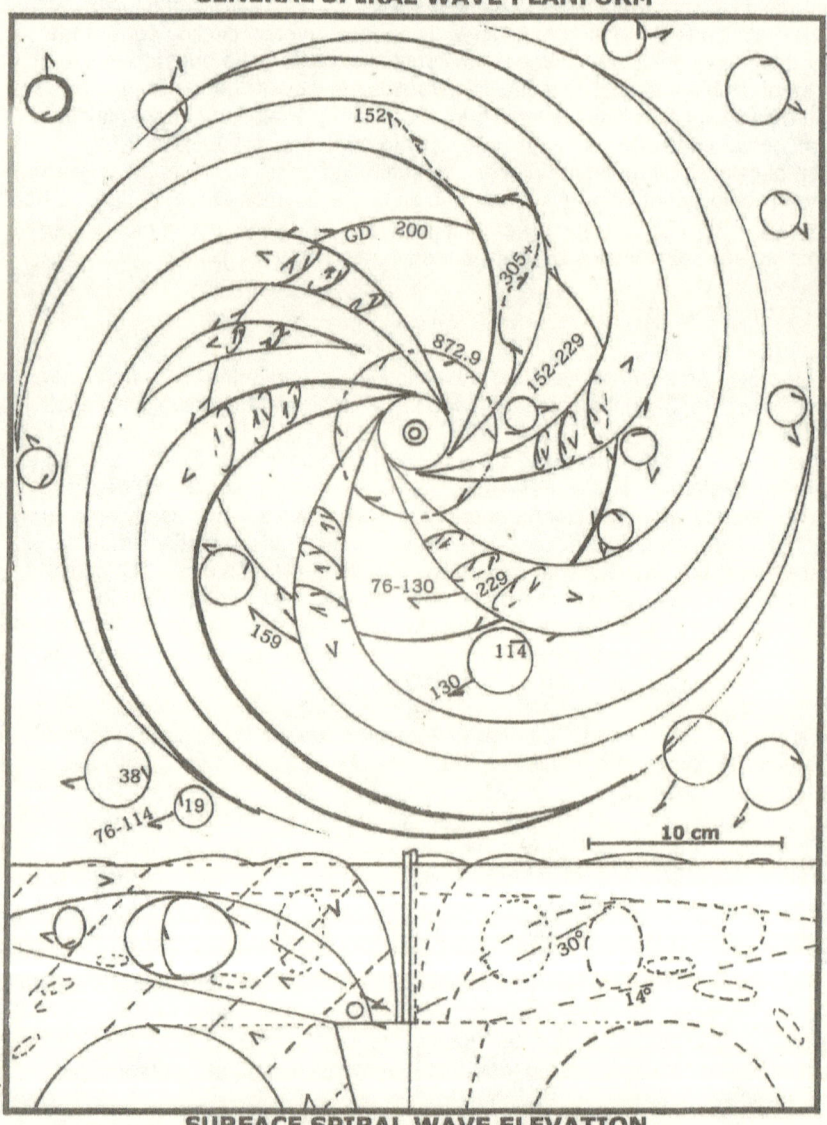

SURFACE SPIRAL WAVE ELEVATION
(PARTIAL VIEW of HEMISPHERIC VORTEX)

Figures 3-2 and -3. These figures compile the flow velocity data from the different flow structures (Figs. 3-1a through f) of a turbulent conic vortex, in both the immersed vortex (Fig. 3-3) and the equivalent surface vortex (Fig. 3-22 bottom), in their relative locations as they appear interlinked together in the lab. This completes the sequence of interacting structures in vortices standing free of boundaries. (See text for details.)

FULL SPHERIC VORTEX – IMMERSED ELEVATION

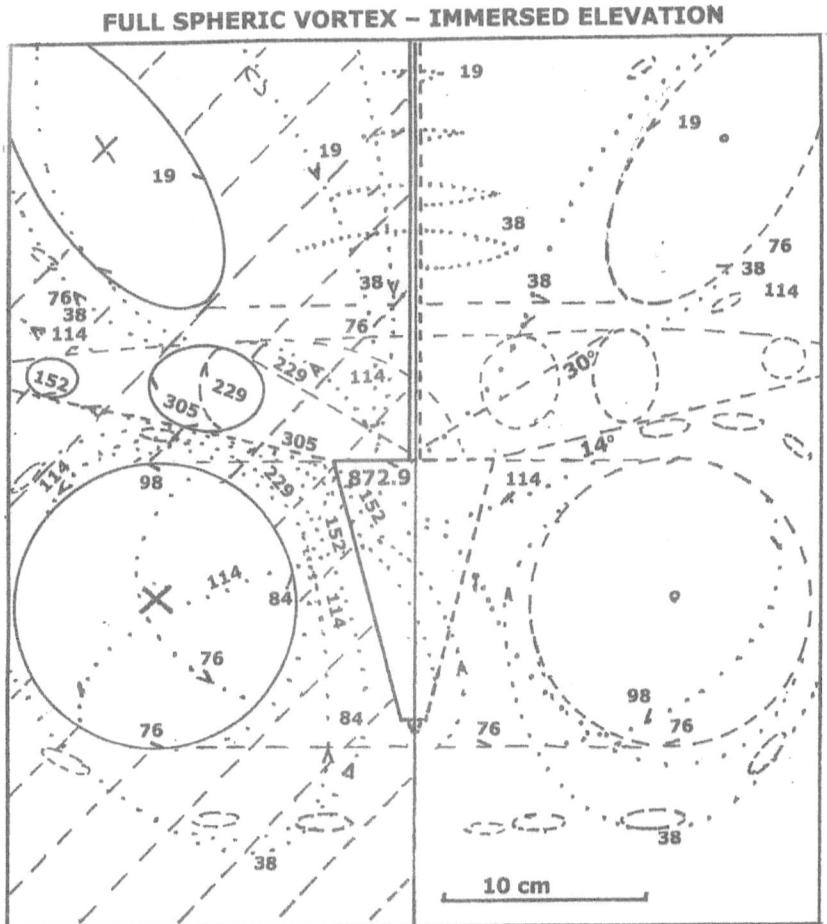

Figure 3-3 Conic Vortex Currents Cross Section Elevation. Flow rates in centimeters per second.

This completes the sequence of actions of interacting structures in a single sphericly active, centrally driven, turbulent, immersed, conic vortex free of boundaries and other nearby similar vortices. However, there is one important limit on the actions that has a very useful equation. The GD shown in the figures of this chapter is the mean plan position of the edge of the disk of concentrated turbulence above the base of the drive cone. This edge vibrates outward slightly as each rotating spiral wave passes above any point in the circular plan of the edge. But the edge itself is clearly visible as the base of the cone is held at its diameter depth in the water or is moved a few percent closer to the surface. This gyre diameter has an empirical equation that is derived in Chapter 4 as a scaling reference for the gyre force equations (or more fully in an expanded text in Volume 2 and Howard, 2009.) The equation is also required here to complete the description of the vortex flow currents with data on where the turbulence ends and the viscous eddy subvortices arise in quantity in an isolated conic vortex. Accordingly:

$$GD = d_c \left[1 + e \left(\frac{\rho}{\eta} \right)^{0.6667} V^{A\sqrt{V}} \right]. \qquad (3-2)$$

where the unit for all lengths, including the gyre diameter and the cone base diameter d_c, is chosen such that the number for d_c is between 1 and 10. The same unit (which may be unusual) is used to calculate areas $A = (2 A_b + A_s) 100^{-1}$, A_b is the area of the base of the drive cone, and A_s is the wetted driving area of the sides of the cone. Then $V = V_P 1000^{-1}$, V_P is the peripheral velocity of the driving cone in cm s^{-1} at any scale, e is the base of natural logarithms, ρ is the density of the fluid in gm cc^{-1}, and η is the viscosity of the fluid in centipoise. (In the cases of weather vortices, the last two symbols are evaluated at scale height, where the density is the mean half of surface density, usually at a nominal 20,000 feet. A good handbook of chemistry and physics is useful. After the calculation the numbers for the two diameters are converted to the preferred conventional units for all other uses.)

To test the general empirical validity of Equation 3-2 it was scaled to an unusually well observed violent incident in Hurricane Ivan. This was a landfall event with an outbreak of 23 documented tornadoes (Watson et al. 2005) centered between Cape San Blas and Marianna, Florida, at about 370 km (200 n mi) northeast of the eye, which was initially at sea with 213 km hr^{-1} and 59 m s^{-1} (115 knot) winds falling to 204 km hr^{-1} and 56.6 m s^{-1} (110 knots.) when ashore near Mobile Bay. This outbreak occurred along a distinct rainband beyond an open space as the band moved well inland across Florida and into Georgia. Though most of the area is rural woodland and swamp, there were at least 6 deaths, and a number of homes and business buildings were destroyed. The most distinctive item about the 23 tornadoes is where they occurred, in the last significant outlying rainband of Ivan in its right front quadrant.

At the time the eye outer diameter in satellite images is 86.1 km (46.5 n mi), and the inner diameter (ID) is 34 km (18.5 n mi). The measured eyewall cloud height is 20 km. The mean diameter of the eyewall is 60 km ± 1%. A 30° cone frustum with this diameter at 10 km mid-height has diameters of 65.36 km at its 20 km high base (from satellite data) and

54.64 km at sea level. The scaled Ivan drive cone, then, is a 20 km immersed frustum of a cone. In 10 km units, the base area is 33.552 square units. The wetted side area of the frustum is 38.265 square units. Then A in the general scaling equation (3-2) is 1.0536. For V, assuming that the stated 56.59 m s^{-1} wind at the sea level diameter of the figurative conic frustum conserves angular momentum in rising to the 20 km altitude with 47.31 m s^{-1}, then V becomes 4.731. Since density and viscosity vary with altitude, a suitable estimate for weighted effects is the mean of handbook values at the nearest altitude (7 km) to the scale height for half the sea level density, with a density-to-viscosity ratio of 0.036.

This scaling process yields a GD limit for highly turbulent (violent) weather in Hurricane Ivan at the stated time of 747.26 km with a radius of 373.63 km. The peak height of the Blountstown-Marianna tornado cell in the center of its rain band at that time is at about 357 km from the center of the eye, 17 km within the limit. No potentially violent portion of that rain band is more than 10% farther from the eye at 393 km. Then, the outer parts of this rainband are 5.1% beyond scaling (Eq. 3-2) over six and a half orders of magnitude from the GD lab data (Figs. 3-1a-c and 3-2), which itself includes ±5% with 7% peak variation at the point of passage of spiral waves through the lab GD (Fig. 3-1c).

This gives an initial validation to the coordinated empirical lab current equations and current flow structural data of this chapter for further employment in mutual force data on centrally driven turbulent conic vortices for scaling the required vortex micro-quanta components of the sub-nuclear particles. Further supporting information is contained in expanded versions of this condensed chapter in Volume 2 (and in Howard, 2009.) Empirical quantitative lab data are now made available on organized primary, secondary, tertiary, and quaternary flow structures in turbulent conic vortices of both the near-surface and fully immersed types. A flow structural basis is laid for formal equations, and quantitative scaling, of empirical force data generated within and between such vortices, over wide ranges of size, viscosity, and other parameters, in Chapter 4.

Like cyclones around central heat engines, lab gyres develop from primary viscous flows around a conic drive (like eyewall winds) with secondary upper and lower toroidal rings of spiral sheets (like rainbands) separated by a centrifugal disk of tertiary spiral-wave subvortices that generate quaternary eddies and vibratorily breaking pressure and shear wavelets out to distinctive gyre diameter (GD) limits of quasi-fractal turbulence. (like tornado bands.) The GD limit varies with central flow velocity over a horizontal scale factor of 10^6, six orders of magnitude, scaled upward in this case.

REFERENCES OF SPECIAL INTEREST
(Full references, tables, and appendices are in Volume 2.)(In Prep, See Website)

AMSLER, C., et al., 2008, Biennial Report of Particle Data Group, PL **B 667**, 1

HOWARD, F. E., JR., 2009: Turbulent conic vortices: Part 2, 3D gyre forces control tornadoes, supercell tops, & hurricane eyewalls.

Chapter 4

Turbulent Conic Vortices -
Lab Forces Source Data, Equations, Single/Dual Macro-Scaling

SUMMARY: This is the second of two chapters on laboratory scale experiments with centrally driven, highly turbulent conic vortices which, except for their scale, would be congruent with the required microquantal components of the QM and PDG sub-nuclear particles as described in Chapters 2 on necessary overall structure and 3 on the measured empirical current flow data and derived equations of these gyres. This chapter adds the lab measured empirical data and derived equations on mutual forces between primarily symmetricly arranged pairs of these gyres. For general force equations the initial conditions of gyres are widely varied in drive cone angles, cone sizes, and fluid viscosity. With pairs of vortices the exponential growth of viscously generated tiny eddies begins to show up in weak inertial and resonant gravity couplings in a lower frequency band. The eddies are true mass energy stores oriented evenly in all directions. Electric charge forces sum from the hemispheric coupling directions of the turbulent spiral waves on the bases of the drive cones. Strong forces sum from the cone point sides of vortices. Weak forces arise from the outwash that escapes between upper and lower toroid gyres. These overall vortices are clearly the great hidden simplifier of the quantum mechanics of sub-nuclear particle interactions with all the forces from a single vortex source. There are also in this chapter further brief validations of the lab data against empirical data at weather scale on centrally driven, violently turbulent natural vortices, as in Chapter 3. In view of these accurate matches with up-scale nature, this chapter makes these empirical force and flow equations ready for broad generalization to scale down in following chapters to the microquanta that necessarily compose the QM and PDG sub-nuclear particles and causally drive all their empirical forces, masses, actions, and other characteristics in a necessitated space medium.

This second chapter of new numerical data on turbulent lab vortices, adds measured lab forces and empirical force/velocity scaling equations to the flow velocity and momentum configurations of turbulent conic vortices in the prior chapter, pointing out briefly correlations with up-scaled, violent, cyclonic weather formations from the surface to the mesosphere for initial validation of equations in natural applications on the broadest base. These researches originated as lab experiments in water with fully submerged and surface conic vortices driven by electrically powered central cones (like weather heat engines rotated by Coriolis effects) to high rotational turbulence for measurement of both the anatomy of water particle velocities in the (primary and secondary through quaternary) flow formations of gyres (Howard 2009) and also the forces developed by single and symmetricly dual gyres. The rarity of such data, especially on dual vortex mutual forces, contrasts with the large body of literature on vortices that are not as directly relatable to turbulently vigorous natural actions, but correlate only with quiescent fluid activity, such as laminar surface gyres, two dimension gyres without secondary flows, isolated smooth cylindric or toroidal gyres, etc.

For direct correlation between chapters the experimental vortices for force measurement are driven by the same motors as in Chapter 3 and arise around the same $\pi/6$ (30°) whole angle, right circular cones and test conditions except where otherwise stated and varied to obtain general empirical equations. As before, this provides, within each gyre, known angular and peripheral velocities on a conic boundary layer surface of viscous rotational flow, on which forces can be measured in summed resultant. The velocity-tracking particles and video camera of the prior chapter are replaced by twin force-measuring pendulums mounting the drive motors and cones at suitable angles. The force-to-displacement scales of the pendulums are recalibrated for each adjustment. For most direct analysis and reduction to equations over a wide range of variables that do match the 76.9% minimum to 81.25% maximum majority of quanta within particles in normal, stable matter, the force experiments on two gyres emphasize symmetricly balanced variables. [The other 18.75 to 23.1% of quanta, depending on the normal, stable elements involved, are in the quarks' linking orbits, one pair of quanta for each quark {with the definition that normal matter contains only up and down quarks.} Quanta in expanded linking orbits of quarks are not systematicly symmetrical with other quanta about a common orbit center, as are the majority of quanta in each particle, including the internally entirely symmetric atomic electrons.]

The principal aim of experiment is derivation from measured vortex forces of empirical equations that enable quantitative scaling of mechanical effects over a very wide range of many orders of magnitude in scale lengths. The general scale length is the gyre diameter (GD) of high centrifugal turbulence in spiral wave disks of centrally driven conic gyres. Because of the change in the nature of the necessary fluid for micro-quanta, some of the empirical equations must be re-generalized for scaling that far down in the next chapter. Axial point thrust (**PT**) on drive cones at GD scales must also be subtracted vectorially to find most other summed forces in each measurement. Amperages of gyre drive motors at 120 Volts indicate energy stored and transmitted by GD scaled volume of experimental vortex fluids (plus unmeasured gear friction losses.)

Force-Measuring Pendulums Used in These Experiments

Each 2 meter pendulum had at mid-length a clamped sliding angle joint between its two flat 5 cm metal bars for control of axial attitudes and water immersion below a fixed cylindrical metal 2.54 cm (1 inch) OD horizontal bar supporting perpendicularly the 0.48 cm (3/16 inch) cylindric axis of the pendulum. (The rolling error introduced by this axis was neglected as at least a factor of 10 less than the vibration induced by vortex turbulence.) At each pendulum foot there was a parallel/perpendicular mounting for direct or geared drive to a $\pi/9$ or $\pi/6$ (20 & 30°) whole-angle cone by a reversible 10 mm electric drill turning extendable 6 mm shafts driving cones under water. Resultant angles off vertical or horizontal of each drive shaft were measured to about 1 degree accuracy before and after runs and corrected for as needed. Voltages were checked and amperages were measured under the various loads, while optical rpm instruments were held near the drill chucks. (Power factors were not measured.) Pendulum deflections to 5° (about 20 cm) between cone centroids of volume (CV) were recorded along horizontal reference lines. In most experiments low friction guides along both sides of pendulums eliminated significant lateral or twisting deflections except when orthogonal lateral forces were recorded; twisting moments were not measured.

For fully immersed gyres with horizontal axes (for which vertical pressure differences necessitate crawl corrections), right angle gear trains were used below adjustable rotary collars at mid-pendulum. Cones were base or point driven. Combinations of reversible 1:1 and 2:1 right angle gears gave drive ratios of X/4, X/2,

1X, 2X and 4X to motor shaft rotation rates under the various loads. Motors were overloaded at 4X gear ratio, but brief testing was feasible. Test periods were held to 10 to 30 seconds per run to eliminate reflected-wave wall interference in fluid tanks.

On each readjustment, pendulums were calibrated with weights and a low friction pulley for radial or lateral forces up to 240 grams force at 10 cm deflection, with spot calibrations to 400 grams, at an estimated average accuracy of 1.0 gram. However, accuracy of force measurement on that calibrated scale was reduced by averaging constantly varying readings due to inherent gyre turbulence. Finally, 5% accuracy in force measurement is rare. Ten percent of the measurement accuracy is typical over the range of 50 to 150 gm, but increases gradually to about fifteen percent at the lowest range near zero and at the highest range near 300 gm (due to turbulence.)

Gyre Diameter Single Scaling Reference

The striking feature of strongly driven, near-surface or fully immersed conic vortices is the central gyre diameter (GD) disk of apparently chaotic turbulence in centrifugally ejected fluid from the base of the drive cone. The turbulence is organized above the primary vortex in about six tertiary, spiral standing waves of spinning water parcels (Fig. 3-2) in a secondary centrifugal disk covered with quaternary subgyres (not shown.) Each spiral wave core is itself a small vortex with its inner end nested with others around the axis of the cone's base. The spiral wave arms subside abruptly from turbulence to smoothly spreading outward waves at the gyre diameter (GD). The diameter varies ±5% in vibration to a peak of about 7% at the point where each wave crosses the perimeter, while the disk planform rotates slowly around the drive axis under viscous coupling to the inertia of the surrounding water mass. GD subsidence of turbulence is due to the viscous transfer of turbulent energy to numbers of small eddy gyres. These float off in the larger currents to fill thinly (in an isolated single vortex) the large central outwash, the low velocity upper toroid, and particularly the lower toroidal sub-vortex with much less energetic small and lower velocity eddy gyres before they are eventually disrupted in the point intake of the currents up the sides of the drive cone. The disrupted eddies are constantly replaced at the GD so that after their number is built up, it is steady. When two driven gyres are closely interacting, especially if co-rotating, and even when they are held static, there is an estimated order-of-magnitude increase in the number of these eddy gyres due to the frictional interaction of opposed adjacent currents, largely from the impact of spiral waves and outwashes, but also from the two lower toroids distorted by each other. (Counting eddies would have required sealed under-water plane or sheet illumination and under-water photography of more highly reflective, neutrally buoyant, smaller seed-lets, which was not attempted but should be included in further research.) - - The water packet velocity tracks of the thick and slightly conic GD disk and the primary vortex below it do not change within the observable error if the cone base at a depth of one base diameter (Fig. 3-3) for near-surface gyres is lowered to fully submerged, forming a completely spheric conic vortex with viscous toroidal flow above and below the GD. This sphere may be moved around slowly in the water. As an essential part of the generation of gyre forces by transfer of momentum outward, the GD is the scaling reference for the vortex currents and forces.

The GD dimension in cm (Fig. 3-2) is a function (Eq. 4-1), of the cone's size and shape, its peripheral velocity around the base ($|V_p|=V_p$), and fluid viscosity and density. (Cone surface roughness was not shown to affect GD significantly unless a coating is used that increases the cone diameter or there are large body irregularities.) Two types of right circular cones were used for most trials, the larger one of lightly sanded and unpolished maplewood, the other of polished or roughened brass. The $\pi/6$ (30°) whole-angle (β) wood cones with 7.9375 ±0.05 cm base diameter (d_c) were truncated at 12.38 cm length for a cap nut on the 6mm drive shaft. The $\pi/9$ (20°) brass cones, with 3.175 ±0.01 cm base diameter, 8.7 cm length, and slightly rounded tip, were received threaded in the base for a 10mm shaft. The respective base areas A_B were 49.48 cm^2 and 7.917 cm^2. The side areas A_S were 195.36 cm^2 and 43.87 cm^2. When directly driven (1X ratio) in city tap water at 20°C, the base peripheral or tangential velocities V_P were 872.9 cm s^{-1} and 356.5 cm s^{-1}. At that standard temperature the approximate fluid density ρ (from handbook tables for distilled water) is 0.998 gm cc^{-1}, and the viscosity η is 1.005 cp (centipoise.) (Right angle gear losses in dual 1X gears did reduce both peripheral velocities slightly, as shown on horizontal figure scales, Fig. 4-1, etc.)

Water temperature variations to 18.75 and 16.2°C with changes of nominal density to 0.999 gm cc^{-1} and of viscosity to 1.11 cp had no observable effect at 2X drive gear ratio on GD or forces within their 6-10% uncertainties (though the 2X peripheral velocity of the 30° cone changed 0.8%, shown by doubled velocity ticks on horizontal figure scales.) Water at 49°C provided (briefly) a viscosity of 0.56 cp with density of 0.989 gm cc^{-1}; and 20% by weight sucrose solution at 18°C doubled viscosity to 2.08 cp with density 1.085 gm cc^{-1}. At 2X drive gear ratio, the $\pi/6$ (30°) cone's GD and peripheral velocity vary significantly with viscosity changes (Fig. 4-1 scale tick-marks).

With two right-angle drive gears the increased frictional load at 1X (2 gear) ratio reduced the $\pi/6$ cone's V_P to 789.35 cm s^{-1} at 19°C. This is the reference velocity for comparisons with X/4, X/2, 2X, and 4X gear ratios. [Some force data are at 1X (Direct) with V_P of 832.6 cm s^{-1}.] The $\pi/6$ 4X series overloaded the motors with V_P of 1232.0 cm s^{-1}. The $\pi/6$ 2X (2gear) series was similarly limited to 1072.4 cm s^{-1}. The $\pi/9$ (2gear) series ran 328.2 cm s^{-1} at 1X, 570.98 at 2X, and 977.77 at 4X.

Within these test conditions (also used for forces) GD in cm varies as:

$$GD = d_c \left[1 + e\left(\frac{\rho}{\eta}\right)^{0.6667} V^{A\sqrt{V}}\right]. \qquad (4\text{-}1)$$

where $V = V_p/1000$, the coefficient e is the base of natural logarithms, and $A = (2 A_B + A_S)/100$. Note that V has its own square root in its exponent and that the base area in cm^2 is given twice the weight of the side area. The unit of length for d_c, A_{B-S}, and GD must be chosen so that d_c is between 1 and 10 units, though they may be unusual units (such as 10km units for hurricanes. But V_p remains in cm s^{-1} units at all scales.)

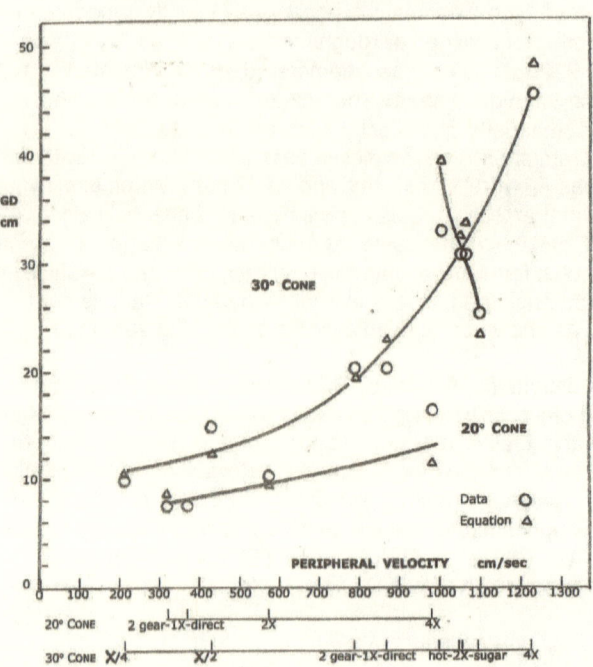

Fig. 4-1 Turbulent Gyre Diameter (GD) Versus Peripheral Velocity of A Drive Cone Immersed At Least 1 Base Diameter or More in Lab Water. Empirical data on turbulent water vortex diameters (defined in text) and derived Equation 4-1 with either of two right circular cones with different base diameters and whole cone angles (described in the text) driven by the same motor. This yields different peripheral velocities of the cone bases. Drive at 1X gear ratio is taken Direct and with two right angle gears at 1:1 net drive ratio showing some loss in peripheral velocity due to gear friction losses. The 30° cone is also geared down by factors of 2 and 4. Both cones are geared up by factors of 2 and 4. At 2X Gear Drive the main 30° experimental cone also has water viscosity halved by heating and doubled by sugar solution. (In hot water the near-surface vortex turbulence cools the water in the vortex so rapidly that valid true hot gyre viscosity could not be obtained. in the 4X data with the 20° cone, the 10 mm drive shaft appears to be excessively contributing to the GD compared to the 6 mm shaft on the larger cone.)

Excessive deviation between measured and computed $\pi/6$ (30°) GD (Fig. 4-1) in 2X low viscosity (hot water) is attributed to very rapid evaporative cooling of the water with near-surface turbulence during this measurement. (V_p and motor current were always checked first for run-up equivalence.) At doubled viscosity (and little change in density) in sucrose solution GD is very low. With low mass in the small volume driven to turbulence and small surface for coupling momentum outward, the cone peripheral velocity is distinctly higher than with lower viscosity tap water. In the next section the current, or power, drawn by the motor is significantly less in the higher viscosity fluid. These effects reverse in hot fluid and prove significant for scaling.

Drive Power & Single Vortex Energy

Motor drive currents and voltage with and without wet loads were measured at 2-3% accuracy, and the wattage differences are plotted (Fig. 4-2) for sets of test conditions. Since gear train losses are not well resolved, the plots are taken as upper limits of power transferred through the gyres to surrounding fluid beyond 2GD under steady conditions. A separate attempt was made to measure power difference due to mutual force between immersed gyres close together at 1X (1 gear) drive ratio with $\pi/6$ cones in either spin sense. The best measurement was <6.6 watts due to maximum attractive force in the vicinity of 300 grams (compared to <30 watts gyre drive power of a single gyre without mutual force.)

After the motor starting surge, average currents during the build-up of the turbulent gyre were estimated as similar to the steady running currents. Averaged stopwatch estimates of build-up time provide approximations of watt-seconds or joules of energy stored in gyres (Fig. 4-2). The trends of the combined estimates and measurements (Fig. 4-2) are similar to the GD data curve (Fig. 4-1), including the 2X gear ratio variations with viscosity due to its inverse relation with gyre volume. A similar effect on mutual force appears in later sections. Again, the measurements in heated water were at a lower than intended temperature and are low.

The energy and angular momentum stored in the immersed gyre are in two volumes, the nearly spherical region of strong primary and secondary flow at about the GD diameter, and the thick subplanar disk, or wide and shallow cone, of centrifugal turbulence in the spiral wave planform within the GD (Fig. 4-1). Both volumes scale with the cube of GD. This is scaled down (for overlay comparison of storage volume and power stored) by the ratio of the two volumes to the drive cone volume reduced in a convenient proportion at which the volume ratio curve (Fig. 4-2) overlays the power stored. This implies a threshold (of kinetic energy per unit volume of fluid near a shear interface) beyond which the disk volume of turbulence expands until it is just above threshold level and the sphere is just below that, for a threshold of about 3×10^{-3} J cc^{-1} or J gm^{-1} in lab water estimated in the 1X (2gear) $\pi/6$ case with about 12 joules total stored and with 16 J s^{-1} (Fig. 4-2 watts) transferred from the sphere to surrounding fluid. Some portion of the energy is in the observed (but un-unmeasured) turbulent vibrations of pressure waves, with the remainder in flow shear waves.

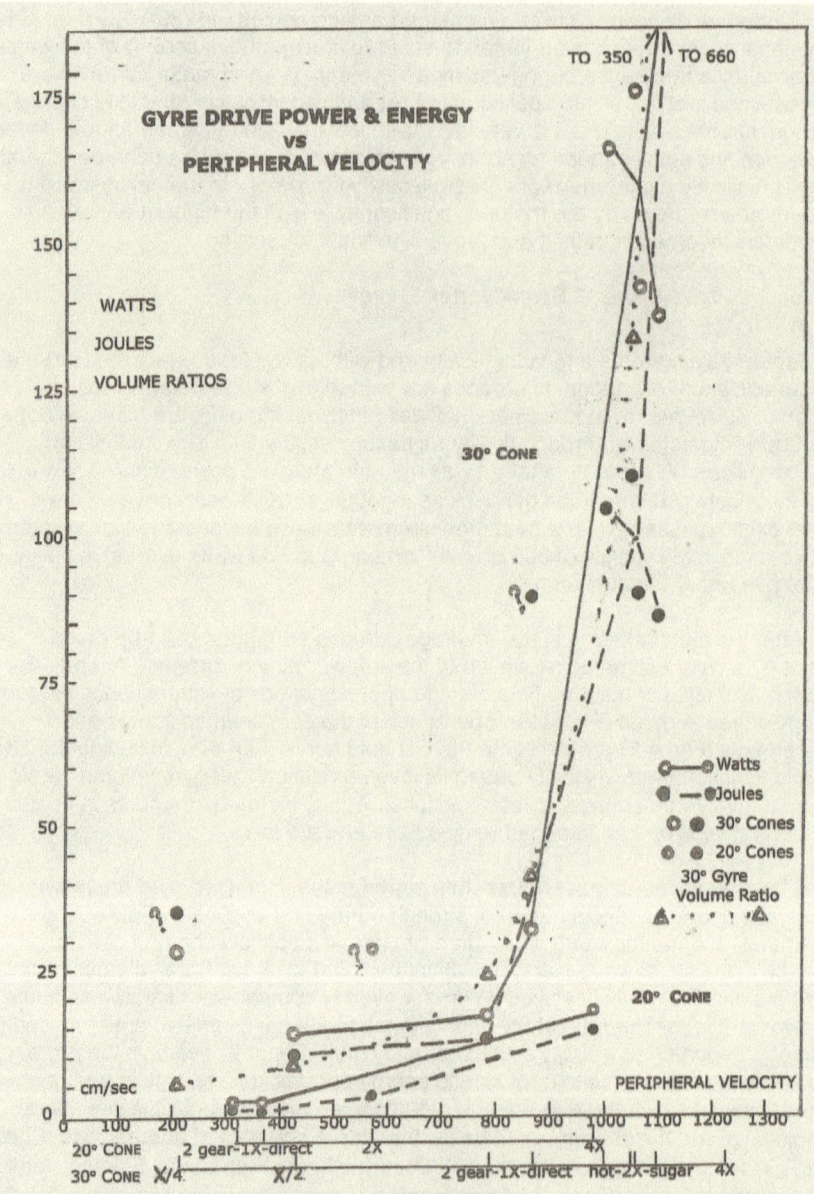

Fig. 4-2 Electric Drive Power & Stored Energy of Turbulent Conic Vortices. Measured power and energy are proportional to the dotted curve for Volume Ratios of water spheres with the GD diameter to a fixed volume reference. Crossed lines at 2X drive are for 3 viscosities; high point is for a single 2X gear.

Point Thrust (PT) Force of Single Gyres & Macro-Scaling

In reading vortex current data figures to this point it could have been noticed that the accelerated centrifugal ejection of fluid from the base and sides of the spinning cone is in two conic streams at about $\pi/12.86$ (14°) and up to $\pi/6$ (30°) toward the base axis. This volume also includes a separable flow in the tertiary spiral waves. Here added and ejected momentum has a rocket-like component of axially reactive, point-directed thrust force, PT (for which the vector magnitude |**PT**| is also used as PT with axial point direction assumed in text.) For every conic gyre this important force must be subtracted from the net force in dual gyre measurements to obtain interactive mutual forces.

PT data from single gyres are plotted (Fig. 4-3) against peripheral velocities of the bases of the two main types of drive cones for the various drive conditions. Data from surface and immersed gyres are generally consistent, with midrange uncertainties well below 10%. The more erratic X/2 data on the 30° cone are for three trials of two single 2:1 gears and a two-gear train with different gear frictions. The force at 2X drive gear ratio for the high-viscosity sucrose solution did not match the GD and power data by being lower than forces near the standard viscosity. Yet the hot measurement for low viscosity definitely confirmed the negative slope of PT (and other features) with viscosity in the equation derived to express the data in general form:

$$|\mathbf{PT}| = \frac{\sin\beta}{10\pi}\left(\frac{\rho}{\eta}\right)^{0.25} |\mathbf{V}_p|^{1.525} \frac{A_S}{9.807 \times 10^2}, \tag{4-2}$$

where β is the whole cone angle, 9.807×10^2 is the conversion factor from dynes to grams force (for convenience in measurements), the area of the sides is in cm^2, and other terms were defined earlier. The sine factor describes vanishing PT for the cylindric limit and for the 180° whole-cone flat-disk drive. (Herein, empirical equations for turbulent gyres contain elements of the Reynolds number with unusual exponents.)

In PT the data for the two cones overlap vertically (Fig. 4-3). The smaller cone at the 4X measured 977.77 cm s^{-1} peripheral velocity produces 17.55 gm PT from the equation (4-2) with the viscosity and density factors very close to 1.00. In the equation the larger $\pi/6$ (30°) cone would have the same PT at 285 cm s^{-1}. At these V_p levels (Fig. 4-1) the GDs for the two cones are also very nearly equal. The fluid volumes (Fig. 4-2) will also be similar. Likewise, the smoothed and extended power and energy measures for the two cones at the same respective V_p levels (Fig. 4-2) are quite similar.

The viscous boundary layer on the larger cone then models the smaller cone's outer vortex at the same GD with equal effects, and vice versa. Any of these features is sufficient to find the modeling equivalence of all. Later equations apply in the same way to each cone's vortex and to its mutual equivalence to both the other cone and to many other cones not actually driven. [This effect appears down to 5° cones (that are as short as the 30° cone} but is disturbed by the relative size of the drive shaft.] Thus, in a later quantum gyre it is not necessarily required to prove existence of an exact driver. One definable quantum gyre scales a range of possible drive cones by 1 cone, GD, V_p, etc.

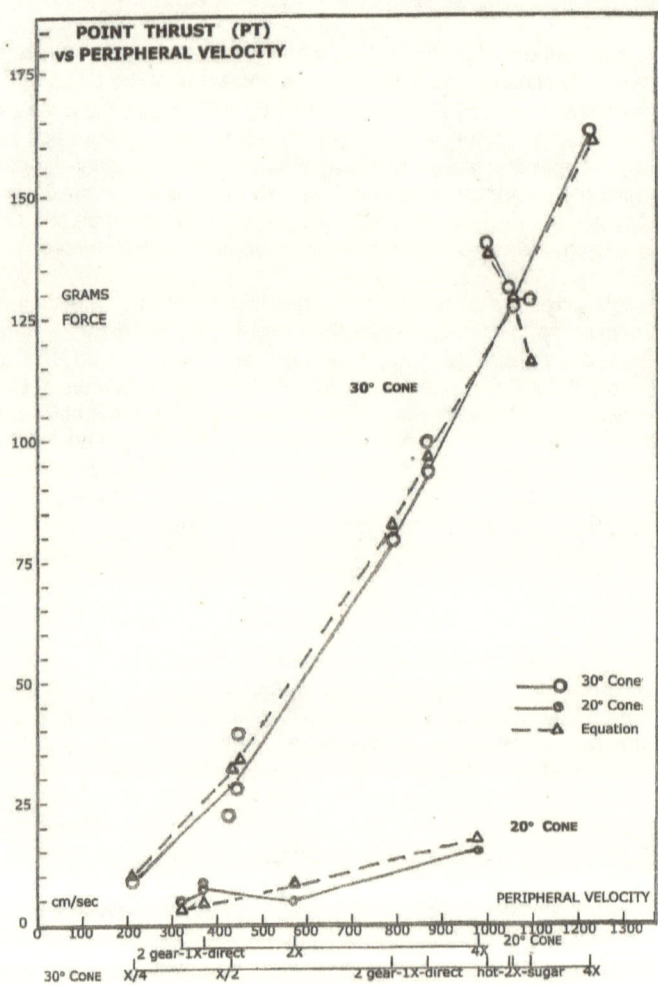

Fig.4-3 Measured Axial Point Thrust (PT) of Gyre Drive Cones at Peripheral Velocities in Water. This newly demonstrated PT force, inherent in all centrally driven conic vortices, must be subtracted from total measured mutual force between a pair of such gyres to obtain mutual radial force, and must also be added for scaled quantal forces within a subatomic MQP paradigm particle. (PT also accounts for descent to the earth of the conic body of a tornado against surrounding up-drafts that generated it at higher altitude in a stormcell or hurricane rainband.)

Similarly, the empirical equations scale gyres mutually over much larger ranges. It was already demonstrated (Chapt. 3) that Eq. (3-1) defines the turbulent weather limit of a tornado outbreak in Hurricane Ivan's vortex circulations driven by heat engines. Full experimental modeling of weather phenomena in this way is not feasible because of Coriolis effects, decreases of pressure and density with altitude, temperature lapse rates, latent heat release, etc. But the fluid mechanics actions are validations of these empirical equations as correctly describing within their uncertainties nature's great common and general system of causal fluid interactions. Consequently, given that a heat engine in a tornadic thunderstorm cell has set up an intense vortex in a gaseous medium, mechanical correlations to a laboratory turbulent conic gyre (Howard 2009), or lower hemigyre, can be numericly scaled for validity, as with the hurricane tornado line.

In Project VORTEX (Ziegler et al. 2001) a tornado appeared at the ground after a mesocyclone at 2-5 km altitude "developed downward," as in many eyewitness reports of tornado funnels moving down from clouds. Since cyclones in general (Rasmussen et al. 2000; Markowski et al. 2002) have upward flows, visible tornadoes must necessarily propagate down against updraft air (often evidenced by large hail.) From PT lab data, a tornado body developed aloft can thrust down against convective upflow by centrifugally accelerated ejection of air angled upward. PT Equation (4-2) can model such cases.

In the same project in a series of very clear, scaled pictures of a tornado below clouds at 2 km altitude, the upper visible funnel was measured photogrammetricly at 1.4 km wide (Rasmussen et al. 2001). The dust cloud at the ground was 400 meters wide. The reported ground damage had the same path width. This gives a blunt, clear air cone of $\pi/6.43$ (28°) whole angle around a funnel cloud. Doppler radar at the time also showed at 4 km altitude a 43 m s^{-1} differential velocity in a tornado about 1 km wide. At the same time there was also a radar mesocyclone of 46 m s^{-1} differential velocity 7-8 km wide at not less than 4 km altitude, which does represent a GD-like structure.

Assigning the 43 m s^{-1} differential velocity from one side to the other on radar at about 1 km diameter (Rasmussen et al. 2001), as the photogrammetric 1.4 km diameter outer funnel diameter below the clouds at a 21.5 m s^{-1} rotation of radar reflective and heavier centrifuged rain drops, and assuming conserved angular momentum in air parcels with inner fog droplets that reflect very little, then at a 0.7 km diameter inner core of a 20 degree cone base 2 km above ground, air particles could have a rotary velocity of about 43 m s^{-1} and at the ground the 400 m diameter could be at about 75 m s^{-1}, which is high enough for the reported tornado damage. In moist summer air with the ground at sea level (rather than the unreported level in Kansas) 43 m s^{-1} V_p would yield a PT of 1.2394×10^{10} gm on the cone base at 2 km altitude, or 10.19685×10^4 gm m^{-2}. With a worst case drag coefficient of 2 for flow interferences the tornado body cone might have an upward drag force of 3.5894×10^4 gm m^{-2} from an outer updraft of 20 m s^{-1}, which is not unusual. The PT is far more than enough to bring such a tornado down. The updraft could increase to about 33.7 m s^{-1} before balancing the down thrust and forcing the tornado to retreat aloft, which they usually do eventually. Though very approximate, the equation scaling estimate is well within reasonable uncertainty.

To validate the part of modeled gyre structure above the cone base: In spite of the known vorticity of large convections (e.g. van Delden 2003; Hoskins et al. 1985), there is usually little discussion of vortices in connection with sprites and other electrical discharges in the transparent stratosphere and mesosphere above thunderstorm cumulo-nimbus (CN) supercells in the American plains (e.g., the historical survey of Lyons and Armstrong 2004). This occurs because the outer visible clouds of 100-mile-wide storm cells do not exhibit much vorticity. One 40 year synopsis of thunderstorm observations discussed mesocyclonic rotation only twice (Fujita 1992). However, a STEPS program picture (Lyons and Armstrong 2004) shows a sunlit wave cloud that is spirally shaped above the turbulence of a large cell. A radar image of a CN cell suggests vorticity during sprite discharges (Lyons, Nelson, and Fossum 2000) (though not all plains supercells produce sprite night glows,) And references in the just prior case above show that study of CN vorticity is increasing in some researchers.

One western supercell was photographed in infrared by a GOES satellite twice in the same part of the night as mountain-top photo records of distant brilliant sprites show the supercell clouds below and faintly glowing "gravity" waves coming out in the mesosphere over the camera. The IR images show the horizontal plan of distinctly spiral wave cloudtops of a definite supercell vortex (Sentman et al. 2003, Fig. 2). These typically 20 km high clouds are in the rising and spreading equivalent (Fig. 3-3) that flows up the sides of the lab cone and out under the upper spiral wave disk as a continuation of the central drive cone base, equivalent here to the storm heat engine in the central cell vortex driving this entire system from troposphere to mesosphere. Much more striking is the 2 hour video record from the mountain top (courtesy of the senior author), with 25 seconds exposure every 30 seconds, showing the nearly circular sectors of outward moving waves of continuous faint airglow well above the brighter 5 to 10 millisecond sprite discharges above the lightning-lit supercell clouds. At about 500 km from the storm the 30 second video frame interval confirms the outward 85 m s^{-1} velocity of "gravity" waves 20 to 30 km from trough to peak and 40 to 50 km in wave length measured by Sentman et al. (2003). These outer waves correlate exactly with outer parts of the upper spiral wave disk in the lab experiment (Fig. 3-2). There is even closer correlation in the images of the origin and departure of the waves going around the elliptical image of the 85 km high central ring of airglow above both the sprites and the cloud tops under them. One excellent frame shows the outer waves and their central source in the report (Sentman et al. 2003).

The lab centrifugal spiral flow coupled to the cone base below it rotates both inside each wave and as a disk of waves around the drive center out to 2GD. The net flow is not only outward, but also upward (Fig. 3-3) between $\pi/12.86$ and $\pi/6$ (14 and 30°.) The gross movement correlates with the glowing spiral wave origin at the storm's 85 km with momentum transfer outward and upward in mesospheric "gravity" waves.

The centrifugal spiral waves visibly originating at 85 km in the Sentman et al. image (2003), extending laterally several hundred kilometers in the low viscosity air of the stratosphere and mesosphere, also (like any centrifugal water pump) effectively draw down air of the lower ionosphere within a scaled upper toroid ring (Fig. 3-3). This provides (with a reduced lifetime at the higher density) ions for the sprite, blue jet, elf, and airglow displays seen above these large storms at night. This correlation redefines generalized signs of an upper level subsidence of the ionosphere in clear atmosphere above CN cells (e.g., van Delden 2003, Fig. 3; Hoskins et al. 1985, Fig. 8). It also validates at large scale the wave flow regimes of the upper half of the fully developed turbulent conic vortex of Figures 3-2 and 3-3 herein. These waves are also frequently recorded by aircraft as clear air turbulence (CAT) above such storms. This type of validation thus extends to the upper, lower, and turbulent spiral wave disc central sections of the single lab conic gyres. That validation applies to flow regimes, forces, and mutual modeling of single gyres, as already covered.

This completes the preparation for taking up next the mutual forces between two immersed symmetric gyres, with coverage of the full range of possible symmetric axial angles at which force is measured. Since the flow regimes do not change, the prior validations will also continue to apply. (The variety of cases of asymmetric axial angles for force measurements become almost infinite in number, and limited amounts of these also have been run to support force estimates between LQ particles. Those asymmetric angles of the axes of two gyres arise within the LQ particles only with the linking orbits of quarks. They occur in great numbers between any two gyres that are in separate particles of any kind. A case of this type will be taken up later with the force between two electrons at the end of the chapter on single electrons.)

PT-Corrected Radial Mutual Forces Between Two Symmetric Conic Gyres

This section measures forces only within a symmetric pair of conic lab vortices. This is done separately for two sizes of central drive cones with same and also opposite senses of base rotation in any pair. In about 900 force measurements, there are nine independent variables, plus near equivalents of two, as well as the prior GD and PT with major interactions. The resulting graphed data are cross-sectional samples of a larger field, but do cover the radial mutual forces of dual symmetric gyres. Empirical equations are derived for numerical scaling of mutual gyre forces felt in opposite directions along a radial line between their centers of drive cone volume (CVs) by each of the two vortices. Two of these equations apply generally across the entire field of possible symmetric variation. About 20 variable coefficient equations (for substitutions in the two general equations) will each apply for various limited sectors of the entire field. At the end of the equations, Table 1 lists which of these ≈20 variants for substitution into the two primary equations apply to each type of case that can occur with turbulent symmetric conic gyres. This approach condenses a great amount of possible discussion and clarifies the sequence of computations for forces between exactly symmetric conic vortices.

In most equations and measurements in this section the prior PT and GD have been measured, the sum of all actual radial forces has been measured, and the PT force vector at its angle from the radial line between CVs has been subtracted before data display or usage. (Within a particle then, the PT must be separately added in again, by vector addition at its proper angle, after mutual radial force is computed.) First, Figures 4-4a and b define the mutual symmetric angles between two vortices.

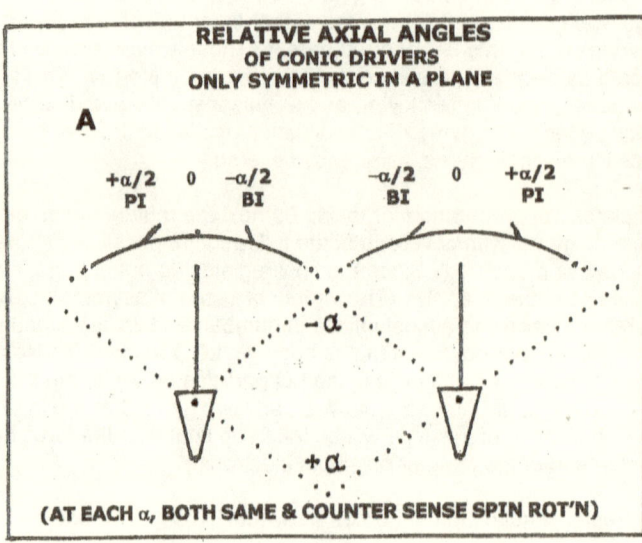

Figure 4-4a Relative Axial Angles of Symmetric Vortex Drive Cones. Points inward angles are positive. Base inward angles are negative. Each axis moves half the complete angle. As shown with parallel axes the angle is zero. This applies to all the empirical force equations derived from the measured data in this chapter.

Gyre pairs in opposite (contra-) rotation handedness are H- for force generation. Gyre pairs in the same senses of (co-) rotation handedness are H+ for force equations.

When the cones are near the water surface with the bases at a depth of one cone diameter, the condition is S+. If the drive cones are deeper so that the entire vortex is fully immersed, the condition for force generation is S-.

The large dots in the bodies of the cones represent the centers or centroids of volume (CV) of each cone. CVs are the references from which distances are measured and on which forces on the vortex are summed up from around the body of the cone. Positive radial mutual forces push cone CVs apart along the line between centroids. Negative radial mutual forces push the CVs together along the same line. Lateral forces are at right angles to the plane in which the cone axes are symmetric. Equal computed forces are on each vortex (but as a fluid pressure force the two forces are not added.)

GD scales of separation of vortices are measured along the line of radial separation between CVs. They are also measured in centimeters in lab scale.

The measured forces on a vortex transferred in the fluid to the cone body are always the total force and thus include the PT force generated by the individual vortex itself. To obtain the mutual force generated between two gyres the angular component of the PT force along the radial line between the two CVs is subtracted from the measured total force. The mutual force measured is then either from the currents along the sides and in the lower toroid, or the side outwash, or the spiral wave disk, whichever is opposed.

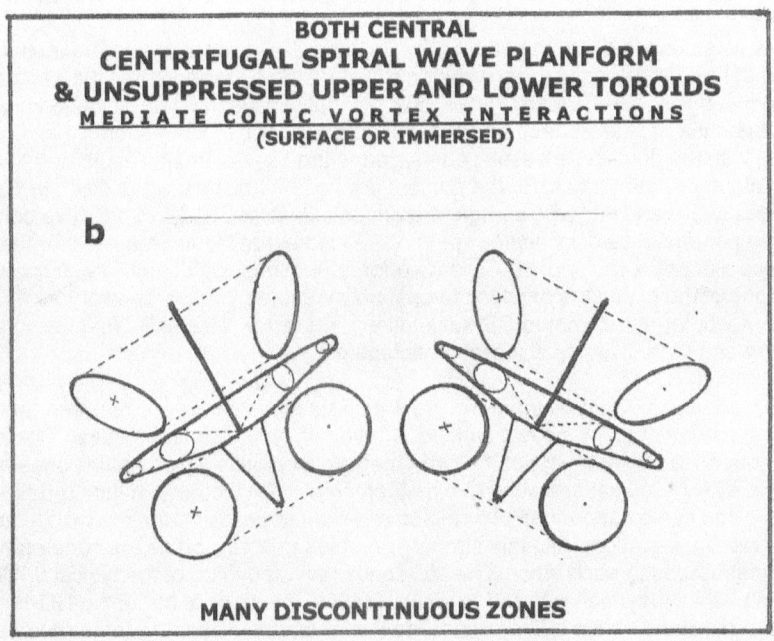

FIG 4-4b Vertical Cross-Sections of Immersed 30° Cones & Turbulent Gyres at 2.5 GD Separation. These sphericly complete sections are in the plane of the symmetric axes and at +60° relative axial angle in left hand circulation, as shown by the current arrow tail feathers in the central circulation of the toroids. The uneven profiles of the currents account for large changes in mutual forces with relative axial angle and GD separation. At this spacing the currents are pushing the gyres apart. With smaller separations these primary current flow envelopes compress each other flexibly. At about 0.7 GD separation of CVs in this and smaller axial angles the two lower toroids will have closed out opposing circulations between the cones in favor af a single larger toroid going around both cones. This single toroid circulation pulls the gyres together attempting to make them merge. The weaker upper toroids do this too with slightly less separation. This reversal of force is a general feature of co-rotating vortices. It does not occur in contra-rotation. (During that process there are also lateral forces from the opposing currents perpendicular to the plane of symmetry that tend to make the gyres circle around each other.)

    In setting up and calibrating lab pendulums for measurement of mutual forces, it is often necessary to make an initial estimate of the maximum limit of mutual force that can be generated with various equipments. That limit will usually be reached just as the two drive cones come in contact at minimum separation. For cones in contact, F_M limits vary proportionately with PT (Fig. 4-3) and mean centrifugal fluid ejection angle (Fig. 3-3). With a 0.9 factor from interference between cones, the peak empirical attraction between 30° cones with 1X direct drive V_P = 873 cm s^{-1} and PT = 95 gm is about:

$$\lim_{R \to 0} |\mathbf{F}_M| \approx 0.9 \frac{|\mathbf{PT}|}{\tan 14°} \approx 3.6 |\mathbf{PT}| \approx 340\, gm \qquad (4\text{-}3)$$

which is consistent with dominance of PT momentum components off the cone sides and base and with the highest force measurements herein at that peripheral velocity.

As was pointed out in the caption for Figure 4-4b above, there will be many changes of the mutual radial force within this limit as the axial angle and the separation between two drive cone CVs are varied. For a simplified introduction of these variations there is the case of near-surface vortices (which do not occur in the subnuclear particles) with the drive cone bases submerged to the base diameter depth and the drive shafts at or near vertical for the earlier viewings of the spiral wave disk. In Figure 4-4c cones of 20° and 30° whole angle are driven in that way at 1X direct drive cone rates (see peripheral velocity scales in Fig. 4-3.) Note that the force scales for the two cone sizes differ by a factor of 10. The data for both cones use the same GD scale for separations of the cone CVs at which forces are measured, but the centimeter measurements for that common GD scale differ by a factor close to 3. (In prior force figures the common distance scale is in centimeters.)

The most important factor in the Fig 4-4c data is the difference between the forces and movements for the two types of rotation of symmetric cone pairs. Contra-rotation (opposite handedness) of the two otherwise completely symmetric cones and vortices in every case causes attractive mutual forces. On the other hand co-rotation (rotating in the same hand direction) causes repellant forces between the two gyres over the wider separations, but this strongly reverses to attraction as the cones (and pendulums) approach each other. The 30° cones reversed force at the typical 0.7-1 GD separation, while the smaller cones reversed force in this case at about 1.5 GD separation (or about 11 cm.) Both cone sizes were unstable near the force reversal as indicated by the reversed arrows for unsteadiness of direction. The larger cones also pulled into cone contact with much greater attractive force than could be measured with the low pendulum weights in use, as shown by doubled down arrows. This over-riding strong attractive force at very close approach, including force reversal in co-rotation, is one of the principal vortex component features that directly explain the prior PDG data on particles with positive or negative charge. The explanation also includes the feature that charge is dependent on the direction of rotation of components which (like cones) are not symmetrical along an axis of rotation. The effect occurs reliably over a very wide range of other conditions. - - This is a near-field (NF) force effect at <1.5 GD separation of CVs. [Transitional-field (TF) is at 1.5-2 GD, and far-field (FF) starts at 2 GD to beyond 4 GD where F_M varies (Eqns. 4-4 & 4-15a) closely as the Newtonian inverse of radius squared.] Co-rotating symmetric immersed conic gyres usually reverse mutual force at <0.7-0.75 GD, where a lower toroid circulation around both cones begins dominating the effect of other currents [not unlike laminar momentum vortices (e.g., Melander et al. 1988; Smyth and Peltier 1994; Riccardi and Piva 1998).]

That process has a very strong effect on the amount of lab mutual force between cones in all the symmetric interactions. This is expressed in general empirical Master Equation 4-4 for the numeric amount of the force vector F_M between two fully immersed, centrally driven, turbulent, symmetric conic vortices. As before, the force is converted from dynes to grams by division by their ratio for lab use, and the density to viscosity

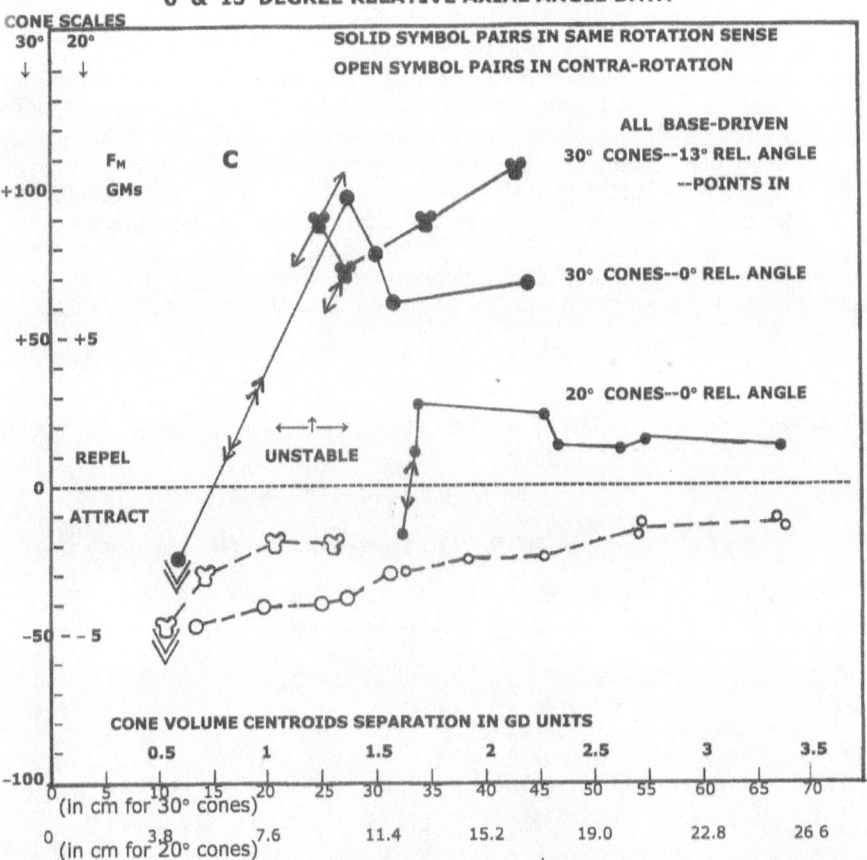

Fig. 4-4c Examples of Radial Mutual Force in Grams Between Two Symmetric Turbulent Near-surface Conic Vortices in Water. Axes are vertical with 0° relative axial angle or near-vertical with 13° relative axial angle. Gyre drives are in co- or contra-rotation (as marked) at 1X direct drive by 30° cones (left), or by shorter 20° cones (right.) The common bottom scale is in GD units of radial separation of cone volume centroids, with separate cm and grams force scales. (See Fig. 4-3 scale for cone peripheral velocities.) Note the reversed double arrows for abruptly unstable force changes in both direction and amount. Note the large double down arrows for contact between the larger cones with force too great to be measured with the light weights in the pendulums in these initial trials. (Weights were added there-after.) See text.

ratio previously defined is for the fluid used. The sine is of the whole cone angle β, the area of a cone's sides is in centimeters squared, and velocity of the perimeter of the cone base is in centimeters per second. There are changes of force levels by functions of relative axial angle, and of the separation of cone CVs , in two variable coefficients:

$$|F_M| = \frac{1}{2} \frac{\rho}{\eta} \frac{A_S \sin \beta}{9.807 \times 10^2} |V_P| f(\alpha)_{TBD} f(R)_{TBD} \qquad (4\text{-}4)$$

wherein the two coefficients are to be determined (TBD) from Equns 4-5 to 24 next by organizing rules in Table 4-1, where f(R) has a sub-Master Equation (4-15b) with further TBD coefficients, but f(α) has no master equation. When simpler, numbers in equations for α° are in degrees rather than in radians, or angles in fractions are in radian measure (as π or π/4 where π = 180°.) [In the lab cone point ends which are not inward (BI vs PI in tests) may be either point driven (PD) or base driven (BD) by crossed shafts.]

First are the empirical lab data coefficients for f(α). Exact Eqns. (4-5 & 4-6) are awkwardly long and are avoided by using points from simplified curves Figs. 4-5a-b:

$$f(a)_1 = 1 - f_Q(y) = 1 - \frac{y}{10^9}[y-(1/6)][y-(1/3)][y-(2/3)][y-(3/3)]...[y-(6/3)][y-(12/3)]$$

$$[y-(13/3)][y-(14/3)][y-(15/3)][y-(21/3)][y-(22/3)]...[y-(28/3)] \{(1/2)-(H/2)\}.$$
$$(4\text{-}5)$$

where y = |α|/(π/6) & .... indicates continuing the prior series steps until the next change.

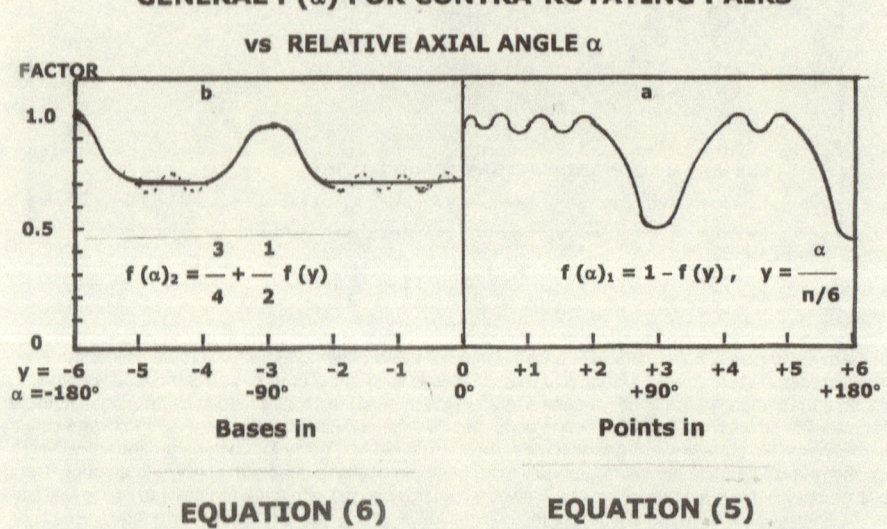

Figs. 4-5a & b. Empirical Equations 4-5 & -6 Curves for Coefficients in Equation 4-4 per Table 4-1. The curves are to be used to replace lengthy empirical series equations in the text. They adjust functions of the relative axial angles of symmetric vortices as coefficients f(α) in the equation for radial mutual force.

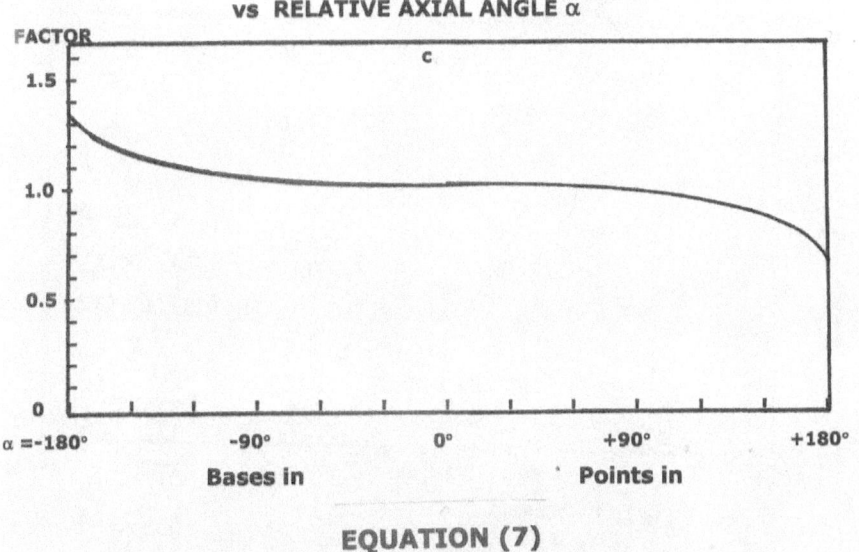

Fig. 4-5c Empirical Equation 4-7 Curve of Coefficient Correction Used per Table 4-1.

Equations 4-6 and 4-6a continue to include the excessive length of Eqn. 4-5:

$$f(a)_2 = (3/4) + (1/2) f_Q(y) . \quad (4\text{-}6)$$

$$f_{BI0}(a) = 2f(a)_2 \quad (4\text{-}6a)$$

$$f_H(\alpha) = 1 - \frac{1}{17} \tan \frac{4}{9} \alpha. \quad (4\text{-}7)$$

$$f_{HV0}(a) = f_H(a)\left[\frac{1}{2} + \left(\frac{1.5}{0.34}\right)(z-1)\right], \quad (4\text{-}8)$$

where $z = (V_p/800)$ generally and when V_p is limited to not less than 750 cm s^{-1} in Equation 4-8 only (not for limit of z in other equations.)

$$f_{HV1}(a) = f_H(a) \frac{1}{\left(\frac{3z - 0.75}{2} + 1\right)} \quad (4\text{-}9)$$

$$f_{HV2}(a) = f_H(a) \frac{1}{\left(\frac{3z - 0.75}{6} + 1\right)} \quad (4\text{-}9b)$$

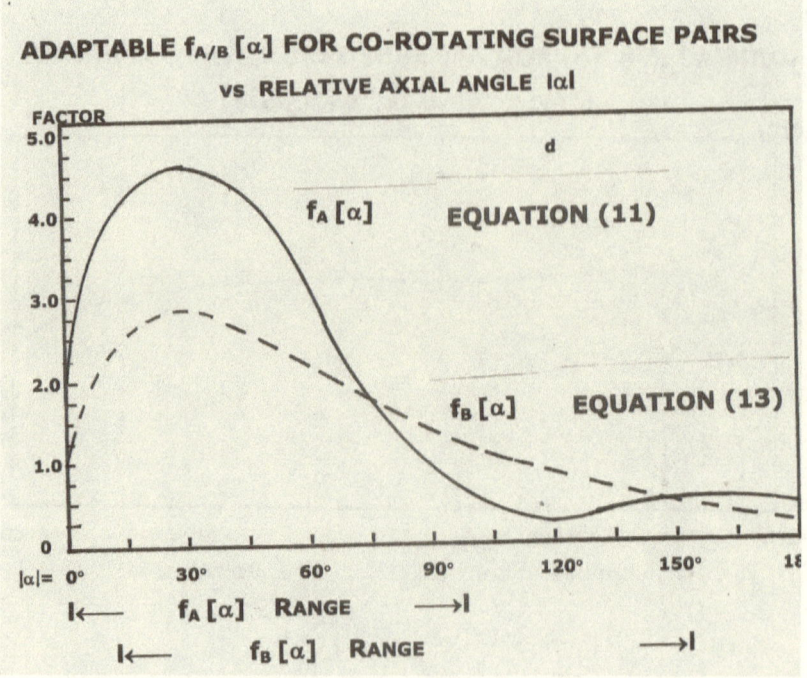

Fig. 4-5d Two Variations of Impulsive Exponential Equation 4-11 & 13. This adapts to some large peak forces between vortices in the lower range of relative axial angles and in a few cases of about 100 to 150 degree, as required below in the listings of Table 4-1.

$$f_{HV3}(a) = 2^{f(w)} f_H(a) \quad , \tag{4-10}$$

where
$$f(w) = \frac{-3\pi}{5z^z} w \ln w \quad , \tag{4-10a}$$

where z continues to include V_p, and w = 2 z − (1/2).

$$C_W = 2^{f(w)} \quad . \tag{4-10b}$$

$$f_A[a] = 2 + \frac{3ye}{e^y} - 5 \sin\frac{\alpha}{9} + f_p(y) \quad , \tag{4-11}$$

where y = α/π/6 (again) from α = 0 to + π□ (Fig. 4-5d), and:

$$f_p(y) = \frac{y}{\pi \times 10^8}(y-1/2)(y-1)(y-3/2)(y-6)(y-7)(y-8).....(y-14) \quad . \tag{4-11a}$$

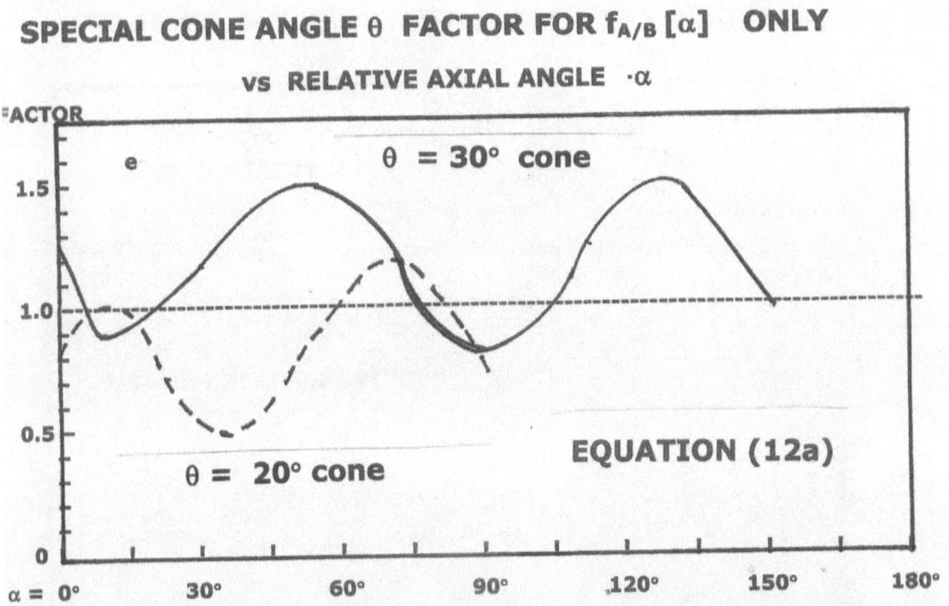

SPECIAL CONE ANGLE θ FACTOR FOR $f_{A/B}[\alpha]$ ONLY
vs RELATIVE AXIAL ANGLE ·α

Fig 4-5e Adaptation of Equn. 4-12a for Whole Cone Body Angles in Special Conditions. It is plotted versus absolute relative axial angle between cones as a factor in Equn. 4-12 when required in Table 4-1.

$$f_{A\theta}(a)_{PI} = f_A(a) f_{BM}(\beta) ,\qquad(4\text{-}12)$$

where, with angles in degrees magnitude (Fig. 4-5e):

$$f_{BM}(\beta) = 1 - 0.35 \cos\left[2a + \frac{100}{1+(0.2\alpha)^2} + 4\frac{\alpha + 7.5(30-\beta)}{\frac{100}{\alpha+1+15(30-\beta)}}\right] - 0.015(30-\beta) + 0.015(\beta-20)$$

(4-12a)

$$f_B(\alpha) = 1 + \frac{2ye}{e^y} - 2.5 \sin\frac{|\alpha|}{9} ,\qquad(4\text{-}13)$$

where y returns and absolute values of α apply.

To finish the f(α) factor in Equation (4-4):

$$f_{B\theta}(a)_{PI} = f_B(a) f_{BM}(\beta) .\qquad(4\text{-}14)$$

All of Equations 4-5 to 4-14 are for f(α) in Equn. 4-4 as shown in Table 4-1 later. The next series of equations are for f(R) in the same way with effect as in Fig. 4-5f.

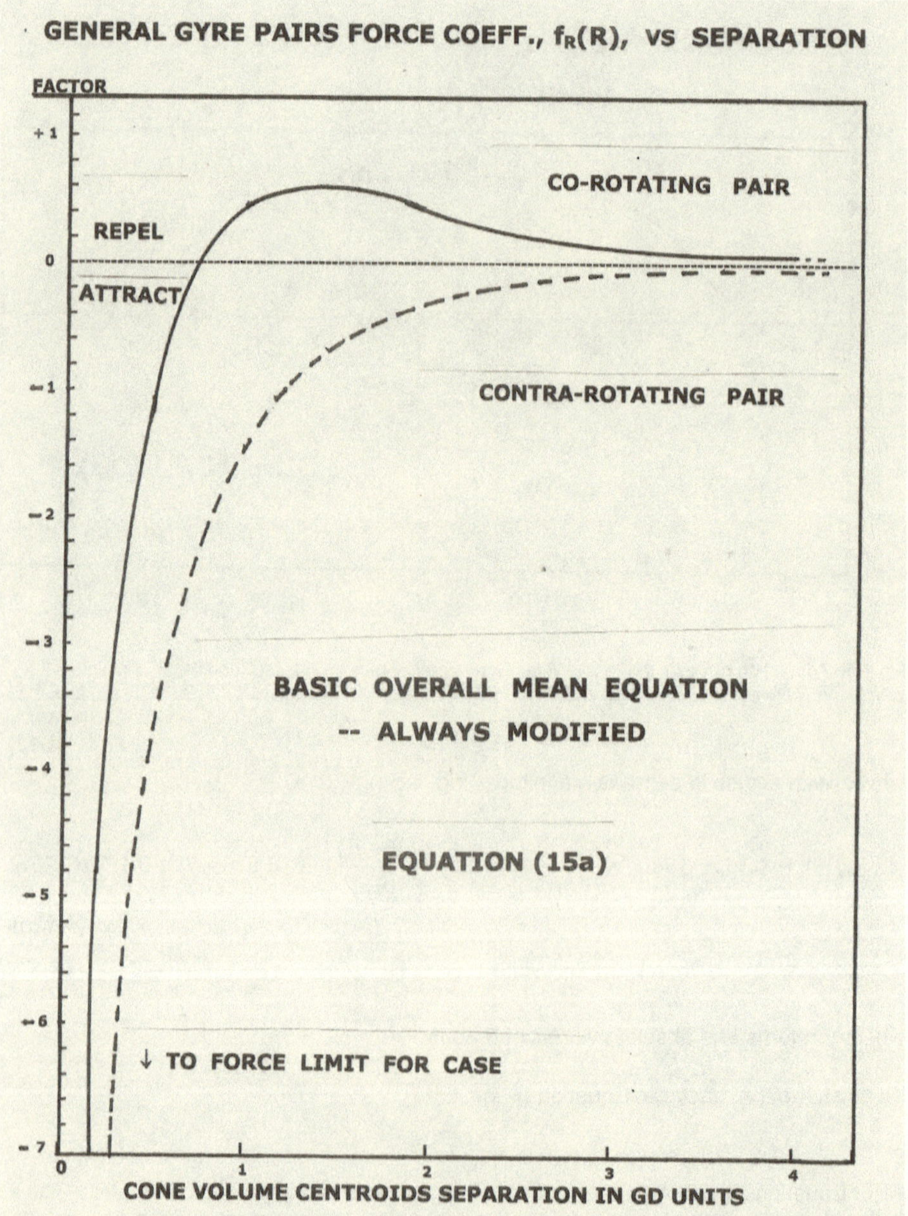
GENERAL GYRE PAIRS FORCE COEFF., $f_R(R)$, VS SEPARATION
FACTOR
+1
REPEL
0
ATTRACT
-1
-2
-3
-4
-5
-6
-7
CO-ROTATING PAIR
CONTRA-ROTATING PAIR
BASIC OVERALL MEAN EQUATION
-- ALWAYS MODIFIED
EQUATION (15a)
↓ TO FORCE LIMIT FOR CASE
0
1
2
3
4
CONE VOLUME CENTROIDS SEPARATION IN GD UNITS

Fig. 4-5f Basic Master Equation 4-15a For $f_R(R)$ in Equn. 4-4, with Coefficient Equns 4-15b to 24. These equations smooth the empirical lab data on symmetric gyres listed in Table 4-1 and displayed in graphs.

Figure 4-5f is the general dual curve of Master Equation 4-15a for the effect on mutual force in lab experiments (per Equation 4-4) of radial separation f(R) between the CVs of two symmetric co-/contra-rotating gyres. The f(R) in these equations below is a function of the simple GD scale measured quantity in data figures. Data R is multiplied by variable other functions of GD and other variables for $f_R(R)$, which compresses or expands the curves (Fig. 4-5f) horizontally and effectively shifts them to left or right on this actual GD or cm scale. The main cause for these GD shifts (GDS) is the variation of contacts (Fig. 4-4b) between two complete gyres with variations in relative axis angles. Other variations are expressed in the general force equation (4-4) and in the relative handedness (H) of the two gyres. H = -1 for contra-rotating gyres and H =+1 for co-rotation of the two gyres. Particular equation variations are specified in Table 4-1 for various ranges of the angle α between relative axis angles of gyres.

For the general Master Separation Equation between symmetric vortices then:

$$f_R(R) = \left(\frac{1}{2} + \frac{H}{2} - \frac{16}{(1+3R)^2}\right)\left(\frac{1}{(1+R)^{R-1}}\right) + \left(\frac{H}{(1+R^2)}\right), \quad (4\text{-}15a)$$

where:

$$f_R(R_{i,j,k,l}) = f_R(RC_iC_jC_kC_l), \text{ etc. next,} \quad (4\text{-}15b)$$

$$C_V = z^z, \quad (4\text{-}16)$$

where z = (V_P/800) starts occurring again.

$$C_\eta = (\rho/\eta)^{2/3}. \quad (4\text{-}17)$$

$$C_{2s} = (2\sin\beta)^2. \quad (4\text{-}18)$$

$$C_{3s} = (3\sin\beta)^{2/3}. \quad (4\text{-}19)$$

$$C_{3sl} = 1/C_{3s} \quad (4\text{-}20)$$

$$C_{311} = 1/(3\sin\beta)^{3\log z} \quad (4\text{-}21)$$

$$C_{313} = 1/(3\sin\beta)^{3\log 3z} \quad (4\text{-}22)$$

$$C_T = \frac{1}{1 - \frac{2}{3\pi^2}\tan^2 0.485\alpha} \quad (4\text{-}23)$$

$$C_{SS} = 5/(3a), \quad \mathbf{(4\text{-}24)}$$

where α is in radians.

This completes the mutual radial force equations for symmetric turbulent conic gyres as shown in Table 4-1, wherein the functional coefficients in Eq. 4-4 are to be determined (TBD) from Equations 4-5 to -24 by rules summarized here. In this Table f(R) has a Master Equation (4-15a) with further TBD coefficients, but f(α) has no such single master equation in which substitutions are made. Numbers in the Equn. line for $\alpha°$ are in compact degree notation rather than in radians, angles not marked with the degree symbol are in radian measure (such as π or $\pi/4$ where $\pi = 180°$), and both 0° and 180° relative axial angle may be sensitive as to whether they are listed as PD (few point-driven cones) vs BD (the usual base-driven cones). [Measured cone roughness effects in the gyre turbulence were too small/limited for adaptation of these equations.]

In accordance with the rules of Table 4-I in use of the prior equations (for which the prefix 4- is omitted to save space), the computed and measured F_M mutual force data for turbulent symmetric conic gyres are displayed in Figures 4-6, -7, -8, where effects discussed above for Figures 4-4d and 4-5f are prominent. Drive gear ratios, V_p and GD for each curve are consistent with prior scales. The frequent abrupt instabilities and repeated force reversals contrast strongly with studies of smoothly merging laminar vortices (e.g., McWilliams 1984; Melander et al. 1988; Smyth and Peltier 1994; Riccardi and Piva 1998.) These distinctive effects may occur in any turbulent vortices.

Note that where space in a data page of one major class of vortices is available, a few data have been specially marked and included with gyres that are of a different sub-class within the main class for the page, such as gyres driven by cones of 20° whole angle rather than 30°, or the smaller number of data plots may be of near-surface trials rather than fully immersed. Near-surface indicates horizontal bases of the drive cones at one base diameter depth. Much of this data is little different from the fully submerged data, but does lend itself to reduced uncertainty of measurements. The large majority of the lab data are at the 10% uncertainty level due to the effects of the necessary turbulence on optical recording and/or direct visual measurement. Additional instrumental error is negligible. Large variations of the data are due to turbulent fluctuations of the forces and currents. Near points of reversal of force between co-rotating gyres sudden large changes of force by turbulent currents are common.

In experiments with horizontal axis cones on right angle drives used to obtain the full ranges of symmetric axial angles, it was found necessary to determine a "crawl" force to correct both mutual radial forces and also mutual forces perpendicular to the radius of separation of cones. The asymmetry of upper and lower hydrostatic pressures in viscous loads on the sides of the cone makes a tilted or horizontal gyre crawl as if the downward side of the drive cone were getting traction on a roadbed beneath it. This is important in coaxial plus 180° (PI) gyre pairs in which lateral forces can act to keep the vortices coaxial (as in the coaxially orbiting gyre PI pairs found necessary in Chapter 3.) Figure 4-9a shows that BI (base inward) pairs that are co-rotating are forced out of coax, but all others are forced into coax alignment as separation comes within GD, and contra-rotatiing PI and BI pairs are always in coax lock. Figure 4-9b shows a few measurements of combined lateral and crawl forces (with F_L equivalent to F_P depending on axis assignments.) Fig. 4-9c displays isolated crawl force data with horizontal drive.

Table 4-1 Symmetric Mutual Force Equation 4-4 Factors Schedule of TBD Usage. This Table systematicly organizes required Coefficient Equations for each lab data-taking condition in applying empirical Equn. 4-4 to smooth, analyze, or use that lab data, esp. by ranges of relative axial angle of 2 equal-size symmetric turbulent vortices in either contra- or co-rotation & points-in PI or bases-in BI also PDriven or BaseDrive, (Both apply at 180° & 0° rel. axial angles.) Upper and Lower Curves indicate either co-rotation or contra-rotat'n. (Both degree and radian angles are used to fit table space.)

Table 1. Equation factor groups by ranges matched to data for symmetric gyre pairs

	Immersed Spheric Vortices (S=−1)				Surface Hemispheric Vortices (S=+1)			
	Contra-rot'n (H=−1)		Same rot'n (H=+1)		Contra-rot'n (H=−1)		Same rot'n (H=+1)	
	Lower curve α°		Upper curve α°		Lower curve α°		Upper curve α°	
	PD-BI	BD-PI	PD-BI	BD-PI	PD-BI	BD-PI	PD-BI	BD-PI
Equ'n	− +	0 30-150 180	− +	0 30-60 90 120 150 180	− +	0-13-150	− +	0 13-90 120-150
(4)F_M	*	*	*	*				*
(5)$f(\alpha)_1$	* *	*	*					
(6)$f(\alpha)_2$	*	*	*					
(6a)$f_{Rad}(\alpha)$								
(7)$f_{ts}(\alpha)$				*				
(8)$f_{srv0}(\alpha)$								
(9)$f_{srv1}(\alpha)$			*					
(9b)$f_{srv2}(\alpha)$			*					
(10)$f_{srv3}(\alpha)$								
(11)$f_{A}[\alpha]$ basic				* * *	* * *		* *	* *
(12)$f_{Aol}[\alpha]_{PI}$				* *			* *	* *
(13)$f_{Is}[\alpha]$				* *	* * *		* *	*
(14)$f_{Isal}[\alpha]_{PI}$				* *			* *	*
(15a)$f_R(R)$ basic				* * *				* *
(15b)$f_R(R_t)$ *				* * *				* *
(16)C_V		*		* *			*	
(17)C_q		*		* *				
(18)C_{2A}	*			*			*	
(19)C_{3a}	*							
(20)C_{3a1}								
(21)C_{3B}								
(22)C_{3D}								
(23)C_T								
(24)C_{SS}								
(10b)C_{VV}	*							

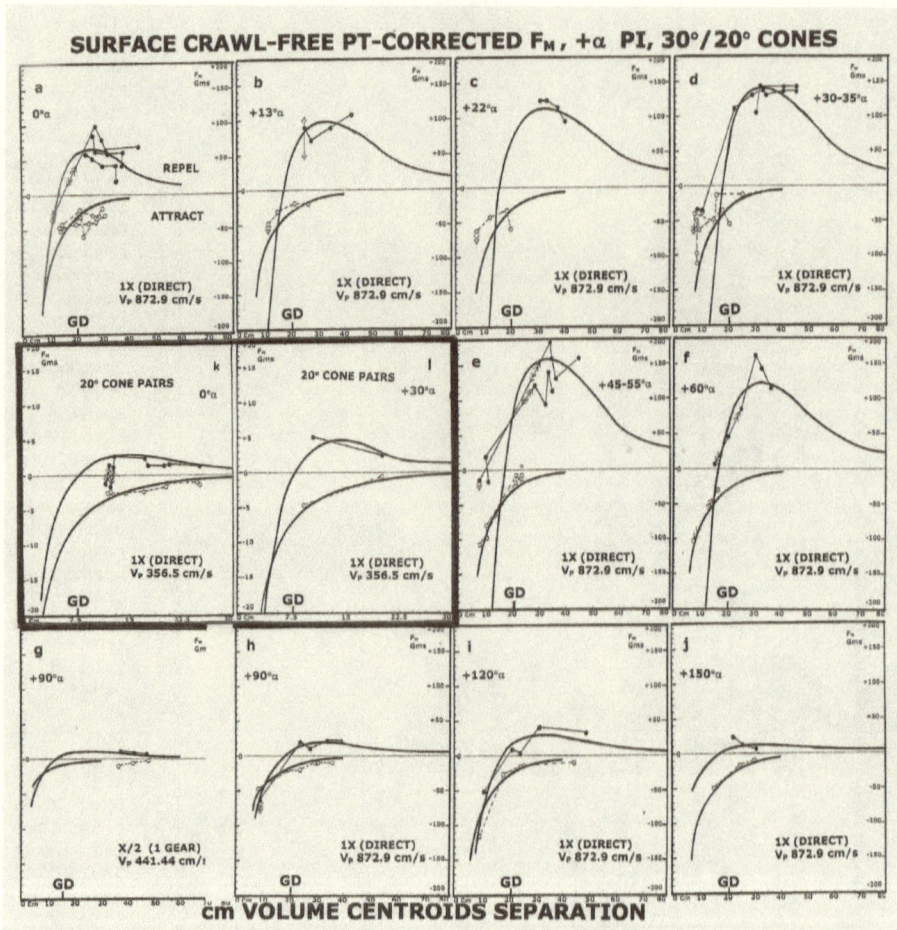

Fig. 4-6 Grams of Radial Mutual Force between Two Symmetric Near-Surface Turbulent PI Vortices, Lab data that it was possible to measure in close approach with two meter pendulum instruments and computed curves per Table 4.1 variations of Equation 4-4 are plotted versus centimeters of separation of drive cone centroids of volume with direct drive by initially vertical drive shafts (in all but one plot.) Note that turbulent Gyre Diameters (GD) were measured in each case for single gyres, as were the force corrections. Ten plots are for 30° drive cones, and two are for the 20° drive cones with smaller (about 1/3) base circumferences and also much smaller cone areas. These differences are fairly well matched by the empirical equations. Upper curves are for co-rotation of the two cones. Lower curves are for contra-rotation. Positive mutual forces repel and increase separation of the cones and gyres. Negative forces make the cones attract each other. Note unstable abrupt changes of force and reversals due to turbulence. Also note the significantly higher forces with direct drive than in the next figure in which every plot shows the effect of frictional gear drive losses of peripheral velocity of cones with either one or two right-angle gears in the drive train in each case. Again the equations matched the force measurements reasonably well with wide changes of peripheral velocity, including two large viscosity changes (See text).

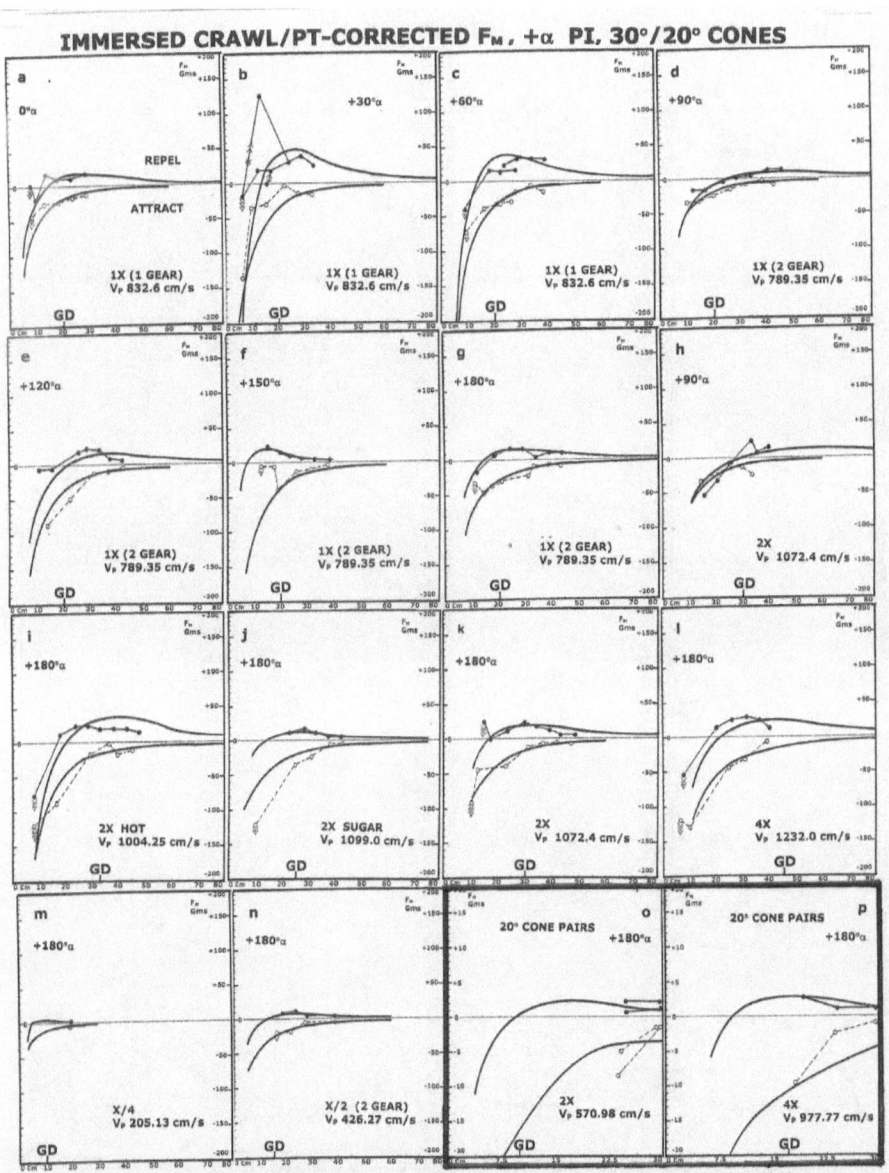

Fig. 4-7 Grams of Radial Mutual Force between 2 Symmetric Immersed Turbulent PI Vortices, See last entry for Fig. 4-6 on other contrasts, plus full immersion of the cone axes to beyond more than two cone base diameters with nearly horizontal axes. Further immersion did not change forces recordably beyond the typical 10% uncertainties due to necessary turbulence. All these tests have 90° drive gear losses.

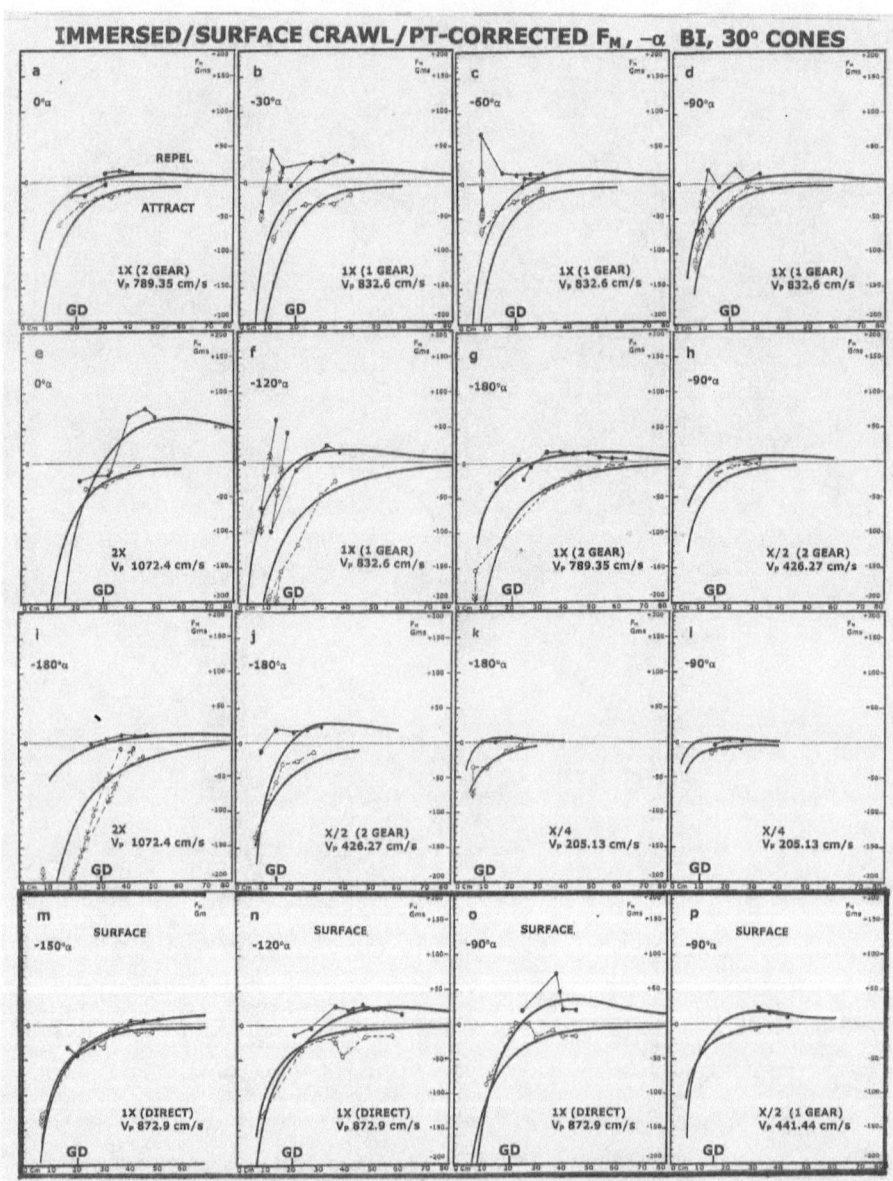

Fig. 4-8 (Continuing Figs. 4-6-7) Lab Data & Equations on Grams of Radial Mutual Force for 30° Conic Symmetric Turbulent BI Vortices. Negative relative axial angles apply to Bases Inward. The bottom row of near-surface runs has three with direct drive vs a dozen immersed runs. Again the equation variations listed in Table 1 provide a reasonable empirical interpretation of the trends of the measured lab data.

Figure 4-8 completes the radial symmetric mutual force data that could be obtained in the lab due to limitations of the pendulum measurement system with large enough cones to enable definite observations in turbulence. (More complex strain gage instruments should be tried in very large tanks with bottom and side windows or immersed cameras, etc., for later lab measurements.) More specialized data follows.

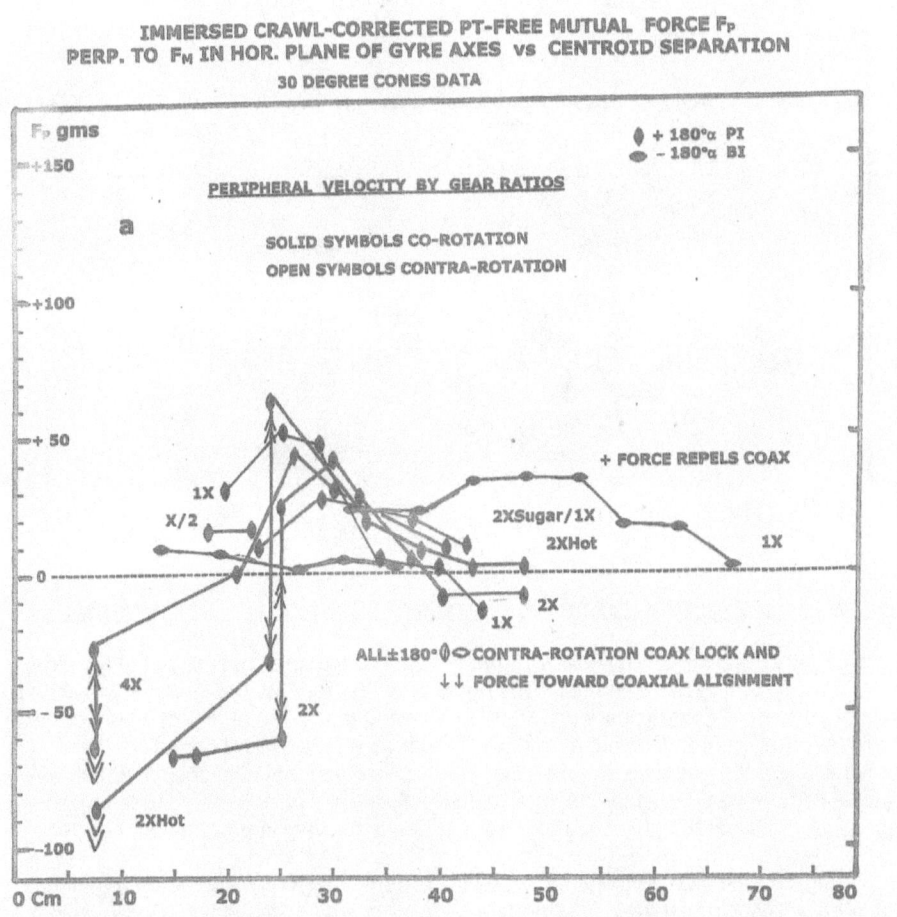

Figure 4-9a Symmetric Coaxial + & − 180° Fully Immersed Mutual Forces Orthogonal to Radial Force. These forces F_P perpendicular to the radius of centroid separation are in a variety of viscosities and drive gear ratios. The relative GDs and peripheral velocities of 30° cone bases can be taken from prior graphs. The contra-rotation negative force details are omitted to insure clarity in the more complex co-rotation data. While the radial mutual force equations correlate with the mean magnitudes of these lateral mutual forces, the two types of force are not always in phase with each other nor fully out of phase vs separation. That aspect of the equations does not apply here. Reversed arrows show some unstable direction of force, and doubled downward arrows indicate forces much larger than could be measured with reliability.

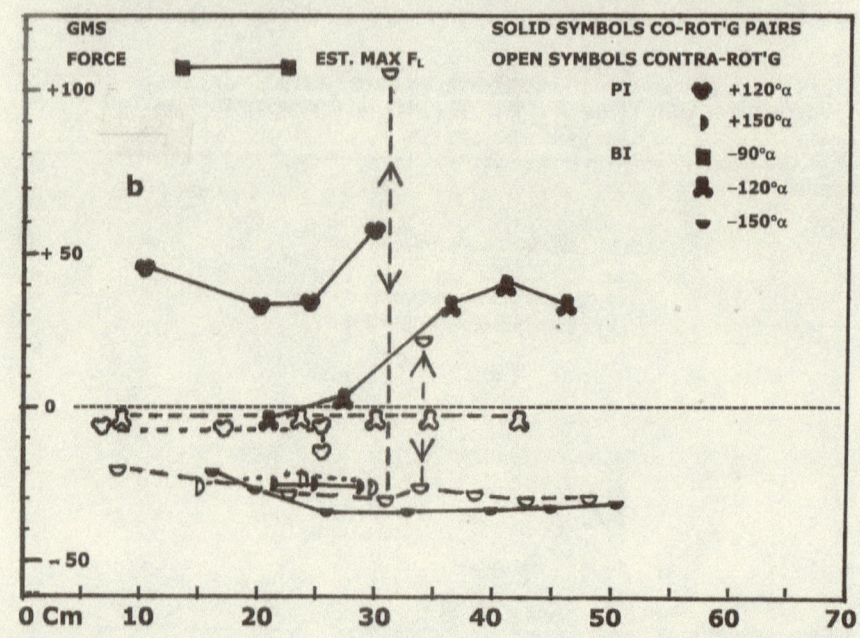

Fig. 4-9b Typical Mixed Crawl and Lateral Force Data Needing Correction by Crawl Force Subtraction.

Figures 4-9a and -9b show symmetric lateral force data that are not fully in the organized system of radial mutual force equations. In Fig. 4-9a the steep down slope to the right of most PI data is very important in showing that in the region of 3 GD, or here 60 cm separation of PI Coaxial gyre drivers, both co-rotational and contra-rotational forces scale in the negative direction that will stabilize and maintain coaxial alignment of coaxial pairs of vortices while they orbit with multiple pairs in stable electrons or protons, etc. (This is typical of many such functions in direct auxiliary support of QM needs.)

Figure 4-9c next contains the crawl force data used to correct the data that was obtained with horizontal axes for the drive cones in the available tanks. (It is possible that much deeper immersion or tanks pressurized to 90% or more of the fluid pressure at the depth of the drive cones would reduce the crawl effect to insignificance in the turbulence of the lab gyres. - - In the scaled micro-quanta, the average space all-directional pressure equivalent, if any, would be uniform over gyre outer currents. Still, a secondary crawl effect might occur if a fast frame-dragging of the medium comes very nearby, perhaps in the wake of a passing heavy particle moving relatively close to the speed of light. That might in concept disturb the even pressure over a few quanta.)

IMMERSED SINGLE GYRE CRAWL FORCE, 30° CONES
IN HORIZONTAL PLANE OF GYRE AXES vs CONE PERIPHERAL VELOCITY V_P

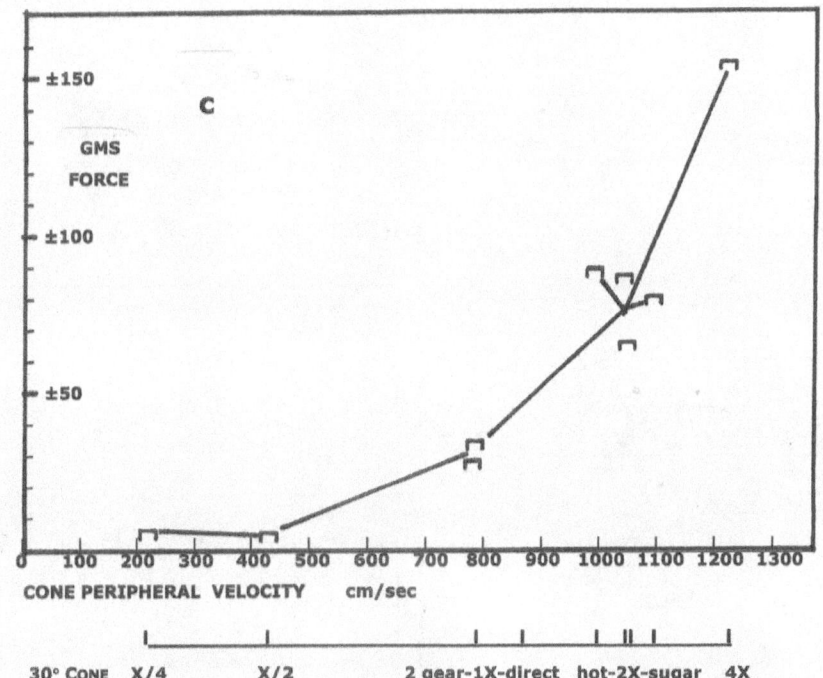

Fig. 4-9c Grams of Immersed Crawl Force of Single Horizontal 30° Drive Cone in Standard Lab Water.

Asymmetric Pairs of Gyres with Mutual Radial and Orthogonal Forces

The prior 30° cones have been set up with co-planar 1X two-gear drives angled asymmetricly with graduated steps in three different ways to measure both radial mutual forces and in-plane perpendicular forces toward or against coaxial alignment of vortices (Fig. 4-10). Illustrative plots are partly consistent with prior data, but also contrary. It takes many graduated steps of angles for valid adaptation of equations to each type of asymmetry. Data stocks are still too sparse for equations versus angles at separations.

However, this short series includes the largest measurement to date of attractive mutual radial force of 324 grams (Fig. 4-10c), at very near the limit of 340 grams for 1X peripheral velocities estimated in Equation (4-3). That measurement occurred with one 30° cone 30° off coaxial Bl alignment with another. This force measurement illustrates most conclusively the distinctive downward pumping action of the spiral wave disk in the

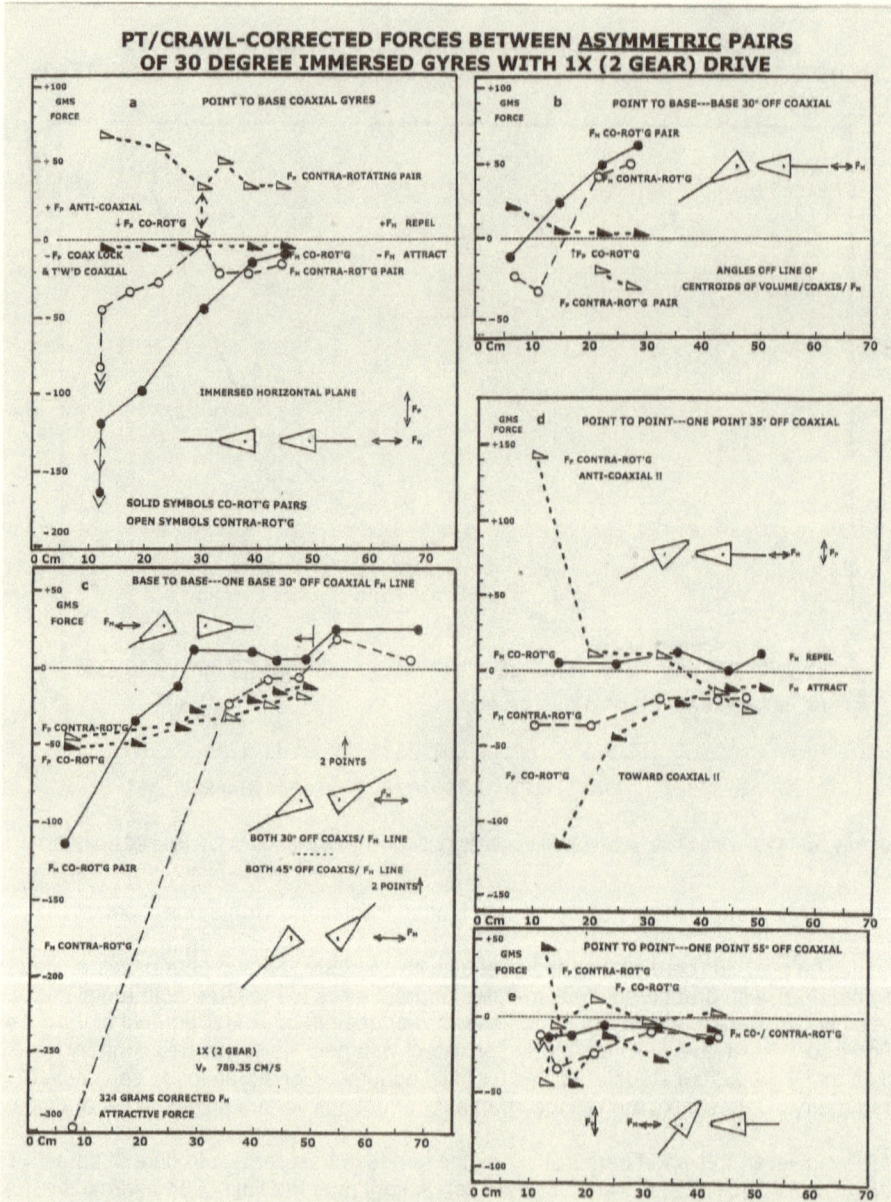

Fig. 4-10 Gram Mutual Forces vs Cm Separation of Asymmetric Gyres. Radial between CVs, co-rotation solid symbol & line, contra-rotation open symbol & long dashes. Perpendicular forces in short dashes. Above zero line repels coaxial position and radially. Below zero attracts coaxial alignment & radially.

central upper hemisphere of gyres as a separate addition to PT jet reaction effects from the sides of cones. [This powerful effect, opposite to PT in this case due to pulling a vacuum against the opposite cone base, was pointed out earlier as a downward inflow from the mesosphere over an observed weather super-cell (Sentman et al., 2003) very much like lab near-surface gyres that pull air down to the cone base (Fig. 3-2). Above the more powerful vortex of a hurricane the downward velocities over a much larger area of the mesosphere would focus a large total downward momentum at low density over the center of the hurricane. This momentum can carry mesosphere air all the way down to the ground for the clear skies and calm weather in a hurricane's eye. (Radar chaff rockets, or windsondes, could test this.) Also, when a hurricane begins moving ashore fast in a high southwesterly circulation ahead of a northerly cold air front near landfall, the hurricane drags the surface eye ahead of the center of the down momentum in the mesosphere and stratosphere (just as a stratosphere wind can offset a supercell anvil top to one side of the cell for cloud-top-to-ground lightning.) Then the offset down current of mesosphere air will hit the trailing side of the eyewall. That momentum component accounts for the opening of a hurricane eyewall near landfall, setting it up for violent pre-tornadic downbursts. Reconnection of the upper downflow to the center of the hurricane would cause observed second eyewalls to form before the momentum of the first is dissipated. These scaled effects confirm the lab vortex model,]

The other forceful asymmetric orientations of gyres in Fig. 4-10 occur, along with many thousands of others, in particles, either with one of a pair of gyres in a linking quark orbit interacting with a single gyre of another pair in any other orbit in its own quark or in another nearby quark sphere (Figs. 2-6 and 2-7), or between any two gyres separated in adjacent spheric particles, such as two electrons (in the next chapter.) These force curves are obviously radically different from the data on symmetric gyres. (Figs. 4-6 to –9) It is estimated that at least four times as much data will be required on such asymmetric gyre duos as on the prior symmetric lab work to support adequate fitting of empirical, and eventually theoretical fluid mechanics, equations. (This needs a well funded program in a state-of-the-art tank lab facility with instrument design and machine/electrical/photo shop support.)

Chapter Conclusions

This chapter numericly (and empiricly) fully validates the structure arrived at in the just previous chapter as the necessarily required, centrally driven, fully immersed, turbulent conic vortex structure for the universal micro-quantal component of the primary sub-nuclear particles of nature. Such a vortex is a structure which occurs naturally at many sizes over a very wide range of dimensional scales covering many orders of magnitude. At the smallest scale of the micro-quantum this structure generates in particles exactly the rigorously required functions previously defined over more than a century of the most advanced and worldwide science of sub-nuclear physics as treated by Quantum Mechanics. Broadly, this quantal vortex is both a discrete, but partially elastic, sub-particle body and an outer assembly of waves and currents in a medium with surrounding waves, all of which interact with uniform quantitative precision. As a conic, centrally driven body, Usually operating in combined spheric orbits of conserved coaxial spinning pairs with conic points inward, the quantum inherently and constantly generates not only an inward self-force of jet reaction point thrust, but also from the inner sides of the main conic current flow aided by the main toroid a close-range mutual attractive force that combines with the PT force in a strong force of binding attraction at a low effective resonant frequency within the primary particles generally. There is further a very weak outward repellant wash force component in a belt around the center of action. Facing outward from the base of each drive cone there is also a hemisphere of much higher frequency electric charge mutual force activity for resonant interaction with other gyres through a tuned set of multiple (usually six) spiral wave sub-gyres of much smaller size and even axial distribution. This charge force has polarity for mutual attraction or repulsion interaction of waves and currents depending on whether the two drive cones are co-rotating or contra-rotating (either at left-hand on the bases or right-hand.) These quanta also possess spin momentum, both from individual rotation and from rotation of the co-axis of a pair around another central orthogonal axis yielding synchronized spheric orbits. However, conserved pairs can in some Usual and other Extreme cases operate in self-spin on the co-axis alone where this fits the local situation of balanced forces. Further, in combination of all the classic and QM forces from a single source, the gyres also generate by internal viscous friction of currents and transmitted waves in their medium a number of smallest conic vortical eddies which both couple to the local medium for a collective mass inertia and also randomly at a very low resonant frequency interact very weakly remotely with all other eddies out to an infinity in a gravity attractive force dependent on the number of the quanta in a primary subnuclear particle and sum of the individual quantal masses from contained eddies times a special frictional effect of the charges of quantal pairs. This gravity effect of mass eddies also depends on the fifth power of the number of quanta in any primary subnuclear particle and further, in hadron particles, on exponentially declining with the number of the primary particles, times the sum of their masses. (It is by far simpler and more accurate to write and read the earlier equations on mass herein and on gravity force between masses in the non-relativistic Newtonian texts of classical physics.) What is measured in the equations and data of this chapter is the sum of all the relevant forces in a specified radial direction from each quantal vortex at the specified separations between centers of drive cone volume. Filtered resonant readings of force waves at separated frequency bands will be necessary to isolate such weak forces as that of gravity from the "weak" force, the much stronger electric force, and the even greater strong force. These forces all operate at the space equivalent of an optimal fixed subsonic displacement wave velocity within water. (In the next chapter it will also be necessary to add another force of entanglement operating at the velocity of at least the equivalent of sound in water, if not at an equivalent of hypersonic shock.)

It still remains necessary to check (in the next chapter) the generality of the empirical equations for scaling down to the subnuclear dimensions and to generalize them further where necessary to describe the natural vortical micro-quanta at true scale. All of these processes then can give the radial force vector on each one gyre from each other single gyre in each LQ particle sphere of typically two to two dozen gyres, each of which requires this separate calculation. For six gyres in a sphere there are half of six times five radial force calculations (to which lateral forces should be added), and five of these sums must be vector summed for each gyre to get the momentary cohesive force on that gyre. Repeating this six times fully describes the linearly additive forces on the gyres' fractions of the total mass of the particle. (But the mass varies as the fifth power of the number of gyres in other calculations. Large particles become very unstable.)

Experimental laboratory measurement of the little known mutual forces between pairs of highly turbulent conic vortices, and their individual forces, finds new explanatory correlations also with the mechanical features of several forms of large scale violent weather. Outstanding in this are: A numerical correlation of the newly described point thrust (PT) force of single conic vortices with the descent of tornadoes from storm cells against the storm updraft. Correlation of the newly described structure of turbulent conic vortices with the causes of mesospheric sprites and gravity waves of airglow observed over supercells at night, of violent disturbances of hurricane eyewalls, and of initiation of tornadoes. Correlations with violent weather indicate an opportunity and approach for further field and laboratory research on the most violent weather vortex forces. (They usually do not have the complication of many interacting vortices.)

REFERENCES OF SPECIAL INTEREST
(Full references, tables, and appendices are in Volume 2.)

Sentman, D. D., Wescott, E. M., Picard, R.H., Winick, J. R., Stenbaek-Nielsen, H. C., Dewan, E. M., Moudry, D. R., Sao Sabbas, F. S., and Heavner, M. J., (2003) Simultaneous observation of mesospheric gravity waves and sprites generated by a midwestern thunderstorm. J. Atmos. Solar/Terr 65(5), 537-550 (& 2002 preprint.)

BERNOULLI, D., (1738) Hydrodynamics

HOWARD, F. E., JR., (2010) Striking Correspondences of the Micro-Quantal Particle Paradigm & Quantum Mechanics, www.particlephysics.info and www.electron-particlephysics.org

Howard, F. E., Jr., (2009) Turbulent conic vortices: Part 1, Lab gyres define tornado band in Hurricane Ivan. Part 2, 3D Gyre Forces Control Tornadoes, Supercell Tops, & Hurricane Eyewalls, www.particlephysics.info and www.electron-particlephysics.org

Acknowledgements. Particular appreciation is due to Matthew Clark, who had his students in woodworking at the Okaloosa County Vocational and Technical School (in Fort Walton Beach, Florida) turn the four large 30 degree cones out of maple wood for these experiments. I must also thank Richard G. Henning of the US Air Force Weather Detachment at Eglin Air Force Base, Florida, Professor Keith G. Blackwell of the University of South Alabama at Mobile, Alabama, and D. D. Sentman of the University of Alaska Geophysical Institute at Fairbanks, Alaska, for providing data and discussions of weather effects which provided helpful background for this paper. Dr. Sentman's courtesy in providing the video of sprite and gravity wave activity over a Nebraska supercell is especially deeply appreciated. I thank Fred E. Howard, III, for a critical review of the text and many constructive comments on the general physics involved, and H. Blevins Howard for photographic instrumentation of lab vortices, special support on computer equipment and software used herein, as well as for monitoring internet weather data sources. I continue to owe much special appreciation to Cheryl Mack, senior librarian, and Christi Rountree of the US Air Force Armament Laboratory Technical Library at Eglin Air Force Base, FL, and likewise to new librarians, Eleanor Baudouin and Michael Jackson, for excellent and patient assistance of long standing.

Chapter 5.
A Self-Consistent Structure for the Electron/Positron -
The Prototype Micro-Quantal Particle

SUMMARY: Last century's many attempts at electron structure failed to match Rutherford's initial atom, which resolved the conflicting necessities in prior data analysis and laid a correctable basis for further insights. The new Micro-Quantal Particle (MQP) paradigm, that resolves the structural conflicts in prior electron analysis, arises in the previous chapters from newly found mass/charge power law equations defining interactive mass-energy structures of the massive particles in the Particle Data Group Summary Tables as necessarily being built up in sphere-like shapes (Fig. 5-1) from micro-quanta in the energetic form of centrally driven turbulent conic vortices (Fig. 5-2). In a simplest possible prototype for <u>u</u>sual particles in Nature as we know it, three negatively charged (by definition) coaxial pairs of uniform gyres in three sphericly concentric orthogonal orbits, with <u>full symmetry</u> between individual gyres at all times, and only outward electric force, solve at once the century-long problem of a feasible causal structure for the electron built on the uniform single vortex source for all the known forces. (Its 3D form is clear in simple, quick plastic foam models.)

Scaling to the MQP mass-charge equations with correlated lab data equations of directly and radiantly interactive vortex forces, momentary electron force-balance iterations through an orbital cycle converge simultaneously on 6 conic vortices that both spin and orbit their center (in a constrained medium) with polarized external charge forces, strong internal cohesion, mass-energies, and apparent elastic radii of electrons either in atoms and low velocity electric circuits or near light-speed in colliders. Instantaneous summed <u>non-symmetric</u> force estimates between the 12 gyres of two electrons (or positrons) also approximate their classic charge force. The three orthogonally orbiting vortex pairs of the wave-particles sum their spin effects at the quantum mechanics quantization angle of arccos $1/\sqrt{3}$ from all three orbit axes. The particles necessarily interact by whole-particle sums of mass and charge shear-wave forces of gyres at light-speed and by tuned entanglement of high superluminal pressure wavelet forces between component gyres in a constrained medium. This establishes a basic natural prototype for the massive PDG particles.

Electrons today are like the indeterminate atom before Rutherford a century ago. He shaped real structure to his firmest data, defining nucleated atoms to a correctable approximation. As Feynman noted, Rutherford's original structure later became much modified. (Feynman et al., 1964) Yet, it was the necessary step from which corrections creating today's atom were discovered. A correctable baseline structure for electron-positron mirror-twins is long overdue.

Last century's inconclusive electron models (Hestenes and Weingartshofer, 1991. Springford, 1997. Dowling, 1997), include vibrating strings and quantum mechanics (QM) "preons" (Salam, 1989. Haisch, et al., 1994). "Structure functions" are devised (Slominski, et al., 2001. Muryn, et al., 2004) as if electrons have undefined "partons". One concept (Waite, et al., 1997) wraps electromagnetic field "vortons" around point dipoles. Proven "entanglement" of electrons and photons (e. g.: Pan, et al., 2000. Mair, et al., 2001) is studied by elaborately disproving implications to electron structure. Yet, Einstein's comment that it would be enough to understand the electron (Mac Gregor, 1992) still applies .

Ideal Analysis Constrains Physical Structure Approach

The structurally broadest prior effort on electrons is Mac Gregor's monograph summarizing his refereed electron papers, thorough references, and related Quantum Electrodynamics (QED). He analyzed conflicting requirements, from charged body spin radii for QED line-splittings of light spectra, to empirical point-collisions, and theoretically destructive self-repulsion. Mac Gregor's analyses determined that only ideal, perfectly rigid spheres of relativistic "mass" m_p, corrected to slightly more than √3 times the early Compton electron radius (r =3.86 x 10^{-11} cm) and equatorially rotating an ideal point unit "charge" (perhaps fractional point "charges") at precisely the relativistic limit, can meet all spin, QM, QED, gyromagnetic, etc., necessities of electrons in atomic orbit at speeds v, and may also carry out point collisions (r<10^{-14}cm) near the speed of light c, where:

$$m_{pv} \equiv m_{p_0} f(v/c) . \qquad (5\text{-}1)$$

In the meantime a vast amount of empirical data on subatomic particles has been gathered by the international Particle Data Group (PDG). In Chapter 1 herein the new MQP paradigm for the massive particles, independently derived (Howard, 2005, 2006) from the PDG Summary Tables (Eidelman, et al., 2004, & prior), was first discovered as an exponential particle mass law (Howard, 2005) for quarks (Q) building hadrons (H):

$$m_p = N_q^y \Sigma m_q , \qquad (5\text{-}2)$$

where y asymptoticly approaches 5 (from <1) with decreasing mass sum Σm_q of the number N_q of the quarks. Extrapolating this QH law downward (Howard, 2005) with y at limit 5 derived a more fundamental lepton/quark (LQ) power law (Eq. 1-2 of Chapt. 1), requiring in electron/positrons (Howard, 2005, 2006) 6 micro-quantal components, each of small mass-energy and 1/6 conserved charge in 3 conserved pairs, ++ for positrons or – – for electrons. In this law, other leptons and quarks have more, equal, or fewer, charged or +– neutral pairs, usually in spheroid clusters predicted from the earlier QH mass curve. Here the initial QH total component mass Σm_q of ≤3 varied masses of charged quark components (Eq. 5-2) becomes ($N_c m_u$) of ≥2 uniform micro-quanta in charged or neutral pairs. Then variable numbers N_c of micro-quanta, each at rounded 10.9525 eV mass-energy m_u, interactively increase (with pair-charge factors) to the PDG mass-energies m_p of LQ particles (Eq. 1-2 of Chapt. 1 in alternative format):

$$m_p = N_c^5 (N_c m_u) \left[(n_{\pm}/n) + (n_0/na) \right] , \qquad (5\text{-}3a)$$

where $n_{\pm}$ is the number of charged pairs of micro-quanta, n_0 is the number of neutral pairs, n is the total number of pairs ($N_c/2$)=$n_{\pm}$+n_0, a=3^x, and x=1 in usual LQ particles like electrons (but x≥12 for extreme neutrinos.) This too permits globular quanta in spheres. Eq. 5-3a in the most frequently observed, Usual LQ particles collects terms to:

$$m_p = (2m_u/3) N_c^6 \left[0.5 + (n_{\pm}/n) \right] . \qquad (5\text{-}3b)$$

which reduces further in atomic and lower-speed (ground-state) electron/positrons to:

$$m_p = 6^6 m_u = 0.511 \text{ MeV, rounded}, \qquad (5\text{-}3c)$$

in empirical equivalence to Mac Gregor's ideal inner relativistic increase of mass. (Mac Gregor, 1992) These (PDG data-based) equations requiring pairs of quanta with spin indicate that they have two senses of spinning of points-inward (PI) conic quanta in coaxial pairs. Each quantum spins individually on the co-axis, and that axis also spins at 90° to the axis around its center so that both quanta spin in an orbit as if around a sphere with their conic bases outward. The necessary interactivity of the spinning conic quanta very definitely indicates that they are made up of uniform currents and sets of waves rotating around the co-axis and setting up circularly radiated waves. To do this

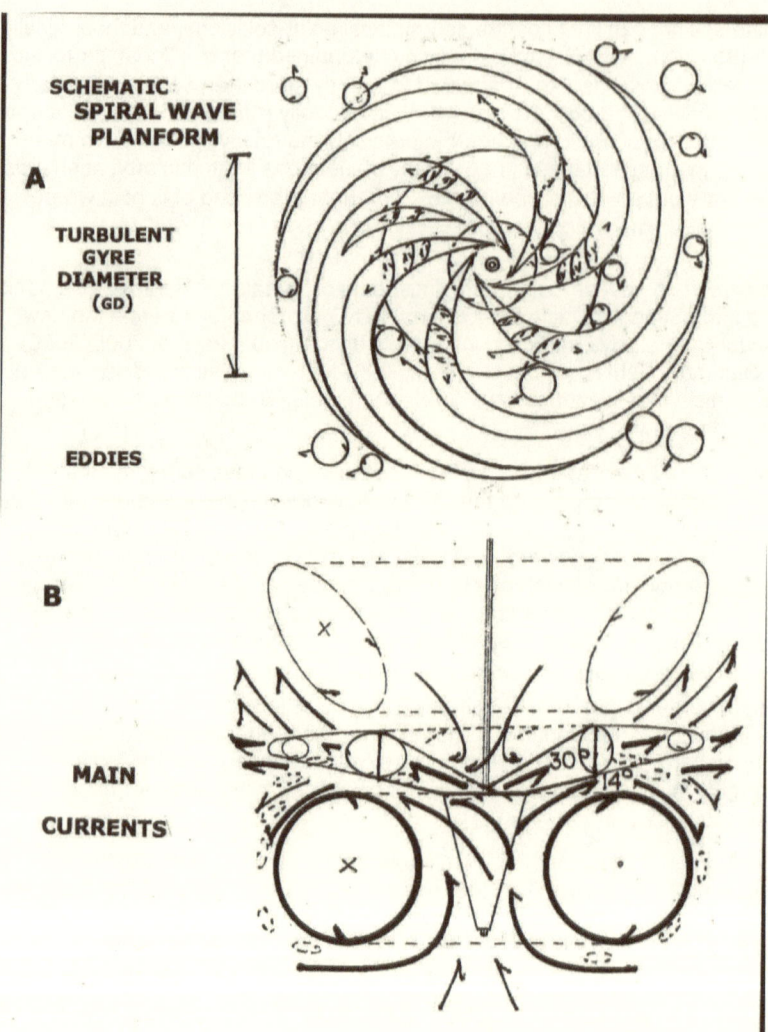

Fig. 5-1 (per Fig. 2-10) Quanta Require Centrally Driven Turbulent Conic Vortices. A. Plan View of Left-hand Spiral Waves in Lab Water. B. Vertical Axial Cross Section. Bold flow arrows indicate measured current flow velocity. In A spiral waves fade into out-going smooth circular waves that die out with travel.

requires conic vortices [as with cone-driven lab gyres (Fig. 5-1).] Pair spin senses affect mass (Eq. 5-3a) in $a^{-1}=3^{-1}$ for the Points-Inward (PI) neutral gyre pairs with unlike-charges versus like-charge pairs, just as PI unlike-spin lab gyres interact smoothly and confluently, but like-spins have viscously opposed shear currents at their wave contacts, creating more eddies (Fig. 5-1) which embody interactive "mass-energy." These eddies

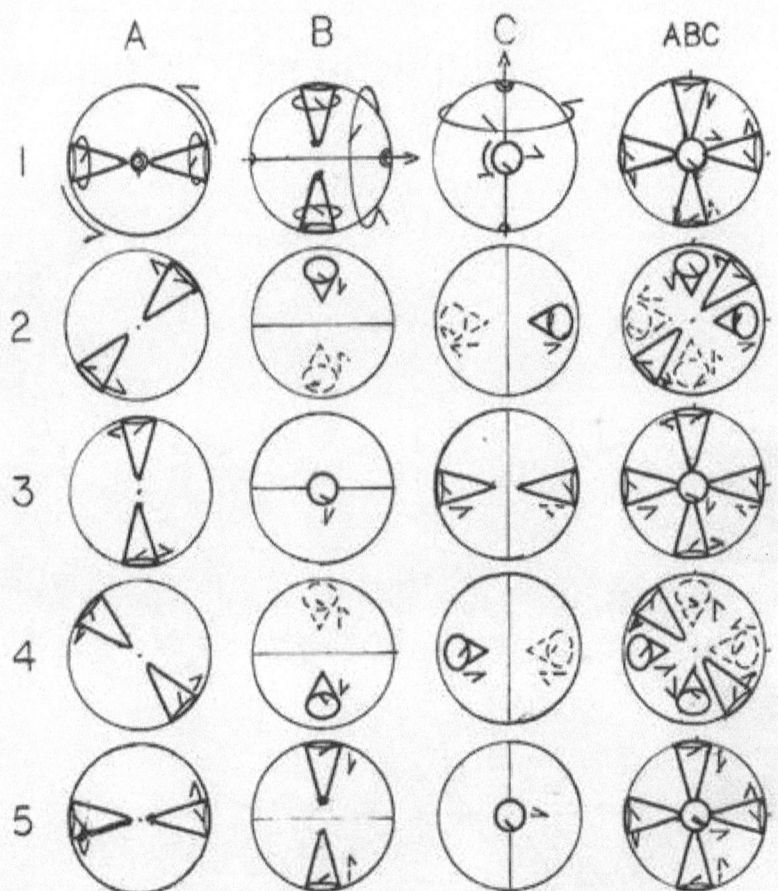

Fig. 5-2. Five Steps of Orbital Rotation Sketch of Three Pairs of Quanta in an Electron/Positron. Each drive cone shares the exponential increase of churning interaction generating mass energy in either type.

arise initially from viscous friction at the GD in each quantum, but increase exponentially throughout the electron by greater viscous friction between quanta from the three pairs of them orbiting together synchronously as shown in Figs. 5-2 & -3. Scaled adding up these contact shears cannot produce N_c^6 power (Eq. 5-3b-c). That requires turbulently radiating gyres with exponentially compounded viscoelastic waves penetrating all gyres in a particle for mass-energy eddies over the particle's volume, arising densely near the sphere surface where interactive currents peak. From there eddies are carried by the micro-quantal currents smoothly throughout the electron volume. Each eddy is itself a low energy tertiary level vortex at random orientation. Together they both couple the

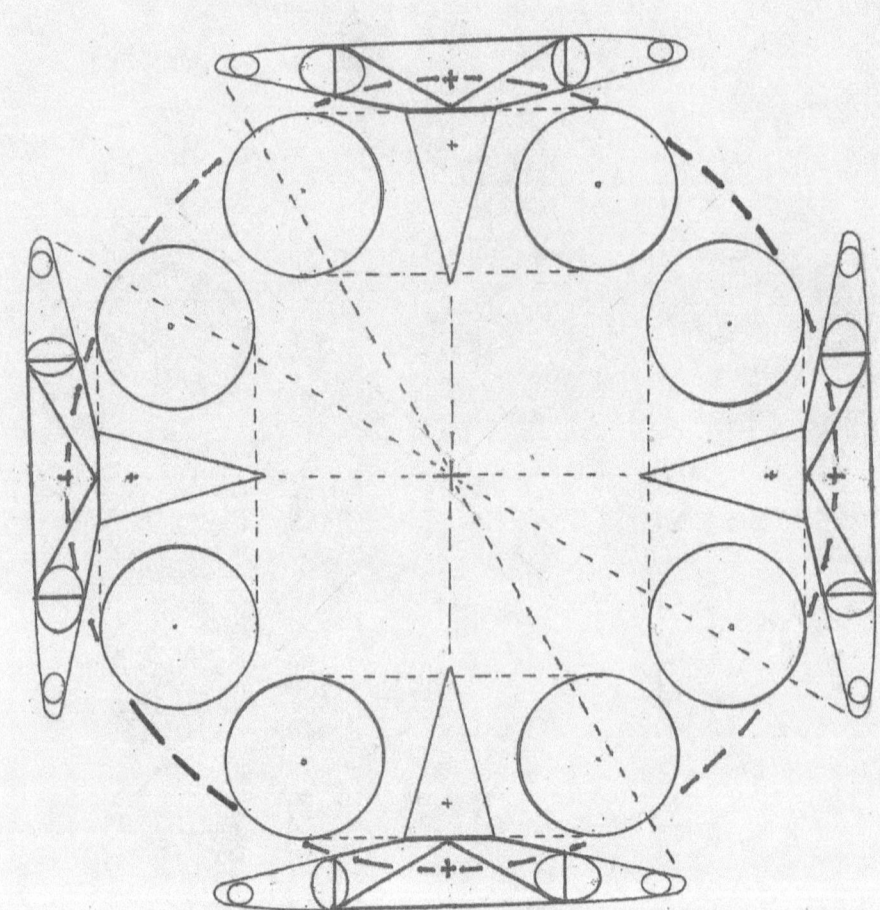

Fig. 5-3 Electron/Positron Axial Cross Section Sketch at Fig. 5-2 Steps 1-3-5. Centroid of each spiral wave sub-conic disk to GD turbulence limit is on the lepton circumference. Tips of full disks cover 60° arcs and come to contact at Steps 2-4 closest approaches near electron S pole. Strongly interacting main lower toroids of stored current energy around drive cone sides, here fully extended circular shells of peak energy cross sections at Steps 1-3-5, distort at Steps 2-4 pushing the toroid volumes in the mid 30° arcs (as dashed in for two quadrants) into the cone side flow increasing eddies & pressure slowing gyre speed V_G going into Steps 2-4 & accelerating gyres out to cross SEq at Steps 1-3-5 at peak V_G. These speeds project on SEq at steady c as required. Also at Steps 2-4, the weak outwash between outer spiral disks and toroids are briefly distorted with a very weak pressure rise. However, quanta around the drive cones are kept in nominal orbiting positions throughout the orbiting cycles by the balance of forces in centrifugal inertia, coaxial correction, and centripetal strong force due to point-thrust and interactions of currents along the cone sides as aided by the toroid interactions at both adjacent range and across the lepton.
entire particle to the local medium in an inertia effect, and also resonantly couple in a comparatively low frequency band to all other eddies in the viscous wave interaction of the attractive force of gravity out to infinity. However, each eddy is very small and low in

Fig 5-4 View Into Electron/Positron of Fig. 5-2 Steps 1-3-5 Toroid Interaction Currents. Seen with the one nearest quantum vortex removed, the five other peak toroid energy contours are briefly clear of each other, but the layers just outside these contours in their four visible near contacts and on the far side of the four quanta in side view are interacting in vigorous conflict with flow components ranging from exact conflict to wide angle crossings. (Simple 3D plastic foam models are helpful in visualizing the relative flows.) Since the axial angles between any two quanta (in the three coaxial pairs of quanta) are always perfectly symmetric in their interactions, the mutual force equations of Chapt. 4 are always applicable in computing instantaneous forces between any two quantal cones for sums of force balances on each one. energy, and the total coupling of force by all the eddies of one electron to another single particle at a significant GD-scaled distance is extremely weak. It requires very great numbers of particles (such as the order of Avogadro's number, 6×10^{23}) to have barely appreciable attractive force of gravity in ordinary terms with a similar group of particles. Eddies continuously die out or escape and are replaced as friction losses to conserved gyre drivers, that are in effect Planckian, ultra-dense, endless spin momentum sources.

The next higher level of force coupling between vortical quanta is between the much more intense sets of about six turbulent spiral waves set up secondarily (in the lab) in very shallow conic disks by viscous friction with a necessary medium off the central conic drive base of each uniform quantal vortex. Each of these spiral waves is itself a long, bent, externally driven, viscous conic vortex (Fig 5-1) which is also wrapped in turbulent shorter and smaller long vortices (not shown for clarity.) Wavelets radiated freely to infinity by these spiral waves from scaled separate interacting vortices resonantly couple their distinctive electric forces, that are polarized from either right-hand or left-hand rotation with respect to the base and axis of each drive cone. (This polarization was arbitrarily designated in an earlier century as either positive or negative electric potential force with respect to a selected standard charge of only fragmentarily known internal characteristics.) Here like-handed polarizations of the spirals in separate charges repel each other through the oppositions of their radiated wavelet currents with the other's internal spirals in interpenetrating interaction off bases. But unlike-handed polarizations attract because of the Bernoulli effect reduction of internal pressures (Prandtl and Tietjens, 1934) on each other's internal spirals when flows in contact move in the same parallel direction. This action continuously orients itself over time for best coupling of interaction. (Note that it is not necessary that the size of the drive cone of a primary vortex be known, as described in Chapter 2; the actual drive cone may be any one of a quasi-infinity of unknown cones with smaller dimensions and higher peripheral velocity but matching outer vortex currents and forces.) The 30° cone described and most computed is a probable upper limit of cone size options yielding the scaled forces.

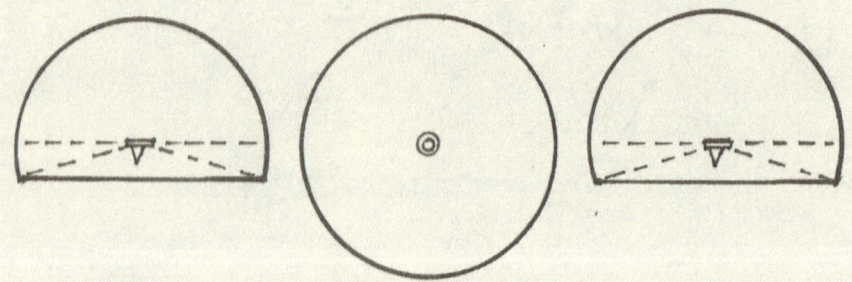

Fig. 5-5 Three-View Sketch of Electric Force Hemisphere Contour Around a Quantal Spiral Disk. Two of these around a coaxial pair of vortices with Points-In create one of three perfect overlaid electric force spheres that do not interact between one lepton's gyres, but only outside every electron/positron sphere. The dashed cone outline is free of force and cannot act on the spiral disk of any of the other five vortices.

The volume of surrounding space in which each spiral wave disk interacts with another is an effective outer hemisphere centered on the volume centroid of the disk with a very rapid decline of interactive forces into the fringe of the opposite hemisphere. With two coaxial vortices in a points-inward (PI) pair the fringe effect is barely sufficient to close out the offset between the two disk centroids (from which the electric force is effectively exerted) to beyond a GD from the spherical locus of the conic volume centroids (at which the scaled electric force is felt and measurable.) In effect at any exterior range with another somewhat dispersed spiral wave disk, the two hemispheres of electric force from a coaxial pair in an electron set up a seamless sphere of force. In the overlap zone, opposite spiral wave edges from the two coaxial quanta complete the spiral force field. The total electron effect then is from three overlapping and separately oriented spheric fields of potential force that are never aligned axially at less than 60 degrees from each other, making the field quite uniform in all directions at all times with no detectable scaled electric interaction from one pair to another within an electron or positron, especially not on average over a very short cycle of scaled orbital orientations.

This particular directional coupling relation between the spiral wave disks of separate quantal vortices orbiting in coaxial pairs in a scaled spheric particle, typified at its simplest by the electron or positron, is extremely critical and important, both to all-directional electric interactions between complete particles, and also to the failure over the entire last century of the body of world physicists to achieve agreement on any possible structure of the electron or other spherical particles. Typically, Mac Gregor was unable to go to a multipart model of the electron because there was no concept of charge force interaction being only hemispherically directed from each of two coaxial gyres. He was constrained by the apparently simple and general prior notion of an impossibly infinitesimal point charge which could not contain a means of generating force toward one hemisphere. In the confinements within an electron's apparent body, the forces between prior fractional point charges would also be quasi-infinitely high, as in Poincare's 1905-06 electron requirement for a distributed charge carried in an infinitely strong solid material (Mac Gregor, 1992). (The general concept of any particle as a wave phenomenon did arise mathematically, but without a structural embodiment or realizable wave structure, and it became almost an antithesis of formed structure.)

The highest level of force coupling is that of the strong force of particle cohesion which comes into play at very short range between the net Bernoulli effects of current components along the sides of the drive cones from cone points to cone bases, aided especially and directly by the jet reaction point-thrust of eddy-carrying currents that are viscously accelerated up the sides and escape at a base-ward angle between the strong lower and faint upper toroids as an expanded but weak circular outwash force. (This correlates with the QM strong and weak forces.) The side-flow coupling is also affected by the direct couplings of PI current components crossing each other between the energy-storing lower toroids of orbiting vortices as they approach each other to 60° spheric axial separation in electrons/positrons. At closest approach these effects are briefly heightened by a flattening distortion of the body of the toroid flow contours, in which the current components around the larger circumferences of the contacting toroids in these leptons are in viscous opposition. Such forces fend off like gyre pairs.

Where the interactions between the eddies are always highly randomized in axial asymmetry and have a net weak attractive gravity force, the two sources of very much stronger forces are always highly organized in internal symmetry within a lepton particle and have vigorously differing repulsive and attractive mutual force interactivities that <u>must be well balanced or directed systematicly to yield a definite summed mutual force.</u>
Folliowing written descriptions of electron current force relations as assisted by structural sketches may be improved by rough foam plastic models omitting the weak upper toroid, but with a small flat saucer representing the spiral disk, a foam cone, a foam cylinder notched and bent into a lower toroid ring, and step-by-step marker pen lines showing currents, plus 5 more sets on 3 axial dowels tied like Fig. 5-4 & -5.

For the measurements of particle forces as in Chapter 4, the forces within one particle are best described by defining symmetric directional angles between any two gyres' axes that intersect in a common center at a separation between the centroids of their drive cones. Each scaled measurement of radial force at short range is of the sum of forces along this line of separation as detected over their various surfaces by each one of two symmetric gyres with relatively broad apertures for reacting to force sums.

<u>Within</u> the electron all the individual gyres are <u>always symmetric</u> with respect to each other gyre. So the organized PI force equations of Chapter 4 are necessary and empiricly sufficient to deal with the <u>internal</u> structure step-by-step. The BI equations are necessary <u>between</u> particles when the gyres are <u>momentarily symmetric</u> and also give a basis for <u>scale estimates</u> when gyres are <u>not symmetric</u>, with added hints in Fig. 4-10. (The unfinished relations of any two <u>asymmetricly</u> arranged gyres <u>between two separate</u> particles do not affect or limit the ability to deal with the <u>interior of one electron</u> or of any similarly <u>symmetric</u> particle. This also applies to the Usual neutrinos and to the spheric orbit parts of the quarks. The gyre pairs in the linking orbits of a quark are necessarily <u>asymmetric</u> to its other interior gyres and usually to all pairs of other linked quarks.)

In Figure 5-2, and in Figures 5-3 and -4 when they are completed with all six gyres, three coaxial pairs of schematic electron vortices balance masses perfectly (for long stable life) in synchronous orthogonal orbits to stir their total mass eddies up to the equatorial c velocity equivalent required by Mac Gregor (1992). Since the LQ power law equations for mass (Howard, 2005, 2006) of Chapter 1 are derived in electron-Volts (eV) from PDG LQ and QH data, their necessitating 6 electron gyres defines mass-energy and 1/6 charge of universal microquanta and systematicly correlates these elements of electron structure with the other massive particles as a prototype particle model scaled in GD units with inherently interpenetrant interaction entangling a particle-wave duality.

A separate determination herein of 6 electron vortices in a scaled force balance confirms the model. Lab data were recorded for vortex flow velocity, size, and force measurements, and reduced to empirical equations for an iterative solution of forces. In model, PI gyres are at $+\alpha$ relative axial lab angles (Fig. 5-2). The exterior conic bases are at $-\pi \leq \alpha \leq 0$ (BI) relative axial angles between gyres in separate electron/positrons. Forcefully interactive shear/pressure waves and currents occur around the sides of gyre drive cones inside particles (Figs. 5-3 and -4) and on the spiral-wave bases and sides between particles. Data were measured (Figs. 4-3, -4, -5, -6, -7, -8, and -9) in 3D PI and BI by camera and pendulum instruments on symmetric dual (and single) underwater (& near-surface) vortices, viscously driven by right circular cones of $\pi/6$ or $\pi/9$ (30°/ 20°) whole-cone angles β, to centrifugal spiral GD turbulence, with toroid rings, and eddies.

Two PI conic lab gyres at high α, either in − − or ++ 1/3 like-charge pair duos in electron/positrons or in neutral +− pairs in other particles, usually force each other laterally to $\alpha = \pm\pi(180°)$ coaxial pair alignment, as shown in Fig. 4-9a. Gyres at low α may resist alignment by lateral force. Similar orthogonal lateral forces make gyres circle each other and drive gyre pairs in orbits. Thus each gyre of a conserved pair stays at $\alpha = +\pi$ (+180°) from its mate and $(+\pi/3) \leq \alpha \leq (+2\pi/3)$ [(+60° to +120°)] symmetricly from other gyres, as synchronized in entangled resonances. Orbital radii vary in seeking force balance within quark particles, but in electrons all forces balance at equal radii. Since PI symmetric like-spins on co-axes are opposed, each gyre pair has no net spin, nor gyroscopic momentum, in electrons/positrons. Pair orbital rotations drive these particles' spins. And like-spin currents oppose in creating maximized mass eddies.

As noted earlier, the spiral-wave sub-vortices (Fig. 5-1) in each pair's base gyre-diameter (GD) turbulent disks generate the paradigm's electric charge forces which act from the volume centroid (CV) of each GD disk <u>outwardly</u> across a solid angle slightly >π(180°) wide (Fig. 5-5) to spread evenly <u>outside</u> the sphere over the orbit as required (Mac Gregor, 1992). With ≥$\pi/3$ (60°) between gyres in a sphere, the spiral wave forces cannot interact <u>inwardly</u> between GD disks in the same electron/positron. These scaled structures have no cause for last century's QM "Poincare'" problem (Mac Gregor, 1992: Waite, et al., 1997) of theoretically quasi-infinite self-repulsive and self-destructive electrostatic force between the parts inside each electron, and so, no necessity for infinite rigidity of the sphere rotating the charge (rather than balanced internal forces.)

Three gyres (in 3 pairs) at their closest +$\pi/3$(60°) approach (Fig. 5-2, rows 1, 3), between primary orbital gyre poles of a sphere matching the predominant gyre spins (Fig. 2-1), move around the spheric octant wherein the unique centroid (Fig. 2-2) is invariantly the primary summation pole S. Thus S is equally distant at θ=arccos(1/$\sqrt{3}$) (54.73561...°) from each pair's primary orbital pole (in Fig. 2-5), as if this orbit structure is the inherent cause of that QM quantization axis angle (Mac Gregor, 1992), which is necessarily relied on in Mac Gregor's analysis for the QED properties of the electron.

At 3/8 cycle (Fig. 5-2, 4th row from the arbitrary row 1 start-point with its given rotation direction), each gyre crosses the S equator at the angle of θ=arccos(1/$\sqrt{3}$) in the same (here CCW) projected rotating direction, but half the coaxials go in opposite polar S directions, with $\pi/3$ (60°, π=180°) between crossings. After a half cycle all cross again in reversed polar directions at twice the orbital frequency (~$\sqrt{3}\times10^{21}$ below.) This Sum Equatorial (SEq) contraction of the gyres, gathering near both S poles, and contraction again to the S equator matches the QM classic electron "zitterbewegung" trembling action (e.g. Haisch et al., 1994; Mac Gregor, 1992; Hestenes, 1991; etc.) In this the three orbits act in balance around the S axis, which becomes the principal action axis rather than being only an accessory for mathematical summing of rotation. This would keep the orbits on orthogonal ABC axes. This is exactly the orbital structure for evenly stirring eddies throughout the electron at the fifth exponential power of the number N of quanta in determining mass energy of the electron/positron. It is also the reference for ++ & -- orbits in synchronizing 3 heavier quarks with C axes parallel and AB balanced.

Thus, in an exactly coherent structural sequence, with the electron/positron as the defining prototype for all particles, none of the electron-positron internal orbits would be on the Sum Equator. For this, all three orbits would be at an angle of 54.73561...° to the SEq plane at a necessarily crossing velocity of c times $\sqrt{3}$, which would project at c on SEq at the crossing (as if it were projected also at the crossing instant on a reference plane perpendicular to the SEq plane and tangent to the SEq at the crossing point.) At two crossing times per orbit the 6 each 1/6 charged gyres clearly sum to the electron's charge on the S Equator at a projected velocity (Fig. 2-3) to match with Mac Gregor's analysis of all the QM and spectrographic requirements exactly at each of those instants - - and between crossings with apparent possible indeterminate deviations that are oppositely balanced around the crossings (but vanishing in that particular reference plane at plus and minus 90° in orbit from the crossing, and also reduced to minimized significance by the modulation of reduced rotation velocity V_R before and after crossing under the influence of the varying summed viscous current rotational force F_R involved in the maximum generation of eddies at the octant of closest approach between gyres with peaked concentration of that component of the force which opposes rotation as noted with Fig. 5-3 on page 116. Three gyre symmetric mutual force confirmation is not feasible in the Chapter 4 lab.) But, even though balanced and forcefully minimized, if not also insignificant, indeterminate deviations may have a logical alternative next.

The main alternative directly or apparently useful structure would unify all three orbits of the three pairs of one sixth charges to one single orbit at the velocity of exactly c in the SEq plane with the 6 each 60° angular phase separations for the coaxial pairs (from Fig. 2-6) located at a smaller radius on the sphere rather than in the expanded link orbit shown. In electron/positrons this would retain many functions except for losing the mechanism for classic "zitterbewegung" and, furthermore, making the roundness of the electron sphere totally dependent on the outer hemispheric field of each spiral wave disk, not on gyre structure within the field. (This would have serious uncertainties as to a planar form of the electron, another type of deviation, in the highest velocity impacts which do actually impact, rather than merely approaching closely and being repelled away in a continuous change of trajectories before impact can occur.) An important prototypic loss would be simple and clear-cut pre-qualification in this lepton of the basic spheric parts of quark structure with essential orthogonal orbits for the somewhat larger quarks. Such orthogonal orbits have the critical feature later of definitively causing perfect balance and stability for very long life with single down quarks in protons and also a slight imbalance (plus lower force-to-mass ratio) and much shorter mean lives with dual down quarks in isolated neutrons. - - Which of these alternate orbit structures best meets the electron necessities may not be entirely clear here, and may not become completely definite to physicists generally, until the forecast future additional extensive experiment and theory fully establish this empirical electron's orthogonal orbit functions.

(There are potentially other more complex empirical structures that may then also be considered, if necessary, but with further questions on achieving the automatic exact synchronizations of orbits and spin sites that occur in these orthogonal orbits. Such not-recommended attempts may explore static spinning, rather than orbiting, of the three electron/positron gyre pairs at three of the four coaxial pairs of octant centroids around the S pole in the fourth, or orbiting for mass eddies, try corrective orbit offsets, harmonic departure from orbit circles, or even periodic reversal at mid-latitudes of the pole-ward component of orbital motion, etc. There are clearly serious drawbacks in all those, but not in the even-harmonic variation of speed in the quasi-circular, near-elliptic orbits.) Since the three orthogonal orbits in the electron/positron are the most complete and comprehensive structure for meeting all the inherent empirical necessities and functions of that essential particle and its anti-particle, and at the same time contribute to the overall development of LQ and hadron particles in the most coherently systematic way, it is the most broadly functioning, existential empiric paradigm base. (The limited alternates are not developed further here.)

However, full establishment of any micro-gyre quantum component of electron-positrons in any overall structure can occur, even empirically, only with one additional inherent special necessity (taken up next) that has been present from the start of consideration of vortices, but has not yet been brought forward and emphasized as a consequence of such gyres, or of any other internally moving electron structure.

A Pervasive Electron/Positron Structural Necessity

For this orbital velocity projected on the S Equator ever to equal c exactly, as ideally required by Mac Gregor (1992), the effective charge center of volume (CV) of the GD disk of each orthogonally rotating gyre (or its infinitesimal fringe, perhaps) must orbit faster than light c in crossing the Sum Equator, at orbital velocity of the gyre $|V_G|=c\sqrt{3}$ cm s^{-1}, as if that is classically possible. Gyre internal peripheral velocity of the driver (V_P) must in any case further exceed the orbital velocity by at least an order of magnitude to maintain vortical activities under the accelerations of orbiting in the three separate orthogonal orbits (or in that discarded single planar orbit that can not always appear exactly spheric in impacts.)

This dilemma of classic relativist impossibility forces concluding that, since early last century, trial electron structures have been pushed into self-defeating compromise with another aspect of the same issue, the impossible infinite singularity of mass at an ultimate speed limit of c. (In the atom Rutherford did not require the c stumbling block! -- If the photon is a true particle, it must not quite reach the limit either!) Elaborate stratagems (Waite, et al., 1997. Mac Gregor, 1992. Barut, 1991) to elude this mass singularity in electrons have not been generally accepted. Most present physics avoids the issue with mathematical point electrons. But many physicists (e. g., Springford, 1997. Dowling, 1997. Hu, 2004), especially Barut (1991), Hestenes (1991), and Mac Gregor (1992), have searched at length for reality-related structure within such a stated singularity limit.

New necessity for real electron structure arises in empirical "structure functions" of particle scattering (e. g., Slominski, 2001 and 2005. Muryn, et al., 2004. Etc.) Many discuss the importance of a physical basis for electron properties (Hestenes, 1991. Barut, 1991. Luty, 1997), and there is another finding (Dugne, et al., 2002) of necessity for composite structure in light leptons, including electrons. Existence of real, invisible structure is strongly indicated.

Similarly, empirical proof of real existence for superphotic action of at least one type should now be considered established as the simplest assumption in reliable entanglement (e. g., Pan, et al., 2000. Mair, et al., 2001. Sackett, et al., 2000. Rarity, 2003. Aspelmeyer, et al., 2003, and subsequent papers) between electrons and electron-related photons, "holes", qbits, etc., as a generalized link between entangled entities, without any established and definite physical mechanism. These real tests also demonstrate extended range for supposedly "instantaneous" entangled couplings. This now bulky empirical evidence clearly demands removal of a false restriction, the now classical impossibility of actions above c, at least in entanglement by and between components inside particles (the 100-year extended controversy on the existence of general entanglement mechanisms to the contrary notwithstanding.)

Furthermore, the classic general rule (Mac Gregor, 1992) of limiting all velocity at c has not been demonstrated for anything other than photons, gravity, and identifiably separate particles such as whole charged leptons and hadrons, especially not for quark components within hadrons, for instance. Generality of the rule for sub-particles is only a cautious assumption. Propagation of gravity force correlates with c and causes the peak of mass effect near c. But there is a difference in the propagation of charge force; charge escapes the c singularity peak with velocity! A major aspect of charge is empiricly not constrained to c, especially not in the same sense as mass is constrained by the mass singularity at c. And in this MQP paradigm of particle structure there are two distinct sub-sources of mass-related and charge-related forces in the single source.

In this paradigm, interpenetrating and radiant pressure-wave forces of gyres from both their V_G movement and the necessarily much higher spinning V_P of the peripheral boundary layer on both the spiral-wave cores and the cone-base structure driving each gyre and its spiral-wave set inherently provide resonantly coupled entanglement mechanisms at >c to resolve the empirically demonstrated necessity of real, if turbulently complex, causes for real entanglement switching effects, if not also for the absence of charge build-up near c. (Cyclic V_G motion frequencies must necessarily modulate the many-orders-of-magnitude higher velocity and higher frequency V_P pressure waves below in many bandwidths of finite Q for quasi-infinite simultaneous numbers of mutual resonant tuning interactions at short range and subsequent stable radiation coupling to remote range. See detailed Appendix 5-A at end of the chapter.)

Still, there is the real and large increase in apparent mass of particles at speeds just below c which require exceptional energy and are thus severely inhibited (as is open observation at >c.) For this, general equation coefficients of mass with velocity must include a typical relativity term (Mac Gregor, 1992), as f(v/c) where v<c, which steeply approaches a reverse-curved asymptotic peak just beyond the confirmed classical masses near c and vanishes above c, with a continuing subluminal to superluminal velocity term, f(v), in the general form:

$$C_{mv} = C_{m0}[f(v) + f(v/c)] \tag{5-4}$$

where $f(v)=f_0(\upsilon)\approx 1$ very closely near and below c (Eq. 5-5e below).

Transonic rockets are partial analogs, with peak fluid resistance from accumulated sound pressure waves propagating on body surfaces in mixed shear/pressure standing waves until broken through with shock pressure waves (Jones & Cohen, 1960). Here, in partial analogy, low frequency mass-effect (eddy) shear waves with very small pressure, initially from the edge-collapse of GD turbulence energy inside a gyre and then further from viscous gyre current frictions throughout a particle (filling its interior), set up radial wave trains outward through and between the outer GD disks of the particle. The waves become temporary components of smeared layers of shear/pressure standing waves on the outer GD disks of the particle. These mixtures of large numbers of weak wavelets at random phases from random eddy axes then propagate outward at the speed of eddy $|V_P|\sim c$ with sphericly phased wave fronts. In addition, the main spiral-wave electric force empiricly propagates at c (Feynman, et al., 1964. Hestenes, 1991. Springford, 1997. Dowling, 1997. Salam, 1989) a smeared shear/pressure component from each vibratory spiral wave core and its multiple sub-spirals, also temporarily in summed standing waves at the moving surface (static Fig. 5-1), before further propagating away with many relatively flat plane pressure and shear wave fronts from many narrowly phased spirals with highly organized axes. These distinctive polarized wave fronts are scattered smoothly over the outward hemisphere, and averaged over each V_G cycle, at two extremely different frequencies, the lower shear component of which is precisely at c. The mixed exterior waves compress the GD faces of gyres of particles moving near c (in iterations below), creating a more condensed interaction of greater mass-energy, and over-inhibiting speed of whole particles to <c. This process of wave components at c is evidenced by Newtonian mechanics, Maxwell's equations, electromagnetics, Einsteinian relativity, and QED, since these theories accommodate particle speed limitation and work primarily with observables of entire particles and waves at $\leq c$. - - But those effects would not reduce the interpenetrating higher hyperphotic speeds of relatively unobservable V_G and V_P vibratory pressure waves on and within substructures of individual component gyres at much higher frequency bands nor interfere with their empiricly hyperphotic, resonantly tuned entanglement phenomena (which are not tuned to resonances involved with emission and absorption of photon energy in later chapters.) Those sensitive and higher Q (sharper) resonant coupling phenomena are also fully compatible with the well-known interactions coupling with their own special resonances at c wave velocities.

In this way, with the combined benefit of Equations 5-2 and 5-3a to –c from Chapter 1, of their generic particle structure consequences from Chapter 2, and of the introductory lab vortex and gyre force data of Chapters 3 and 4, resultant GD disks of electric charge, in gyres with mass energy, orbit forcefully in massive electron, other lepton, and quark (LQ) spheres (and links) above velocity c (to $c\sqrt{3}$ in the near-ground state.) At the proper angle to the SEq plane, this meets Mac Gregor's (1992) basic analytic requirement for components of electrons and provides structure which will throughout the LQ particles herein inherently interact with all the past known and forecastable future forces in the empiricly observed reactions and extreme variations of particle lifetimes which support the Usual apparent complexities of Nature near the surface of the earth. (Effective completion at that level can be forecast to demand decades of additional concentrated and well-supported lab research and theory development.) This assembly of simplicities also begins herein to resolve many Extreme peculiarities of cosmic and local Nature (such as entanglement, micro- to macro-neutrinos, and pre-quarks) that are empiricly unavoidable and impose extended requirements on the hadrons through the LQ particles, on their components, and on their vacuum medium in which "charge" and "mass energy" can causally occur.

(A great deal of new physics that will undoubtedly take some centuries to develop fully is inherently necessitated, initially by the rest of the electron's empiricly scaled structure, as it continues in numericly quantitative detail within the prior requirements.)

Structural Scaling of the Electron/Positron

The force equations of Chapter 4 do not scale down twelve orders of magnitude to the micro-scale of particles without being further generalized empiricly to fit the known hard data on the half-diameter or radius of the electron as analysed by Mac Gregor (1992) from a century or more of repeated efforts to measure it by all of the available methods. In that analysis, employing the resources of quantum mechanics, relativity of whole major particle movement (as if in a rigid crust), and quantum electrodynamics (QED) from the established work of many outstanding others, in a long series of his published and refereed papers that were condensed into the cited monograph, Mac Gregor found that only a particle with slightly greater than $\sqrt{3}$ times the previously accepted Compton radius could have all the properties measured and accepted by physicists generally for the electron. This new empirical investigation does not further challenge Mac Gregor's summary of that century of the best-qualified world-wide research on the electron, but accepts it as the recent bench-mark. [Quantum Chromo-Dynamics (QCD) likewise establishes some necessary quark and baryon requirements that must be accounted for in their structure and in the meson debris of baryons and pre-quarks.]

Accordingly, the equations taken at this point from Chapter 4 lab data are given new within-chapter numbers in a modified sequence due to their new generalization adaptations here to extend the equations coherently down to the micro-scale range of sub-nuclear particles and leptons, as needed for the electron/positron (and prototypicly for the other particles.) Otherwise, the equations themselves recognizably match those of Chapter 4, and include them in the broader generalities, while continuing to be consistent with Table 4-1 Schedule of Applications in relative axial angles and other characteristics. Wherever the equations and table are to be applied to other particles, the same generalization adaptations as applied here are again required for equations with the same terms. Gyre quantal mass (Equation 5-3a converted to grams) at gyre orbit speed at electron radius determines the pseudo-centrifugal force that must balance with these gyre forces scaled to electron size for the gyre structure to be validated.

Vortex forces and mass scale to electron size through general Equations 5-2 to -4 from Chapter 1 and adapted Equations 5-5 to -17 for lab 3D measurements between symmetric underwater PI gyres from Chapter 4, where peripheral spinning velocity V_p is at the scaled velocity contour in the boundary layer on the lab drive cone [832.6 cm sec^{-1} at 1X drive gear ratio with whole cone angle $\beta=\pi/6(30°)$ and base diameter $d_c=7.9375$ ave. cm.] The lab GD reference length of 20.3 cm diameter of central base turbulence (Figs. 5-1 & -3) (Eq. 5-5, initially scaled up 10^6 for tornado and hurricane trials without change) is generalized in iterations here for scaling down 10^{-12} by Equations 5-5a-e:

$$GD = d_c \left[1 + e\left(\frac{\rho}{\eta}\right)^{0.6667} V^{A\sqrt{V}}\right], \quad (5\text{-}5)$$

where $V = |V_p|/1000$, e is the base of natural logarithms, and $A = (2A_b + A_s)/100$ with drive cone surface areas in cm^2 (or km^2 in weather.) When not varied 4:1, fluid viscosity η in cp and density ρ in gm cm^{-3} each approximate 1 in simple lab water [and $(\rho/\eta) \cong 1$ in a simplest space medium.] The exponent of V here generalizes:

$$C_{AV} A^{Q_A} V^{Q_V} = A\sqrt{V} \quad \text{(at lab to weather scale, GD} \geq 5 \text{ cm)}, \quad (5\text{-}5a)$$

where:
$$C_{AV} = 1 + [\{\log (Q_A)^3 / (f(\upsilon))^{0.00889}\} - 0.097315715]^3, \quad (5\text{-}5b)$$

$$Q_A = \{1 - [(\log V) / \{1.4002\pi^3 (f(\upsilon))^{0.015}\}]\}^2, \quad (5\text{-}5c)$$

$$Q_V = [0.5 - [(\log A) / \{2e^3 (|\log| |\log V|)^{1.3366}\}]][2 - \{f(\upsilon)\}^{0.0122}]^{1.0001}, \quad (5\text{-}5d) \text{ and:}$$

$$f(\upsilon) = f_0(\upsilon) = \exp(\chi^\psi), \quad \text{[conventionally e to that exponent]} \quad (5\text{-}5e)$$

where $\psi = \pi^\pi$, $\chi = (\upsilon/e)$, and υ is the exponent in $|V_p| = c^\upsilon$, ($0 < \upsilon \to 3.3+$) with c in cm sec^{-1}. [In Eq. 5-5d the last factor $\to 1$ when $|V_p| \ll c^{2.5}$. Per Eqs. 5-5c-d, first exponents A and V (Eqs. 5-5 & -5a) exchange second exponents from lab to $|V_p| = |c|^{1.5}$, as $f_0(\upsilon)$ adjusts scales for a causal medium (Appendix B), which apparently is the new factor that makes the equation fit from the lab to the particle empirical conditions complex.]

Initial Force/Size Estimates & Final Balance

For initial electron estimates of a suitable range of forces at which to begin calculated iterations that will approach a target size for the gyre GD and other values in the electron within a reasonable number of iterations, it is advisable to begin with a calculation (with handbook equations) of what the PT centripetal force must be to hold a gyre's quantum of mass from Chapter 1 in an orbit the size of Mac Gregor's electron at the gyre orbital velocity just determined above, without involving at this initial point the repulsive forces from the other five equal gyres whose effects must be added later. (This action is so far above c velocity that sub-c relativity does not apply.) From that PT range estimate it is possible to approximate other initial values. The ratio to gyre self-force $|PT|$ (Eq. 5-7, Fig. 4-3) of the average lab net repellant mutual forces $|F_{Mave}|$ between cone CVs (Eq. 5-6a) from the lab graphs (Fig. 4-7c) at the limiting axial angles for gyres and trial sizes, can estimate (Eq. 5-6b) an initial range of feasible gyre GD to particle radius ratios for stable particles in any scale. (Similar $|F_{Mave}|$ is later computed at scale.) Each lab (or computed) mutual force point $= |F_{MRGD}|$. Projection on a particle radius to a given gyre $C_{FM} = \cos[(\pi - \alpha)/2]$. Weighting is by number $M_{ovo} = 1, 2, 4$ of other gyres in each angular relation for each repeated quarter cycle quadrant arc in:

$$\frac{1}{2} \sum_{quad.arcs 1}^{2} \left[\frac{\sum_{\alpha states 1}^{4} (|F_{MRGD}| C_{FM} M_{ovo})}{4} \right] = |F_{Mave}| \quad (5\text{-}6a)$$

$$\left[|\mathbf{F}_{\text{Mave}}|\Big/|\mathbf{PT}|\right] \cong 0.45 \pm 0.25 \quad \text{(Estimated feasible.)} \tag{5-6b}$$

Inverted to 1.43 least ratio, this is close to the particle radius at cone CV of 1.375 GD, the most compact ratio for limited distortion of toroid flows at $\alpha=+\pi/3(+60°)$ closest approach between gyres (Fig. 2-6). (This survives iterations.) The electron quadrant cross section at $\alpha=+\pi/2$ (+90°) (Fig. 5-3), shows the resulting GD CVs at 1.6875 GD radius. Equating this to the analytic electron radius (Mac Gregor, 1992) of 6.6962×10^{-11} cm, sets a scaled GD=3.9681185×10^{-11} cm reference target for loop iterations (Eqs. 5-5 to -16) with $\beta=\pi/6(30°)$ (matching most of the lab data.) Then iterations with error reduction must converge to balance forces on gyres at GD null error (Eqs. 5-5 to -5e).

Self-force for each gyre is jet reaction axial thrust toward its cone point, PT (Fig. 4-3), from axial fluid intake at low velocity (Fig. 3-3) and radial ejection ~10× faster, angled toward the base between lab-measured arctan 1/4 (14°) and arcsin 1/2 (30°):

$$|\mathbf{PT}| = \frac{\sin\beta}{10\pi}\left(\frac{\rho}{\eta}\right)^{0.25} |\mathbf{V}_p|^{1.525} \frac{A_S}{9.807\times10^2}, \tag{5-7}$$

where A_s is the cone side area in cm², 9.807×10^2 is the conversion factor from dynes to grams force (for sea-level, lat 30°N calibration convenience.) Other terms were defined.

The size of orbits of GD CVs at V_G are then balanced in re-corrected iterations over orbital cycles for PI repellant mutual forces F_{Mave} between cone CVs of all gyres [at separations R>0.7GD (Fig. 4-5f, Eqs. 5-8 to -17 below)], against a gyre's stronger PT (Eq. 4-3) [less the gyre's ordinary pseudo-centrifugal force (F_{cf}) per handbook equation for its individual mass at the previous average ground-state V_G] at convergently determined gyre peripheral velocities V_P and null error GD dimensions (Eqs. 4-1 to -1e). When balanced by successive iterations with V_P corrections, the electron couplings of forces keep it strain-free from relative movement between orbit locations and effectively rigid (as in Mac Gregor, 1992) under constant ground state conditions. It will seek new balances under any external accelerative forces on gyre GDs (transferred to cone bodies.) Gyre quantal mass (Equation 5-3a) at V_G and electron radius determines F_{cf}.

Equation 5-7 and Fig. 4-3 show that the viscous boundary layer of a $\beta=\pi/6$ (30°) cone at lower V_P models PT of a smaller $\pi/9$ (20°) cone at higher V_P. Fig. 4-1 (Eq. 5-5 to -5e) likewise shows reciprocal modeling of GD at the same two V_Ps [where mutual forces also match (not shown.)] The two fluid vortices are then equivalent; and matching either property is sufficient for equivalence (with significance below.)

At any V_P, the mutual force $|F_M|$ (Eq. 5-8) in grams, along a radius of separation R in GD scaling units between two co-/contra-spinning symmetric drive cone CVs which are symmetricly spaced (Fig. 4-4a & b) at their axial angular separation α, is controlled by two variable coefficients to be determined (TBD), vertically (in strength of force at a GD separation) by $f(\alpha)_{\text{TBD}}$ (Eqs. 5-9 to -11), but predominantly (in both graph coordinates) by $f(R)_{\text{TBD}}$ (Eq. 5-12) (Fig. 4-5f), with further variables (Eqs. 5-13 to -17).

[The electron/positron at atomic orbit speed or below uses only the central negative slope of the upper curve (Fig. 4-5f) internally, but uses the left-hand high positive slope of that upper curve internally when compressed to below 0.7 GD between cone CVs by the external forces of acceleration to collision speeds near c, or uses all of both curves in external interactions.] As generalized here and in iterations for scaling:

$$|\mathbf{F}_M| = \frac{1}{2}\frac{\rho}{\eta}\frac{A_S \sin\beta}{9.807\times 10^2}|\mathbf{V}_P|\left[\frac{|\mathbf{V}_P|^{0.525}}{\pi^u}\right]^{ut} f(a)_{TBD}\, f(R)_{TBD}, \qquad (5\text{-}8)$$

where $u = 1-(1/z^8)$, $t = 1+[(6/z^{0.45})(ev/e^v)]$, $v = (z-1)/(z^{0.667}-0.6)$, $z = |V_P|/800$, and [in Eq. (5-10) below] $z' = [3/(0.5z + 1/z + 1/1.5z + 5.703/z^{16})^{0.5}] + \log\log z$ with its first term vanishing at electron scale. Other terms were defined. This yields the lab equations at $z = 1$, where $ut = 0$.

The following vortex lab equations apply by simplified electron/positron rules below (from Table 4-1 and Chapter 4) due to almost unique (among all particles) lack of internal oppositely charged components (Howard, 2005).
In same rotation at $+5\pi/6 \leq \alpha \leq +\pi$ (+150° to +180°):

$$f_H(a) = 1-(1/17)\tan[(4a)/9] \ . \qquad (5\text{-}9)$$

For $\alpha = +\pi/2$ (+90°), with scale change by z':

$$f_{HV1}(a) = f_H(a)\,\frac{1}{\left(\frac{3z'-0.75}{2}+1\right)} \qquad (5\text{-}10\text{a})$$

For most remaining ranges of $+\alpha$ with scale change by z':

$$f_{HV2}(a) = f_H(a)\,\frac{1}{\left(\frac{3z'-0.75}{6}+1\right)} \ . \qquad (5\text{-}10\text{b})$$

But for α near $+\pi/3$ (+60°) and lower:

$$f_B(a) = 1 + \left(\frac{2ye}{e^y}\right) - 2.5\sin(|a|/9) \ , \qquad (5\text{-}11)$$

where $y = |\alpha|/(\pi/6)$, and e is the base of natural logarithms.

H=+1 for co-rotating gyres F_M (Fig. 4-5f, upper curve) with its reversal as $R\rightarrow 0$ (which does not occur within the electron at low velocity as in atomic orbits):

$$f_R(R) = \left(\frac{1}{2} + \frac{H}{2} - \frac{16}{(1+3R)^2}\right)\left(\frac{1}{(1+R)^{R-1}}\right) + \left(\frac{H}{(1+R^2)}\right), \qquad (5\text{-}12)$$

Here, $R = R_{i,j,k,l} = R_r C_i C_j C_k C_l$, where real R_r is in actual lab GD units before shrinking/stretching the GD scale for other variables with $C_i C_j$... that are each equal either to 1 or to any coefficient (Eqs. 5-13 to -17) listed later for the case (in GD's principal effect). For lab V_p effects in all cases (as in Table 4-1):

$C_V = z'^{z'}$, with scale change by z': $\qquad (5\text{-}13)$

For $\alpha = +\pi$ (+180°) in co-rotating immersed gyres with scale change by z':

$$C_{311} = 1/(3\sin\beta)^{3\log z'} \quad (5\text{-}14)$$

For $\alpha = -\pi$ (-180°), with scale change by z':
$$C_{313} = 1/(3\sin\beta)^{3\log 3z'}. \quad (5\text{-}15)$$

For $\alpha = -5\pi/6$ to $+5\pi/6$ (-150° to +150°), Eq. 5-15 is used with:

$$C_T = \frac{1}{1 - \frac{2}{3\pi^2}\tan^2 0.485\alpha} \quad (5\text{-}16)$$

For iterative forces, $+F_{Mave}$ at $|V_P|$ (Eq. 5-6a) sums α states in Eq. 5-8 with varying deltas and smoothing over each arc for $|F_{MRGD}|$ using:
For the α arc near $+\pi/3$ (+60°), Eqs. 5-10b, -11, -12, -13, and -16.
For the α arc around $+\pi/2$ (+90°), Eqs. 5-10a, -12, -13, -15, and -16.
For the α arc near $+2\pi/3$ (+120°), Eqs. 5-10b, -12, -13, -15, and -16.
For the α point at $+\pi$ (+180°), Eqs. 5-9, -12, -13, and -14.

Then for any gyre, the projected outward repulsion force along the particle radius to the gyre $+F_{Mave}$, the pseudo-centrifugal force $+F_{cf}$ at V_G below, and the radial inward $-PT$ at $|V_P|$ (Eq. 5-7) converge iteratively from trial inputs of $|V_P|$, d_c, radius ratio, and the listed generalization factors, to a force balance from all the simultaneous equations with GD error null for particle size. ($|V_P|$ is so high that $|V_G|$ Doppler effects are neglectable.)

The 10.9525 eV interpenetrating wave mass energy of each gyre m_u as if alone, before stirring the frictions of the entire electron volume, appears in classic pseudo-centrifugal force F_{cf} at $|V_G|=c\sqrt{3}$ at the mean electron radius of the cone CV and the GD disk CV, or at 1.53125 GD. [At $c^{1.5}$, $f_o(v)$ (Eq. 5-5e) is 1.00000004; it is negligible at V_G.]

Iterations are cut off at $|V_P|=|c|^{1.5}$ (in cm s^{-1}) with stable convergent balance of forces between gyron micro-quanta (Eqs. 5-5, -7, -8, etc.) within 1.2% and very slowly converging to: $F_{Mave}+F_{cf} = -PT$. Here 77% of PT of -1.126206 dynes centripetal force holds each gyre mass in orbit, and 23% balances average mutual repellant forces from other gyres. Each gyre's $|V_P|=|c|^{1.5}$ flow contour is at $\pi/6$ (30°) cone $d_c=1.5872474\times 10^{-11}$cm. The computed (Eqs. 5-5 to -5e) GD = $3.968625094\times10^{-11}$cm, with a scaling deviation of 0.0128% above the GD error reference. This forms a sphere (Figs. 5-3 and -4) with 6 charge centers (CV of GD) at the required electron radius (Mac Gregor, 1992) of 6.6962×10^{-11} cm to yield electron properties at projected c on the spin equator of an electron at typical whole-particle speed <0.01c in an atomic orbit or slower.

Ultimate Vortex Driver Convergence & Energy

The iteratively convergent structural scaling solution (Eqs. 5-5 to -16) above also is required to converge simultaneously (Eq. 5-5 to -5e, discussion of 5-7), in the same scaled vortices, on an alternate, more nearly ultimate drive cone for the case of an electron forcefully accelerated to approximate c velocity at impact in an electron collider rather than the usual mathematical point electron. The same cut-off cycle finds a GD deviation of 2% higher than the ratio to Mac Gregor's maximum limit of electron radius in these conditions, for $d_c=1\times10^{-15}$ cm on a $\beta=\pi/60(3°)$ drive cone at $|V_P|=c^3$. At four orders of magnitude smaller than the ground state, this d_c gives an even more suitable ultimate vortex driver for all gyre conditions, with an extreme match to the current contour of the 30° cone boundary layer under ground state conditions as described earlier for the inter-changeability of small drive cone diameters with increased lab V_P.

This alternate driver thus enables the $<10^{-14}$ cm empirical electron radius in a typical electron collider (Mac Gregor, 1992). In process, exterior acceleration forces inward on the electron from the rear of motion direction (in initial stages of acceleration) added on cycle average to the PT force compress the gyre and particle dimensions slightly until CV to CV separations of drive cones decrease and approach 0.7 of the effective electron (or other LQ particle) component GD. There (Fig. 4-5f) typical mutual repulsive forces F_M between like rotating gyres decrease, reverse to attraction, and steeply increase toward a maximum as the entire electron and its compacted gyres shrink proportionately until the pseudo-centrifugal effect balances forces again in a much smaller particle radius with a very much higher V_P [causing a large increase in gyre interaction energy beyond the usual charge-mass power law level (Eq. 1-2 or -3) for particles in ground-state atoms.] Since this entire internal action is well above c velocity, there is no necessity for cause of internal increase of mass through an abstract sub-c relativity. The observable increase of mass can be due entirely to the necessity for increase of local eddies by internally compressed frictions from continuing currents in much smaller volume and increasing externally layered shock waves from rising external velocity to near c of the entire electron. This calls for a large increase in internal gyre interaction energy as expressed by internal effects on maintaining GD with increasing V_P (Eq. 5-5 to -5e) beyond that present under the usual mass power law conditions for electrons at rest or in atomic orbit. The more ultimate drive cone which can supply gyres with such electron internal energy levels whenever required then is the core of the universal structural micro-quantum in the empirical mass of all LQ particles and thence in the hadrons such as the baryons that make up the nuclei of atoms. (Such an internal energy increase, from the source in Appendix 5-B below (the ultimate source of all "Higgs"-like mass), would cause or equate to matching the now classic or empiricly observed relativity effect on particle mass. However, an empirical mass equation from the known cause in the collider regime has not yet been looked for within the paradigm beyond the general statement of Equation 5-4. This joins other specific and necessarily required objectives for decades of continuing research in the current new century.)

When accelerating forces reduce, if whole electron velocity subsides by collisions toward ground state, the gyres and particle must re-expand in a new balance of forces like the ground state, drawing on the un-ending energy flow of the empirical electron (Mac Gregor, 1992) driven through its conserved component vortex pairs by their unexpended drive cones (in a medium as described in Appendix 5-B below.) There may exist a not-yet-determined continuing state or series of states of near-balance-of-forces throughout the changing particle conditions over time both going into an accelerated condition and coming out in deceleration. Alternatively, if an impact is energetic enough, the electron structure may at that time be driven into being part of a larger structure with accession of greater numbers of conserved gyre pairs yielding exponentially increased frictions generating massive eddy increases, or, as shown later with impacting baryons, that may be a single cycle pre-particle stage unstably re-organizing into any supportable number of smaller groups of pairs with very high external velocities below c and exponentially decreased frictions generating replacements of any eddies that are radiated. [Only in a black hole or quasar is there evidence in their occasionally observed axial jets of sufficient energy concentration to break otherwise conserved pairs of quantal vortices apart into single conic gyres that escape the black hole concentration either at c or in accelerating superluminally, only to recombine eventually in collisions with unbroken pair matter and release the observed photons (in a renewal cycle discussed in a late chapter after baryons, mesons, and neutrinos.)] Again summing of multiple solutions is required as at the end of Chapter 4.

Force Between Whole Structured Electrons and Conclusion

Instantaneous forces are calculated at the ground state conditions (Eqs. 5-8 to -17) between the gyres of two summed-coaxial, fixed electrons at classical (Mac Gregor, 1992) center separation for 1 dyne repulsion (12 GD separation within 1%). Here another vortex lab data equation (5-17) is required at $\alpha=0$, with scale change by z':

$$f_{HV0}(\alpha) = f_H(\alpha) \left[0.5 + \{(1.5/0.34)(z' - 1)\} \right], \quad (5\text{-}17)$$

Since for each gyre's force summation from the opposed electron 2/3 to 5/6 of instantaneous α positions for gyres in the other electron are balanced in offset from symmetry, and lab data herein are largely only symmetric, limited force extrapolations between scaled vortices for non-symmetric duos added to those computed by equations above for the approximately 1/3 symmetric duos of gyres (at 10% 3 sigma error estimate) yield a first-order repulsion between these entire electrons of 1.1195062 dynes, a 12% high deviation. Until a well funded additional lab program has measured the mutual forces between similar <u>non-symmetric</u>, centrally driven, turbulent, conic gyres in duos over wide ranges of separation and between <u>symmetric</u> triplets at the 60° closest approaches in electron structures, this calculated estimate must be sufficient.

Thus, iteratively convergent solutions of scaled electron component equations (Howard, 2005) from lab force data (with minimized extrapolations added only externally between two electrons) and from PDG accredited empirical data on particle masses and charges (Amsler, et al., 2008 and prior) demonstrate a systematic structural scaling method and a scaled electron/positron structure to function correctably within Mac Gregor's analysis (1992) of the structural requirements for the particle's essential properties. This demonstration necessitates [in a constrained fluidic space medium (Appendix 5-B)] a scaled electron/positron structure of 6 lab-defined, turbulent, centrally driven, conic vortex micro-quanta orbiting in coaxial pairs shown in Figs. 5-1, -2, -3, and -4, under the combined empirical constraints of Chapters 1, 2, 3, and 4 above, in necessarily required consequences of the accumulation of the PDG data, as inherently clarified in the relation of charge to mass in Chapter 1. This necessary and sufficient, correctable and extendable, electron/positron structure of 6 uniform, turbulently interactive vortices as systematicly scaled microquanta function physically by their currents and waves in a constrained fluidic medium to generate mass energy with gravity force, strong cohesive force, conserved charge force, weak outwash force, organized spin effects in the particle, and exterior wave radiations of these types as well as superphotic entanglement force. This constitutes a prototype paradigm for the massive sub-atomic particles, including flexibly changing empirical size under high accelerating forces, reclassifying many of them ranging from neutrinos to hadrons, defining a single unified structural source for all the known forces, and demonstrating the proper empiricly derived generation of particle mass by charge structures in the prototype particle for all other particles (as further demonstrated herein.) This empiricly replaces the QM "Higgs" particle mass generation process. Essential requirements met also include the structure for an empiricly necessary resonant reception and radiation, at speeds far greater than the speed of light, of entanglement between sub-particulate superphotic coupling components, without which causal or QM electron structure is impossible and has not been previously definable, as has been well demonstrated over a century of attempts at the highest level of scientific expertise (but without the recent new benefit of coincidentally discovered Equation 5-2 and its consequences herein.) This relaxes or voids the now classic relativistic limit on physical reality among sub-particulate structures in consistency with the present large body of empirical data on proliferations of reliable entanglement. The Poincare' problem of quasi-infinite self-repulsion within electrons and their other (Mac Gregor, 1992) 20th Century structural impossibilities are also voided. Consequently, this chapter correctably resolves a 100 year search for reality-related baseline electron/positron structure suitable for further completion and refinement over the coming century [and also resolves other QM Standard Model uncertainties on the particles such as that they are definitely structurally combined currents, waves, and particles, with additional consequences in later chapters.] - - Beyond a few radii outside particles forces associated with mass and charge dominate the constantly observed and usual gross effects of particles. The electron distributes these basic forces with ideal evenness, electric charge force being nominally summed from three even, uniform, and omnipresent (or substituted) gyre fields at any instant. With initiation of gross particle motion, relatively moving charge distorts its net effects field, and accelerated charge, especially in acceleration at right angles to prior velocity (as in a wire coil), creates temporary distortions of the more fundamental invariant charge electric force that may be recognized as magnetic or electro-magnetic force from a definitely exterior point or from the immediate gyre diameter of each electron component outward. This primary and secondary relation between charge and magnetics may lead to a new definition of magnetism.

If critical experiments Mac Gregor (1992) specifies prove his electron radius to be too large or too small, that corrects the present departure point, but does not obviate this general formulation. The present particle scaling method from a combination of new lab vortex configuration and force data and PDG empirical particle mass/charge data would still apply.

APPENDIX 5-A

Entanglement - - Another Aspect of Vortices in Electron Structure

Within vortical structures there is a direct mechanism for the phenomena of entanglement (e. g., Pan, et al., 2000; Mair, et al.,2001; Sackett, et al., 2000; Rarity, 2003; Aspelmeyer, et al., 2003). Gyre orbits in electron structure are interlocked in resonance by the direct forces of their entangled superphotic currents and waves (in the main text) very much like the natural synchronization of two similar pendulum clocks hung close together on a wall that couples their vibrations. Sensitive resonances of separate force wave couplings at short range have bandwidths of inverse Q in which very narrow widths of linked orbit frequencies (ω), harmonics (H), and dopplers (d) occur in N associated gyres. These are unavoidably frequency/amplitude modulated (FAMed) with sidebands (SB) by all lower state ω and H, FAMed with SB on similar gyre V_p spinning ω and H, and this is again FAMed on ultimate driver ω and H, and shared at the highest pressure wave velocity at driver $|V_p|$ (in the main text) for a randomly tuned signature of multiply resonant mutual entanglement. Thus, almost uniquely tuned entanglement failure probability for the vortices of two electrons in a He atom, for instance, would approach the very small:

$$P_F \cong 1/[\omega_O \, \omega_d^2 \, f(Q_O) \, \omega_{P1} \, \omega_{sb1}^6 \, f(Q_1) \, \omega_{p2} \, \omega_{sb2}^{14} \, f(Q_2) \, N_\omega^{25}] \qquad (5\text{-}18)$$

Once outside the electron particle, the higher speed pressure waves of entanglement must act as a true radiation that can maintain entangled resonant coupling between particles, qbits, etc., or (by the cited empirical observations) be repeated as an additional coupling by photons that have been radiated in the entangled state. These couplings would be sensitive to faint signals at relatively large distances in analogy to amplifiable radio signal tuning; but as with radio signals, reliability of reception would be subject to many other factors, including susceptibility to eventual detuning of a first entanglement by engagement in any new close unshielded coupling with the vortices of other particles after wide separation of the first entangled particles.

APPENDIX 5-B

Necessary Space Medium and Energy Storage in Gyre Drivers
Driver Energy Minimum

Energy storage E_D is required by such a quantal vortex drive cone as discussed herein for maintaining constant drive over long time periods in spite of continuing output energy losses only to the repeated mass enlargements over time for a gyre in a large quark (Howard, 2005 & 2006) in a large reactively varying hadron (Eqs. 5-2, -3) for example, without including the additional equivalents to empirical relativistic mass increases or lifetime entropic losses in mass, or entanglement wave radiation [which energy output could be a contributor (along with the smaller neutrinos in Howard, 2009) to some of the "dark energy" noted by the PDG (Amsler, et al., 2008).] Such a minimal long-term single-gyre energy output from drive cone energy stored (without apparent

detectable sign of reduction of uniform cone peripheral velocity) could exceed:
$$E_D > 10^4 \times 3.9295 \times 10^4 \times 7776 \times 10.9525 \text{ eV} = >3 \times 10^4 \text{ GeV}. \quad (5\text{-}19)$$

For such energy storage (Eqs. 5-4 and 5-5e), at $|v|=c^3$, $f(\upsilon) \approx 6.7 \times 10^{15}$. Applying this in:
$$\rho_v = \rho_0 f(\upsilon), \text{ and } \eta_v = \eta_0 f(\upsilon), \quad (5\text{-}20)$$
with the vacuum or space density $\rho_0 < 10^{-30}$ gm cm^{-3} of the universe (Amsler, et al., 2008), then $\rho_v < 10^{-14}$ gm cm^{-3}, for driver cores, which is still very small. But at $c^{3.3}$ $f(\upsilon) \approx 1.3 \times 10^{51}$, and $\rho_v < 2.64 \times 10^{21}$ gm cm^{-3} with η_v equivalent, creating a limit for pressure waves c_p in the constrained medium described by $f(\upsilon)$, and providing for a yet smaller inner core of the driver d_c as the only impenetrable, undeformable part of the electron's interpenetrating waves/currents. (The cone driver spinning centers must compress to enormous density and rigidity for momentum to last 10^{11} years and far more without apparent diminution.) Such a medium class constrained by the paradigm could within a correctable Planckian limit for peripheral velocity contours of driver cores with speed below c^4 store such large amounts of rotational energy in such a very small volume as the described ultimate conserved gyre driver cone.

Nature of a Space (Vacuum) State

Equations 5-5b-e and -19 indicate an active (rather than inert) and numericly constrained space medium in this paradigm, meeting the necessity of not only an ultimate vortex driver speed of the drive cone periphery near order c^{3-4} for electron energy, but also pressure wave entanglement phenomena near the same speed range, and continuous radiation of electron mass and charge shear and pressure wave effects in much lower speed and frequency bands (as an inherent entropic source of some only partially recovered, dispersed energy, a dark energy of detached waves.) As an overall result, macroscopic fluid effects that are detectable among complete atoms and molecules, or their aggregates, are only results of sub-particulate vortices in a fluidic inter-/intra-particle space, which, with the viscous boundary layers on drive cones of conic vortices in particles, may also be described as velocity contours in a fluid in usual fluid mechanics terms. However, this must include fluids which become very dense and viscous at very high velocity zones under the influence of Eq. 5-5e, usually with very high shear because of the small size of the high velocity zones. Where waves are involved, the presence of shear waves requires elasticity or a viscoelasticity that transitions to almost pure elasticity with vanishing viscous losses at higher speeds and frequencies of small amplitude. This condition is like a non-Newtonian or nonlinear gel with weak interlocking of its ultimate constituents rather than their being completely smooth, rounded, and ball-like in nature. Also, fluid shear currents and wave disturbances co-exist with and cross-couple into pressure wave components with different natural speeds, and vice versa, varying with different fluid boundaries or non-fluid bodies in a manner distantly similar to seismic waves in variable volcanic magma and earth mantle. Higher speed entangling propagations at distance act as longitudinal pressure waves. Herein shear/pressure wave/current forces combine in approximately scalable net mutual forces summed between lab gyres (Howard, 2009 & 2009).

There is at least one conceptual medium which could accomplish this compatibly with the scaling equations. Separated shear and pressure wave velocities here might result from the tieing and untieing and entangling and disentangling of a probability range of incompressible knots made by the random actions of a single type of uniform worm-like strings at Planck length order. (Here there is an analogy to knots of the common earthworms after they have been gathered from fertile earth and confined. Microscopic nematodes have a more appropriate length to diameter ratio, but their flexibility in bending arc radians per diameter scaled length may be too low for good knot tying and entangling from knot to knot without tying two knots on one length.) Knots would be packed closely with very short mean free paths (less than a mean knot radius) for knots in slightly compressing pressure waves. In lateral coupling of pressure waves/shear waves and currents, shear viscosity and effective density of coupled knots would be dependent on ratio of wave passage period per knot diameter (inverse velocity) to most probable period of knot-to-knot entanglement change. Broad lateral couplings with high apparent densities would occur with rapidly moving currents and waves. High velocity stress at low shear wave amplitude would delay disentanglement briefly for viscoelasticity or indefinitely at high shear amplitude in ultimate driver states of Planck-like ultra-microscopic dimensions. This would also result in extremely high viscoelastic rigidity and density of a Planckian class not perceptible in, but many orders-of-magnitude above, ordinary inter-atomic states, as in the discussion of Equation 5-20 in the first topic of this appendix.

(Supplementary Information to this chapter is available in prior extended papers on-line at electron-particlephysics.org &/or in Vol. 2.)

REFERENCES OF SPECIAL INTEREST
(Full references, tables, and appendices are in Volume 2.)(In Prep, See Website)

HOWARD, F. E., JR., (2010) Striking Correspondences of the Micro-Quantal Particle Paradigm & Quantum Mechanics, www.particlephysics.info and www.electron-particlephysics.org

Howard, F. E., Jr., (2009) Turbulent conic vortices: Part 1, Lab gyres define tornado band in Hurricane Ivan. Part 2, 3D Gyre Forces Control Tornadoes, Supercell Tops, & Hurricane Eyewalls, www.particlephysics.info and www.electron-particlephysics.org

FEYNMAN, R. P., LEIGHTON, R. B., & SANDS, M., The Feynman Lectures on Physics, Vol. 2 (Addison-Wesley, Reading, MA, 1964).

HESTENES, D., WEINGARTSHOFER, A., Eds., The Electron, New Theory and Experiment (Kluwer Acad., Dordrecht, 1991).

SPRINGFORD, M., Ed., Electron, a centenary volume (Cambridge Univ. Press, Cambridge, UK, 1997).

DOWLING, J.P., Ed., Electron Theory and Quantum Electrodynamics 100 Years Later (Plenum, New York, NY, 1997).

SALAM, A., in The New Physics (Ed. DAVIES, P.) 481-492 (Cambridge Univ. Press, Cambridge, UK, 1989).

HAISCH, B., RUEDA, A., & PUTHOFF, H. E. Inertia as a zero-point-field Lorenz force. Phys. Rev. A 49, 2, 678-694, (1994).

SLOMINSKI, W., SZWED, J., Phenomenology of the electron structure function, Eur. Phys. J. C 22, 123-127 (2001).

MURYN, B., SZUMLAK, T., SLOMINSKI, W., & SZWED, J., Measurement of the electron structure function. Nucl. Phys B 126, (Proc. Suppl.) 11 (2004).

WAITE, T., BARUT, A. O., & ZENI, J. R., in Electron Theory and Quantum Electrodynamics 100 Years Later, (ed. DOWLING, J. P.) 223-240 (Plenum, New York, 1997).

PAN, J.-W., BOUWMEESTER, D., DANIELL, M., WEINFURTER, H., & ZEILINGER, A., Experimental test of quantum nonlocality in three-photon Greenberger-Horne-Zeilinger entanglement. Nature 403, 515-519 (2000).

MAIR, A., VAZIRI, A., WEIHS, G., & ZEILINGER, A., Entanglement of the orbital angular momentum states of photons. Nature 412, 313-316 (2001).

MAC GREGOR, M. H., The Enigmatic Electron. (Kluwer Acad., Boston, 1992).

HOWARD, F. E., JR., Elementary particle mass sub-structure power law. Florida Scient., 68 (3), 175-205 (2005); & Erratum, Appendix Table C3, 69, 2, 148 (2006). See www.electron-particlephysics.org for correction in place.

HOWARD, F. E., JR., Sub-structure laws of particle masses and charges---a new systematic classification of subatomic particles. Florida Scient. 69, 3, 192-215 (2006). (Also at www.electron-particlephysics.org.)

AMSLER, C. et al. (Particle Data Group) Biennial Report of Particle Data Group for 2008., PL B 667, 1 (2008). Download by Sections, Signed Notes, Summary Tables (accredited data), Particle Listing Tables (detailed & not yet accredited data), etc., at www.pdg.lbl.gov; as well as close equivalents below and biennially for subsequent years:

    YAO, W.M., et al. PDG Report for 2006. J. Phys., G 33, 1 (2006).

    EIDELMAN, S., et al. PDG Report for 2004. Phys. Lett. B 592, 1 (2004)

    HAGIWARA, K., et al. PDG Report for 2002. Phys. Rev. D 66,010001 (2002)

HESTENES, D., in The Electron, New Theory and Experiment. (eds. HESTENES, D., & WEINGARTSHOFER, A.) 21-36 (Kluwer Academic, Dordrecht, 1991).

HOWARD, F. E., JR., The elusive neutrinos--a paradigm for these leptons among leptons and quarks. (2009) See www.electron-particlephysics.org

SECTION IV

Re-Classifications of Constrained & Expanded Proliferations of Larger Particles - Unexpected Outcomes

Chapter 6

A Consequential Step Toward Quarks, Protons, and Neutrons, But With Divergent Neutrinos and Widely Scattered Charged Leptons - - A Fast Overview

SUMMARY: Chapter 5 on the structure of the definite electron, as distinct from a necessary but variously defined muon neutrino, opened the prototypical door on structures of the other non-nuclear and nuclear particles. To here the book follows the original sequence of discovery. But the electron (or positron) is too ideally simple to indicate a clearest single path through organized interrelations between other particles. Research from this point became too multi-branched and irregular to follow with clarity. The evidence on the Neutrinos as a whole depends on decays in the Baryons composed of Quarks. The Quarks begin small, with one more and two more orbital pairs of vortex quanta than the prototype electron. But they next become massive and complex to a natural limit with many decays (involving obligatory neutrinos plus radiated photons), and they yield numerous proliferations of baryons that break up into other "exotic" mesons. Only after review is it clear that the jump in masses from the electron to the other Charged Leptons, the relatively rare muon and tau particles, is the most systematic next step beyond electrons into the tangled sub-atomic or nuclear particles. These the PDG is laboriously organizing and recording (over the last century) in pursuance into and beyond a QM Standard Model of particle physics that the world of science generally can continue to use as a fundamental basis for human knowledge of Nature out across the Universe. [A grasp of such a cultural scope is a worthwhile objective at an earliest reasonable age for as many people as feasible. Essentially, science here is adding a constructive layer that understands and explains a rich cosmos beneath superficial appearances of random aimlessness. The difficulty is that there are so many constructive steps between particles that are each necessary and must be sorted out. New types of display charts help organize the structures and their most basic data with eventual clarity.]

Until Chapter 5 the necessary resultants of a simplifying equation dealt primarily with the characteristics of all particles, with their common structural frameworks, and their discernable controlling, or apparently comprehensive, Laws within Nature in general. This led into details about one specific kind of particle, the one most prototypical and most prominently ever-present, the low mass or very light particle, the electron. Its nature and its essential necessary structure had for a century been resistant to understanding individually. It is so tiny as to be almost undetectable except as floods of multiple trillions of them all together in a spark or lightning or electric lighting powered through copper wires. At the prior very smallest ordinary view, they are only cloudy shells of small numbers of them that orbit around each type of the atoms and link those together in the actions of chemistry. The structural grasping of a single simplest key particle brings this exploration of particles face-to-face again with all the other kinds of particles in their initially confusing variety. Up to this point these have been treated as broadly varied with some limited, but recognizable, characteristics which separate them into obvious classes or types that a very small fraction of the world's population have been working with for a very few generations. With which one, or which group, to continue after the electron is not at all clear. Certainly an electron was the single most commonly available particle in daily life to start with, and atoms themselves are the topic that would follow in that approach. But atoms are not the particles to which the initial equation first opens a clarifying door. Atomic nuclei might be the next most available, but only one, the proton, of the hundreds of the known nuclei, is a single particle.

The Table 1-1, a and b, of the Lepton-Quark (LQ) particles and the two Primary Baryons (from all of the hundred or so in many families that are made up of three QM quarks) should include the most suitable follow-up particle after the electron. (At least that would rule out the Mesons, which occur largely by the break-up of baryons and so cannot be taken up until the baryons are well covered.) The Baryons then, including the Proton, are the main components of the very commonly available nuclei; however, the numerous baryons are composites of other components, the quarks, and baryons are certainly not basic in the same sense of simplicity as the electron-positron. - - Table 1-1 leads next to the Neutrinos, which start the table with its simplest item, the electron neutrino, followed by the muon neutrino, which is quite similar to the electron in simplicity. The Neutrinos appear at first glance to be good candidates for next follow-up to the electron; but here it must be remembered that behind Chapter 1 and its important Equation 1-2, the Charge-Mass Power Law, there is a separation and simplification of the intricate question of how far into the cosmology of the universe the astrophysicists are taking the very smallest micro-neutrinos. There is a further question of how many very massive neutrinos are to be discovered in the region of the tauon neutrinos, as implied by Table 1-1a. Neutrinos also frequently appear in the PDG data tables as necessary consequences of the "decay" of baryons and mesons. Those considerations must clearly postpone a concentrated neutrino discussion, such as the logical next follow-up to the electron-positron would involve. - - Consequently, there are only the Quarks, that cannot be taken up without the Baryons which are the quarks' primary existence, and the electron-positron's Charged Lepton family of the Muon and the Tauon. These two are well into Table 1-1a's list of increasing masses, and so are not really very simple either. Still there are only two of the remaining Charged Leptons (CLs) after the electron-positron. (It may become more acceptable that there are only two of them after full coverage in the next chapter on Baryons with their orbit interferences and unstable imbalances of orbits. These effects do add to the CLs' limits to the present three known members, excluding their anti-particles, in eleven orders of magnitude of particle masses in Table 1-1a.) Otherwise, there are left now only the six established types of PDG QM quarks (that begin with 4 pairs and 5 pairs of quanta in the smallest two, just beyond the electron's simpler 3 pairs.) The quarks extend, by almost six complete orders of magnitude, to the most massive LQ particle, the top quark. For each of these six quarks the MQP paradigm's tables of Baryon families establish that there are two different mass values for each of at least the smallest five quarks. Compared to this complication, there is no question then that completion of the Charged Lepton Family is the simplest continuation beyond the structure of the electron.

In this the selection process for the electron follow-up becomes an over-arching or top-layer Overview of the LQ and Hadron particles without fully reviewing or exploring the many detailed families of the Baryons and Mesons (which will come much later.) (If the first-time reader finds this brief overview too complex to follow and grasp, Table 6-1 provides it in a road-map sketch of prior particle high points with later book sequence. It can be photo-copied as a book place-marker that follows the reader for repeated orientation, best with a copy of Table 1-1 on the back.)

Table 6-1 both condenses this overview and adds some significant factors conveniently, such as the Primary lighter particles and the frequency of occurrence of predominant particles that are the Main ones of their type versus Rareness of most others. There are also relative masses (or weights in Earth's gravity field) and free versus bound states, as well as variations in common names or symbols for the particles.

OVERVIEW OF LQ & HADRON PARTICLES (in both the PDG QM & the MQP Paradigm)
(a Key Simplification of particle relations)

Hadrons = Baryons & Mesons   (In an equation with variable y exponent)
BARYONS in atom nuclei or free $m_p = N_q^y \Sigma m_q$,   Eq. 1-1
  Mesons = Broken baryons, free   "   "   "
ELECTRONS (e⁻) around atom nuclei or free $m_p = N_c^5 (N_c m_u)\left[(n_\pm / n) + (n_0 / na)\right]$,   Eq. 1-2
  Other Leptons = CHARGED & NEUTRINOS   "   "   (wherein y is fixed)   "
  QUARKS in baryons & mesons (never free)   "   "   "

UNIFORM MICRO-COMPONENTS in all LQ particles, charged ±, always paired ++, +-, or -

Free pairs to spheres Small stable	Free spheres Tiny 1 very stable, 2 not	Never free singles joined spheres with expanded link orbits Always unstable (except one small triplet) -->	Free or bound 3 ring-link'd quarks 1 triplet very stable (1 stabilizable)	Free (bound?) 2,4,or 6 line-link'd Never stable
LEPTONS		**QUARKS**	**HADRONS**	
N N	C L	Q ALWAYS CHARGED t	B (BOTH	M
E E	H E	U (letter symbols) b	A EITHER	E
U U	A P	A c	R CHARGED	S
T T	R T	R s	Y OR	O
R R	G O	K d (in order of mass)	O NEUTRAL)	N
I A	E N	S u	N	S
N L	D S	6 ea, 2 masses ea, lighter **Primary**	S **NUCLEAR** (100s rare)	
O	3 ea	2 MAIN Small 4 rare larger	2 MAIN, (100s rare)	
S	e⁻ (tiny)	**up** u s **strange** (lighter)	**p**roton (very stable, charged +)	
3 in PDG?	muon μ	UP STRANGE (heavy)	**n**eutron (proton stabilizes, neutral)	
1 heavy?	tauon τ	**down** d c **charm** (lighter)	(Both made up of Main	
in decays?	(heavy	DOWN CHARM (heavy)	Primary Small u-d Quarks,	
many micro?	&	b **bottom** (lighter)	and these 2 Baryons make up	
(in stars,	rare	BOTTOM (heavy)	about 99.9% of visible Atomic Nature	
collisions,	CLs)	t **top** (lighter)	that all Life lives in --	
novae.)		TOP (heaviest)	plus mainly **e**lectrons.)	
		(Only 1 of 2 tops observed, only in decay)		

BOOK CONTINUING TOPIC SEQUENCE (BY COLUMNS above) after electron Chapt. 5

many	(e⁻MAIN)	MAIN	rare	MAIN	rare	rare
Last	1ˢᵗ (next)	2ⁿᵈ	4ᵗʰ	3rd	5ᵗʰ	6ᵗʰ

Table 6-1 Overview of the Lepton-Quark (LQ) and Hadron Particles. This overview of particles to this point emphasizes the families of particles that are known to be few in types and/or have one to two Main members that are very frequently found in Nature. Where there are numerous rare types or families, that may appear later, the list is severely abbreviated. Note that at the bottom of the family columns is a topic sequence in which the families and groups are taken up later in this book; two Charged Leptons are next. Still, even the simple two remaining Charged Leptons need an introduction of useful tables that condense data for quick comparisons of particle structural differences throughout the rest of the book (and also give a broader simplest aspect of Chapter 5.)

First, continuity from the electron/positron structure into the other Charged Leptons needs a fresh look at just what was accomplished in the electron prototype of Chapter 5. It was essential in building that structure to rely on Mac Gregor's (1992) re-evaluation of approximately a century of effort by many others, and to use significant parts of his independent data in defining the MQP paradigm of electron structure (with its mirror image in the positron.) Compared to QM the paradigm is not now independent even at the empirical level of simplicity which is inherent to it. The paradigm cannot become complete and independent until it is feasible to instrument, measure, and scale (with empirical equations, if not with more complete fluid mechanics theory) full data on two added dimensions of mutual lateral forces between symmetric (and non-symmetric) vortices, along with the practical parts of such data in Chapter 4. (Lateral forces keep the radially bound vortices orbiting at full velocity. In the electron that velocity was specifiable through reorientation of Mac Gregor's necessities.) The amount of additional work and much more elaborately instrumented data, in a much more capable facility, is at least twice that accumulated over a decade and a half in developing the paradigm to date. (The costs will be greater by orders of magnitude.) In addition the current data and equations need confirmation and improved empirical accuracy. And there are also the other data needs already specified, for any two non-symmetric gyres, for three symmetric gyres, and for frequency filter instrumentation to complete the separations of summed lab forces where necessary (or desirable.) When these data and their scaling equations become available, then this electron structure can be fully calculated (and well simulated) in an independent mode. Also the other particles will become independently analyzable in full. In addition, library research to date has found only portions of the coordinated completeness of view on the other particles that Mac Gregor's monograph brings to the electron. All work herein on other particle structures then must lean heavily on the electron prototype structure where possible and be even less complete. However, all these other particles must necessarily function in the same environment as the electron. Furthermore, It is not reasonable for the other particles to require other basic components than the electron's necessary uniform vortices when those fit everywhere very well. The mechanism of the mutual radial forces of symmetric gyres applied to the Mass-Charge Power Law's gyre masses explains a great deal about necessary structures for PDG QM particles, and also about QM particles that the MQP paradigm demonstrates to be unnecessary (like the gluon, for instance.) - - So within these limits the paradigm proceeds in auxiliary support of QM's well-known extremely complex difficulties, rare over-complications (like the gluon), and very rare, but occasionally significant, shortfalls (such as the lack of an empiricly feasible, basic electron structure, and others to come herein.) It is also possible, or predictable, that QM will adapt elements of such new support in more theoretical QM versions.

Next, in proceeding from the electron's prototype, a simpler way is prepared of condensing the many schematic drawings of structural forms that must be kept in mind after the general forms are understood. In reference, Figure 6-I shows a stripped-down sketch of only the conic drivers of the electron's six vortices at the check points for their orbital synchronization. With only six components, such a view still takes significant time to check it over visually, and this chapter's largest particle would require almost an order of magnitude more of such information in about the same size figure. This partial

143

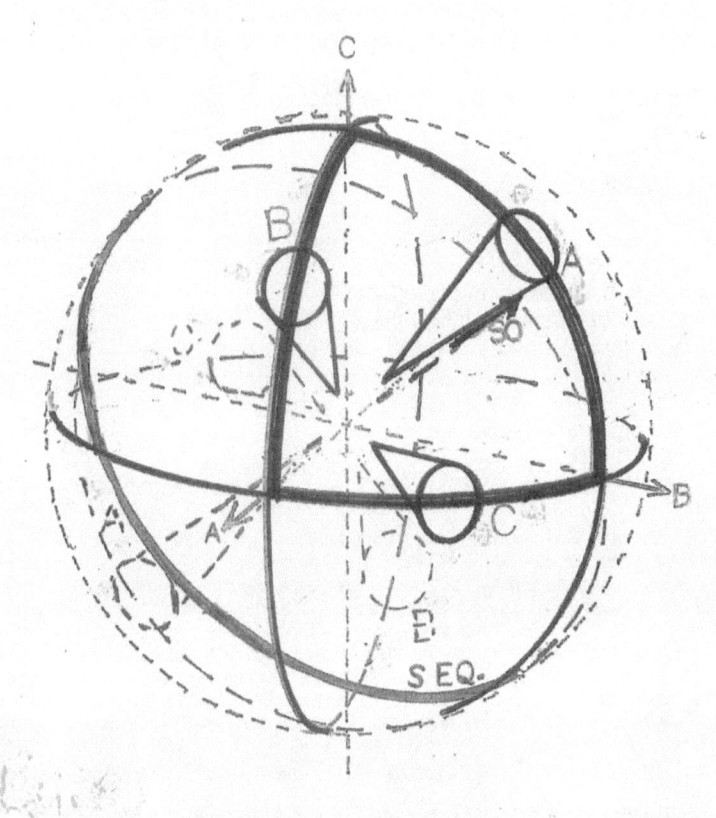

Fig. 6-1 Electron 3D Schematic of Gyre Drive Cones At Their Synchronization Points. The 6 component gyres are orbiting in 3 coaxial pairs which are in orthogonal orbits in a sphere. At the critical points for synchronization three of the paired singles are centered on arcs around the Primary Octant of the sphere with the SO summation axis through the centroid of that octant. (See Figure 2-2 on Page 52 to clarify.)

view with its six negatively charged spinning gyres in three coaxial pairs also exactly duplicates the sync point outlines of the cones in the best fit (under the Charge-Mass Power Law) to the prior PDG-specified limit of the muon neutrino mass with its three neutral coaxial pairs, each with one positive and one negative spin rotation (Table 1-1). Even if detailed the viewer may be uncertain which particle is intended. Table 6-2 next demonstrates the very simple equivalent table of the names of the orbit sites and static spin sites that are occupied by the charged gyres of the electron and the mu neutrino in a way that can be understood and compared at a glance in the two vertical columns for the two particles. Groups of particles can be compared in this way on a single page that would otherwise require many pages of words or much more figure detail. In this chapter such tables show the full meaning of the many occupied sites available in one LQ particle in Figure 2-6 or in a Proton's three quarks in Figure 2-7. (Until it is well understood a photo-copy of Fig. 2-6, Page 56, will be useful in the chapters to come.)

Sphere Pair Orbit Sites around S0 Axis in Electron (e⁻) or Muon Neutrino (Nu Mu, ν_μ)
(In Fig. 6-1)

	Particles (separate)	e⁻	ν_μ (near upper mass limit in QM)
Co-axial Pair Orbit Sites			
A		- -	+ - (Each in 3 orthogonal sites
B		- -	+ - Each in Perfect Balance
C		- -	+ - Each Very Stable)
A'	(Other sites are empty in these 2 cases;		
B'	these are the smallest and simplest Usual		
C'	or most commonly co-organized particles.)		
++			
- -			

S Eq 1 (7.5° from reference plane)
S Eq 2 (67.5°   "   ")
S Eq 3 (127.5° "   ")   (Static spin sites are omitted.)

Table 6-2 Electron and Muon Neutrino Pair Orbit Sites Compared. Each is perfect as three pairs in three orthogonal orbits balanced around the S0 sum axis. The balance is also maintained in each particle since all of its pairs have the same mass, though the neutral pairs in nu mu have only 1/3 the mass of the charged pairs in electrons (Chapter 1.) Orbit sites A'B'C' each lag the ABC orbit sites in the ABC orbits by the same amount, and if A'B'C' only are occupied, they effectively are ABC (unless 2 other pairs are in the usual synchronizing sites of ++ and - - orbits or unless another 1 pair is in one of the S Eq orbit sync sites of Fig. 2-6 in Chapter 2. Either of those would put all the synchronization angles back in place.) The world-wide Particle Data Group (PDG) accredited the measure of the electron's stable mean lifetime at 4.6 x 10²⁶ years, far longer than the calculated life of the Universe. This structure indicates long lives. The PDG for many years accredited the mass of nu mu as less than slightly more than 1/3 the mass of the electron. The MQP paradigm agrees on exactly 1/3 and calculates the quanta pairs per Equation 1-2.

    In using this kind of table it is necessary that the exact number of the proper kind of pairs required under Equation 1-2 to generate the mass of the designated particle be shown with each pair in its proper site to obtain the best fly-wheel balance that those pairs can sort themselves into under their own forces (or an excited equivalent with higher energy and mass.) In such a process of forming a particle it develops that there is a natural priority to form groups of three similar pairs first, or three dissimilar pairs, if that is required to create the measured mass and charge. Then groups of two pairs and static single pairs, with constant attention to balance (which does not occur with both neutral and charged pairs in a group; nor is perfect balance obtainable with mixed plus and minus charges of pairs in ABC groups in particular. In matching observed lifetimes of each particle this kind of mixture does not appear to cause instability in the ++ and - - orbit group in every case. That is the reason for their names.) Since the S Eq orbit sites are on the Sum Equator of the Sum Axis (S0), they automaticly balance on that axis pair-by-pair and may also be mixed in charges. But when more that one pair must be in the S Eq sites, it is because the total number of pairs and the resultant mass of the particle is so high that the mutual force-to-mass ratio is so low that the particle is unstable with a fast mean life in micro-micro-seconds from that cause, and it cannot be made more stable with any grouping of the pairs of quanta. The sequence priorities of orbit groups can be a significant part of the balance and lifetime question in many cases, but not all cases. Unlike electrons the muon and tauon Charged Leptons (CLs) are both heavy enough that balance in sorting pairs into orbits is understood best by following what appears to be natural sequence priorities as indicated in Table 6-3.

Empirical Sequence Priorities in Occupying Particle Orbit and Static Spin Sites

```
Pair Orbit and Static Spin Sites                                          Priority
    A   |
    B   |   Best if 3 Like-charged pairs      Then well balanced            1
    C   |
    A'  |
    B'  |   Best if 3 same    "   "     "      "    "    "                  4
    C'  |
    ++  |   2 charged or 2 neutral pairs       "    "    "                  2
    - - |
S0 Static Site & Opposite Octant, 1 charged or neutral pair Auto-Balance    5
    S Eq 1 |                                All 3 sites Auto-Balance
    S Eq 2 | In Quarks one of these 3 Link Sites has a charged Pair         3
    S Eq 3 |  In Leptons Priority drops to 6 for 1-3 charged or neutral pairs
S1 Static Octant & Opposite Octant Centroids|
S2 Static Octant &   "        "       "     |  Best if 3 neutral pairs      6
S3 Static Octant &   "        "       "     |  Then well balanced
     (In Heavier Particles Static Sites may have 2-4 equal pairs in Cylinders)  7
     (In Heaviest 2 Quarks an outer ABC Shell may be necessary)             8
          BALANCE PRIORITY IS HIGHER THAN ANY SEQUENCE
```

Table 6-3 Empirical Sequence Priorities in Filling LQ Particle Orbit & Static Spin Sites. Frequently in heavier particles these priorities cannot be met and particles become more unstable in mean lifetimes.

These tables are not invariant in details of format, especially in the sequence of sites below the ++ and - - orbits. In this case, though the S0 summation axis is low in site priority as a single static spin center at the centroid of the Primary Octant of the particle sphere, it is set among the true orbit sites for emphasis as the Sum Axis for all sites. In quarks and hadrons it is much more convenient for the three Sum Equator sites in a single orbit to be at the bottom, since they are then the specialized Link Orbits. In baryons each quark has one S Eq pair in a site number distinctive from the other two. With these general understandings the table of pair orbit sites for the Leptons that have been shown to this point are quite clear. Omitting the alternate tau neutrinos shown in Table 1-1, as well as the quarks beyond the common up and down (included for the simple range overlaps), the positions of the three Charged Leptons (electron, muon, and tauon plus the positron as the electron's anti-particle) in the structural development of particles from micro-quantal vortex components is clear as shown in Table 6-4 next. (The reader will also want to be familiar with the prior PDG tables at www.pdg.lbl.gov which are biennially up-dated by the international PDG with the aid of US govt. labs.)

Charged Lepton Pair Orbit & Static Spin Sites with Usual Neutrinos & Small Quarks

Sites	Mu n'trino 0.1703 PDG Lim 3 pairs	Electron 0.511 (-1) 3 pr	Positron 0.511 +1 3 pr	Lt Up quark 1.914 +2/3 4 pr	Lt Down quark 5.11 -1/3 5 pr	Tau n'trino 18.169 PDG Lim 6 pr	Muon 107.19 Heavy,-1 8 pr	Tauon 1744. V.Heavy,-1 12 pr
A	+ -	- -	+ +	+ -		+ -	- -	- -
B	+ -	- -	+ +	+ -		+ -	- -	- -
C	+ -	- -	+ +	+ -		+ -	- -	- -
A'								+ +
B'								+ +
C'								+ +
++				+ -		+ +	+ -	+ -
- -				+ -		- -	+ -	+ -
S0				+ +	+ -	+ -		+ -
S Eq 1				+ +				
S Eq 2								
S Eq 3					- -			
S1							+ -	- -
S2							+ -	- -
S3							+ -	- -

ALL BALANCED (& INTERNAL QUARK-TYPE CONFLICTS AVOIDED)
(Individual quark sites are listed *as if* quarks could exist alone.)

Table 6-4 Charged Lepton Pair Orbit and Static Spin Sites with Usual Neutrinos and Small Quarks. The electron, the muon, and the tauon are the negatively Charged Leptons. (The positron, of course, is the electron's anti-particle, which is very rare in Nature, but frequent in many colliders. Other anti-particles are not usually listed except in Mesons.) While all are balanced, and the very light electron has a PDG measured mean life of greater than 4.6×10^{26} years, the rather heavy muon's mean life is about 2.2 micro-seconds (millionths of a second), and the very heavy tauon has a mean life of about 0.29 micro-micro-seconds (millionths of a millionth), very unstable. Their site distributions shown are not the only possible distributions, but are low energy ground states. The masses listed are from MQP calculations.

The masses shown in Table 6-4 are from MQP paradigm calculations (which have not yet been done at the full accuracy of the electron-calibrated masses to eight or nine significant figures, but are sufficient for tables of broad scope in general actions.) The electron and positron are anti-particles of the same mass. A point to note here is the two and a half OM between the masses of the electron and the muon and the full additional OM to the tauon (frequently called the tau.) It is quite apparent that some process other than chance distribution must be occurring to prevent appearance of a balanced CL over such large intervals of numbers. (However, the quarks, which take full advantage of the S Eq orbit and its three phased sites, have four dual quark masses over this range in Table 1-1, with a fifth quark only a fraction of an OM beyond.)

It is very peculiar and without immediately apparent cause that there is a gap of two and a half orders of magnitude in mass above the electron before the next Charged Lepton is found, and another order of magnitude before there is another. And those three masses are the only CLs. At the lowest level this is a consequence, in part, of the fact that particle components are not increased as singles, but in pairs of the two possible charges, and all odd numbers of components are cut off at the start. Next is the fact that the charge on each component is only one sixth of the charge on free particles. The charged leptons must contain like-charged groups of three pairs or six quanta. That is the smallest quantity at which like-charged pairs of coaxial components can balance in all directions equally with both a plus space location from a particle center and a minus space location from center on each of three planar orthogonal space dimensions. This is a consequence of the local or infinite static orthogonal reference space in balanced completion at three dimensions within which orbital spherical space occurs. The pairs are a balanced conservation through spinning vortex forces of pairs of gyres with dual directions of charge rotation. This constrains all particle charges and their number spacing within simple sphere orbit limits.

Next, the cause of rare occurrence lies in the Charge-Mass Power Law of Masses with its exponent of 5 for the Number of components times the Sum of the component masses, or in the 6th power of that Number times the mass of a uniform single component, or, more truly, in the very high level of mass-energy building interactivity between such energetic components that it requires such an exponent to describe their interaction mathematically. (This is the characteristic which requires the turbulent vortices to provide that intense level of component interactivity building friction eddies of mass energy as well as their charge by alternative directions of rotation. It is especially limiting when after at least three pairs of components are like-charged for a charged lepton, at the next step for every additional like-charged pair, it must be neutralized by an oppositely rotating pair.) Clearly, few of the numbers in their orders of magnitude of mass can possibly be the right number for a charged lepton to occur on.

Altogether, only a limited even mass product of the component Mass-Charge Power Law which balances dynamicly, without having an orbit conflict, can provide a Charged Lepton, if it also has the workable number 6 of un-neutralized like charges in 3 pairs. (For quarks in baryons and mesons there is an equivalent limit for two or four like charges.) There is good reason also why the CLs do not organize themselves in the S Eq sites. This comes out in the angle clearances of Table 6-5 below (wherein there is also very good cause for the quarks' expanded orbits which eliminate conflicts and do allow as many as 12 quark masses in a wider range of masses.

Obviously, not every large number could apply to a particle mass. Neither could those that are not multiples of the sum of component masses. (A less definite number limitation comes from the fact that neutral pair totals that are two thirds lower than higher charged pair totals divisible by 3 must also be distributed in the grouped sites that give fly-wheel balance around the S0 axis at required charge levels for completion.)

In Table 6-4 the Muon and Tauon must occupy the S1, S2, and S3 static sites to find a balanced condition at low spinning and consequently frictional mass-creating energy that will continue for micro-micro-seconds of mean life. This structure of sites appears to be a least-energy ground state (which is not entirely certain at the present accuracy of the known forces.) Table 6-5 shows this as permitted, but this may be the

Spherical Angle Clearance in Degrees between Orbit and Static Spin Sites in a Sphere
(With **15°** Required Minimum Boundary Flow Clearance for Least Drive Cone Sizes)
<u>SEq ORBITS</u> ARE <u>SPECIAL CONCERNS</u> IN **ALL QUARKS**, **CLs**, **& NEUTRINOS**

Site	A	B	C	A'	B'	C'	++	--	S0	S1	S2	S3	S0'	S1'	S2'	S3'	SEq1	SEq2	SEq3
									(Static Oct'nt Sites)				(Cylinder Doubles)				(**BOLD UNDERLINE**		
"													(RARE-Appndx 6-A)				= ORBIT CONFLICT		
																	IF NOT EXPANDED)		
A	*	60	60	45	34	34	60	40	35.3	r'pt	r'pt	r'pt	r'pt	r'pt	r'pt	r'pt	32	20	78
B		*	60	34	45	34	75	75	35.3	"	→		(Static Cyl Doubles				78	26	30
C			*	34	34	45	20	30	35.3	"	→		require **22.5**° or				30	72	26
A'				*	60	60	60	72	35.3	"	→		2x orbit angle rate				**12**	**13**	49
B'	(data to right				*	60	33	33	35.3	"	→		sync below only				39	25	87
C'	repeats below)					*	22	30	35.3	"	→		to avoid conflict.)				65	42	**12**
+ +							*	26	19+	35	19+	35	**19+**	35	**19+**	35	70	22	**6.5**
- -								*	19+	19	18	19+	**19+**	**19**	**18**	**19+**	35	27	65
S0	(Static Octant Site Pairs)								*	70	70	70	*	70	70	70	27	27	27
S1										*	70	70	70	*	70	70	27	27	27
S2											*	70	70	70	*	70	27	27	27
S3												*	70	70	70	*	27	27	27
S0'	(RARE Static Cyl Doubles only												*	70	70	70	27	27	27
S1'	in heavy quarks and CLs usually;													*	70	70	27	27	27
S2'	each must sync at 2x orbit angle														*	70	27	27	27
S3'	rate if below **22.5**° site clearance.)															*	27	27	27
SEq 1- -7.5°	(In quarks after Prequark stage, only one																*	60	60
SEq 2- -67.5°	per S Eq site number per baryon																	*	60
SEq 3- -127.5°	expanded orbit circle or quasi-ellipse.)																		*

(In Bottom and Top Quarks and some isomers of Charms, and in large Pre-stages, 100% larger spheric shells occur very briefly with any of above sites at HALF the required angle clearance, usually seeking balance in the Priority order. In that case, S Eq expansions go to 3x radius shells and are very unstable.) Table 6-5 Actual Spherical Angle Degree Clearances between All Orbit & Static Sites. Most of these columns are readily understandable from the prior information. The S0' to S3' Cylinder Double increases if the four static coaxial pair sites in the eight octant centers (like the S0 pair site on the Sum Axis in the center of the Primary Octant from Fig. 2-2 and its opposite octant dotted in on the rear of the sphere) are needed to provide sufficient sites on the sphere for some of the very large quarks & CLs in Table 1-1 (from the Charm quark with 22 or 24 component gyres in 11 or 12 coaxial pairs to the Top quark with 52 or 54 gyres in 26 or 27 coaxial pairs.) The single pairs on an axis then are doubled by having each of two pairs in parallel cylinders with the S0 axis or one of the three other octant center axes between the two cylinders, making interactive vortical use of the large, otherwise empty, spaces in the centers of the eight sphere octants. This increases the necessary clearance for these sites from 15° to 22.5°, and as evident in the table, both the S0' and S2' static spin sites conflict with both the ++ and - - orbits unless the pair of cylinders is driven (by their enclosed rotating gyre frictions) to rotate also around the common centroid axis between them at twice the angular rate of the conflicting orbits in order to synchronize with those orbits. This occurs naturally from the vortex current frictions, because the adjacent gyres at each end of the cylinders are bound tightly to each other at less than 0.7 GD separation (Figs. 4-5f, p. 98, & 6-2, p. 150), and is facilitated when the adjacent gyres are of the same charge and hand of rotation, but are tilted slightly away from each other and do not interact with each other's spiral-wave charge hemispheres at the sphere surface [that is, the cone pair mutual attitudes balance slightly away from exactly coaxial toward having sides of adjacent drive cones parallel (testable.) The clearest figure of cylinders comes up later with small neutrinos.] - - When this is not sufficient to occupy all interacting gyres, the 100% larger outer shell sphere can occur. Neither occurs with the **small** quarks' 4-5 pairs of gyres & SEq orbit expansions.

cause of the 1.5 MeV and 1.4% deviation in Table 1-1 between the calculated and the measured masses of the muon CL. In the muon (Table 6-4) three of the eight pairs of vortices are in these static spinning sites. Only five orbiting sites are able to pass closely twice per cycle. Though they pass more closely than the average orbital-to-orbital separations, the reduced mutual kinetic energy for such a large fraction of closer interactions between the quantal vortices may account for the lower PDG measurement than the calculated and scaled interaction mass. (More precise and specialized lab measurements would be needed to prove this source of this deviation.) In the case of the tauon the 1.9% deviation is in the opposite direction. The measured mass is higher than calculated. The paradigm presently has no accounting for the mass energy effect of excited states in isomers of the ground state of particles. The deviation should not arise in the same way as in the muon, since only three of the tauon's twelve pairs are in static spin sites, nine rather than five pairs are making closer frictional passes, and the measured mass is higher. It will require much more complex scaled lab measurements than those now available or feasible to account for excited states. (A well instrumented state-of-the-art marine facility could do it readily, though at a cost.)

(References for this chapter are those for the cited sections of prior chapters.)
Lorentz, H. A., (1909) The Theory of Electrons, Dover Publications, New York (1952)

APPENDIX 6-A

Extreme Neutrino Pair-Structure in Static Spin Sites with Cylinder Double Pairs

In Table 6-5 above Cylinder Double Pairs are listed as occupying the static spin sites at the centroids of the sphere octants in particles. In Figure 6-2 next the clearest sketches of this structure show in the central subfigure (from the chapter on neutrinos) two pairs of vortex drivers strongly attracting each other at much less than 1 Gyre Diameter (GD) separation *as if* they were in *virtual* cylinders coaxial with S0. The confining attractive force between the indicated vortices at that close separation is real; the cylinder walls are only conceptual or imaginary, but the vortex pairs both spin around each other and spin individually. If an orbiting pair is passing nearby in an orbit with less than 22.5° clearance (but more than the minimum 15° clearance), the effective cylinder rotation of the double pair must, and will, synchronize itself at twice (or four times, or 6 times, or equivalent fractional multiple times) the relatively constant angular orbital rate of the orbiting pair. Any doubled cylinder pairs are in a high energy state, not the usual ground state. (Since the tauon only has single pairs on the static spin sites, they are not shown in a high energy state; but if there is excess energy in the pre-process of tauon formation, the static pairs could attract three other pairs into the higher energy condition and might account for the tauon's excess mass. Empiric equations for high energy effects in a sphere are not yet developed. This is a good starting point.)

150

Fig. 6-2 Extreme Neutrino Pair-Structure in Static Spin Sites & with Cylinder Doubles. In subfigure E_{OB} the subdued lines of other structure permit a relatively readable view of the effective, but imaginary cylinder with a doubled pair of vortex drivers representing the four real attracting vortices. Note that each driver with its vortex maintains approximate coaxial alignment with its twin vortex and does not change in this while spinning both on the approximate coaxis and with it around S0. Since the center of the strong attractive force between closely spaced gyres is at the centroid of volume of the drive cone, and is thus well below the widest part of the cone at its base, and since this force is balanced against the point-thrust force of a gyre along the axis and pair attraction, there is a tendency for the coaxis to be slightly curved.

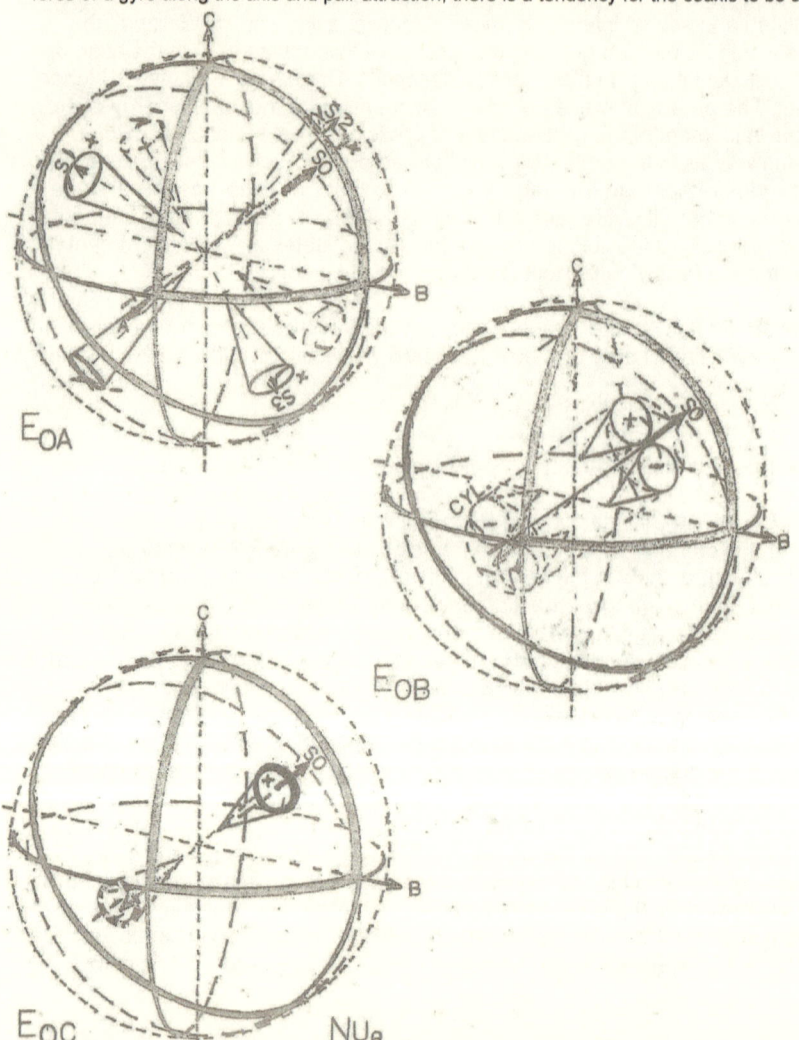

E_{OB} puts the coaxis at the driver points a little closer to the other pair's axis than at the bases. There is a <u>small points-in angle off the parallel for the two pair axes within the four vortex drivers.</u> (See text.)

APPENDIX 6-B

Branching Off of <u>Chemistry Emerging</u> from Its Basic <u>Particle Physics, etc.</u>

There is one baryon particle, the proton, which is also the complete nucleus of an atom. Having worked out the basic MQP Paradigm of the interactions of the subatomic and subnuclear particles to the point of defining a basis in mass, structure, and forces for the electron and its constituents, and at this point being engaged in expanding that paradigm through the baryons, including the proton, the paradigm offers an apparently endless extent. This includes emergent complications such as the portion of particle physics called chemistry, nominally chemistry of atoms, which is separately far more elaborate through the organic biochemistry of life, for instance, than the paradigm could become. The paradigm must presently avoid the door branching out to such a thoroughly well-founded field as the consequences of orbiting electrons around baryon nuclei. Even the simplest hydrogen nucleus, the proton, is beyond the paradigm scope.

There are many other such topics, some of which should be introduced as quickly as possible. The present paradigm obviously operates within the limits of instantaneous electric forces on the basis that magnetism is produced by time and space variation of electric forces with relative motion and variation of electric field of force, which are temporarily neglected. This may be thought of as emerging generally in that now ancient and honorable foundation, Maxwell's Equations. Lorentz came very close to establishing the broader principle (Lorentz, 1909.) Einstein includes too difficult a version for ordinary practice. The structure of the photon noted herein may lead to a further empirical resolution. Until such time classical Physics, Optics, is worthily prevalent with Relativity applied where necessary.

APPENDIX 6-C

COMPARATIVE DATA TABLES ON THE STRUCTURES OF THE QUARKS

Up to this point the discussion of quarks has been dependent on the introductory Table 1-1 as a part of the generation of quarks as a necessary structure to fit the empirical requirements on subnuclear and subatomic particles. This information has been added to as needed in the expanding view of these particles, their structures, and functions. In this chapter the small up and down quarks have been emphasized because of their outstanding importance in Nature. In the coming chapters on baryons and mesons and their well known "proliferation" into the hundreds of particles observed in cosmic ray showers in the atmosphere and around reactors or collision experiments, with classified families of series of these building blocks of the major structures of Nature, it will be important to have a grasp of all the quarks. This includes those that are very massive and rather rare compared to the small quarks, as noted in Table 6-1.

Accordingly, organized Table 6-6a (next) correlates the structural consequences of all the prior quark discussion, including the most massive quarks. Note the locations and motions of mass and charge generating forces per equations of Chapters 1 to 5.

Pair Orbit & Static Spin Sites in Dual Quark Masses
(Individual quark sites are shown <u>as if</u> quarks could exist alone.)

	Up-quarks		Down-quarks,		Strange-quarks,		Charm-quarks		Bottom-quarks		Top-Quarks	
Mass MeV	1.914	2.871	5.11	8.032	82.467	107.19	1.1665	1.395	4.0218	4.4106	170.99	172.1
Charges	+2/3	("MeV")	-1/3		-1/3		("GeV) +2/3		-1/3		+2/3	
Sites, Pairs	4 pr		5 pr		7 pr		8 pr		11 pr		12 pr	14 pr	16 pr	27 pr	26 pr
A			+ -	- -	+ +	+ -					- -	+ -	+ +	+ +
B			+ -	+ +	+ +	+ -					- -	+ -	+ +	+ +
C			+ -	+ -									+ +	+ +
A'							+ +						- -	- -
B'													- -	- -
C'					- -	+ -					- -	+ -	- -	- -
+ +	+ -	+ +			+ +	+ +	+ +	+ +	+ -	+ -	+ +	+ +
- -	+ -	- -			- -	- -	- -	- -	- -	+ -	- -	- -
S0	+ +	+ +	+ -	+ -	- -	+ -	+ +	+ +	+ +	+ -	+ +	+ +
S1							+ +	+ -	+ +	+ -	+ -	- -
S2							+ +	+ -	+ +	+ -	+ -	- -
S3							+ +	+ -	+ +	+ -	+ -	+ -
S0' sync					+ -	+ -	+ -	- -	+ -	+ -	- -
S1' "							- -	+ -	+ -	+ -	+ -	+ -
S2' "							- -	+ -	+ -	+ -	+ -	+ -
S3' "							- -	+ -	+ -	+ -	+ -	+ -
SEq 1	+ +e		- -e				+ +e				- -e	+ +e	+ +e
SEq 2	+ +e		- -e				+ +e				+ -e	+ +e
SEq 3			- -c		- -e				- -c		+ -e	+ +e
A Shell 1									+ -	+ -	+ +
B "									+ -	+ -	+ -
C "										+ -	+ -
A' "										+ -	+ -
B' "										+ -	
C' "										+ -	+ -
+ + "										+ +	+ +
- - "										- -	- -

S0 " etc. (only if an orbit conflicted)
{not a limit; lim →[(16 pr)/shell] + 3 due to SEq orbit expansion from only 1 level; shell may warp baryon)   e= orbits ellipse-like, c= circular expanded orbits
BALANCED IF USUAL (& INTERNAL QUARK-TYPE CONFLICTS AVOIDED)

Table 6-6a Pair Orbit and Static Spin Sites for the Quarks. This is not the only possible plan of orbit and <u>static site distribution, which depends also on quark-to-quark interference per Table 7-1c & excited states.</u>

       This table of the distribution of a form of ground-state (unexcited) gyre pair orbits and static spin sites for the dual mass quarks is also strongly dependent on Table 6-6b next. That table is a restatement in more readily readable form (for present purposes) of much of the data in Table 1-1, which has to do with the necessary generation of the masses of the LQ (lepton-quark) particles by a required microquantal component, later determined to be necessarily a turbulent conic vortex.

MICROQUANTA PAIRS for DUAL QUARK MASSES

	LQ # PAIRS	LIST OF PAIRS	(Chrgd/Total)	NET Charge	Calc.MASS	
e neutrino upper limit	1	+ − (a=3⁵)	0/1	0	2.885	eV
Mu neutrino PDG lim	3	+ −, + −, + − (a=3¹)	0/3	0	0.1703	MeV
Electron	3	− −, − −, − − "	3/3	−1	0.511	"
Up Quark A	4	+ +, + +, + −, + − (or 2 ++, 2 + −)		+2/3	1.914	"
B	4	+ +, + +, + +, − − (or 3 ++, 1 − −)		"	2.871	"
Down Quark A	5	1 − −, 4 + −	1/5	−1/3	5.11	"
B	5	1 + +, 2 − −, 2 + −	3/5	"	8.032	"
Tau neutrino PDG lim	6	1 + +, 1 − −, 4 + −	2/6	0	18.169	"
Strange Qu'k A	7	3 + +, 4 − −	7/7	−1/3	82.467	"
B	8	1 + +, 2 − −, 5 + −	3/8	"	107.19	"
Muon	8	3 − −, 5 + −	3/8	−1	107.19	"
Charm Quark A	11	6 + +, 4 − −, 1 + −	10/11	+2/3	1.1665	GeV
B	12	4 + +, 2 − −, 6 + −	6/12	"	1.395	"
Tauon	12	3 + +, 6 − −, 3 + −	9/12	−1	1.744	"
Bottom Qu'k A	14	4 + +, 5 − −, 5 + −	9/14	−1/3	4.0218	"
B	16	1 − −, 15 + −	1/16	"	4.4106	"
Top Quark A	26	10 + +, 8 − −, 8 + −	18/26	+2/3	172.1	"
(2005 STATUS) A−	27	7 + +, 5 − −, 15 + −	12/27	"	170.986	"
		(POSSIBLY NEAR PAIRS LIMIT)		WIDE STEPS		26

Table 6-6b Microquanta Pairs for Dual Quark Masses, Charged Leptons, and Neutrino Limit Masses.

Entanglement Within and Between Quarks vs Particles

Another factor of unknown, but foreseeable, force influence, especially between the SEq orbiting quanta of quarks in baryons, is that they soon become differentially tuned in entanglement, in addition to being initially grossly synchronized, with various elements of their own quarks and with the link orbits of adjacent quarks. The stable baryons become well entangled, but the unstable mesons may not in the average or mean survive long enough to acquire a significant amount of this further stabilizing influence. The particular orbits of greatest relative effect and the time constants of this tuning to near or precise resonance is required to be distinct from the more gross tuning of entanglement between entire particles, such as electrons and their photons. The playing out of these harmonizations will require extended research, first to establish a velocity vs the velocity of light for entanglement force waves between separated particles. In the model of entangled synchronization of clock pendulums on a wall within their tuned coupling envelopes, or the model of bandpass of the tuned transformers of intermediate frequency amplifiers in superheterodyne radio receivers, the velocity of coupling must be significantly greater than the highest velocity that is coupled. Here the V_P of the ultimate gyre driver must be well beyond c, and the V_e beyond the sum of both.

Chapter 7

Effects of Interference Constraints on Quark, Baryon, (Meson), Charged Lepton, (and Neutrino) Proliferations

SUMMARY: It is controlling to both the proliferation of baryons and the quark structures that compose that proliferation (as well as to the easing of limitation so that there are not fewer of both) that quarks do not generate unitary charges equal to those of the electron and other charged leptons but lesser charges of 1/3 and 2/3 the lepton charge. This less demanding level of charge for each quark with two mass options allows them to build numerically four times the total particle quantity of three for the CLs with twelve Quark masses over the same large range of quantities. However, the quarks are further limited by their necessity to add up to baryon rings with unitary charges of 0, 1 (to match electrons), or at most 2, from their three fractional charges of dual polarity. (Disrupted baryon fragments also have the meson class option of briefly totaling to unit charges from either two, four, or six quarks and anti-quarks in equal numbers to yield observed meson masses.) The resultant baryons then give all the observed masses and, with some significant re-classifications, generate the observed families of baryons with systematically stepped masses. These result from step-by-step replacement of each quark's lighter mass by its heavier mass. This is initially confused by excessive numbers of accredited PDG baryon candidates for the limited numbers of possible lowest-mass step leaders, here named isotons. Finding that excess PDG particles are only high energy isomers in the step groups develops from requiring smooth curves for all Exponential Law y exponents versus step particle masses as found for baryon type leaders in Chapter 1. There are only minimal such constraints on building electrons, positrons, and up to mu class neutrinos.

Smaller Quark and Prototype Baryon Particle Structures

Small Quarks were designated in Table 6-1 as the next general topic, with Main Baryons to follow them. Since quarks and hadrons can only come into existence together (as will be discussed again later with Precursors), and neither quarks nor hadrons in general can exist without the other, the discussion of either type of particle cannot progress very far without taking up the other to about the same level. An entire subchapter about any Quarks without an equal introduction of some Baryons would also be difficult to understand, if not pointless. The main lighter mass forms of the two small quarks and the main two of the three prototype baryons provide the basis for the atomic nuclei of most by far of the visible material Universe. They are a very important combined topic. And since the two lighter small quarks are so well related in particle size to the electron, it is definitely clearest to begin with the two quarks as they are shown in prior Table 6-4 in coordination with the Charged Lepton masses.

As Table 6-5 illustrates, it can be only fragmentary to open up a necessary area of background on any quarks, such as spherical angular clearances between orbit sites inside each of the two small quarks, without including some of the scope of those relations which shows how that type of feature must relate to the necessities of large quarks. The small quarks are not really different in their essentials from the larger quarks. It is only that the masses and the numbers of component quanta lead quickly with mass into difficulties of structural balance and instability that constrain what natural interactions of necessary components can construct. In proceeding further, it will soon become evident that even the second one of the two small quarks, the down, is severely limited in the ways in which it can interact with other quarks constructively. This shows that larger quarks will rapidly become more limited. Therefore, it should not be surprising that these two quarks enable such a very large fraction as over 99% of the visible nuclear cosmos that has been built up by the very small particles. Their general limits and broadest implications then arise from their simplest basic interactions.

All the quarks are treated herein as PDG recognized particles from Chapter 1 onward. Quarks of all 6 types are plotted with little discussion from PDG data as known QM particles in the graph of Figure 1-3 on page 32. Their specific masses as even more generalized quantities of unspecified components with charge are developed with two obligatory numerical masses for each quark under the Charge-Mass Power Law (Equation 1-2 of page 34) and are shown in Table 1-1 on page 35. These generalized masses are emphasized in Figures 1-4 and -5 on pages 36 and 37 and again in Figures 1-6 and -7 on pages 43 and 44. The orbital structures of the Up quark and the Down quark are accommodated very generally in the build-up of Figure 2-6 on page 56. The full structure of the two quarks with the lightest masses as linked spheres of specific orbits of micro-quantal vortices is shown without discussion in Table 6-4 above, near the end of the previous chapter. Quarks, as necessities identified by QM, have been kept in correlation earlier herein. The apparent minor structural nature of the four and five components of the two smallest quarks in tables is the source of their importance due to basic stability. In this they follow the prototypical electron and provide prototypes for the unstable heavier quarks, which reach the internal sphere limits of Table 6-5 on pairs.

In Table 6-5 there are in a single sphere of orbits a small number of interferences such that the apparent minimum 15° clearance for the full width of the drive cone of a vortex is not available. The A' site center has only 12° clearance from the site center of SEq 1 and 13° from SEq 2. C' has only 12° clearance from SEq 3. The ++ orbit has only 6.5° clearance from SEq 3 (as in Table 6-5.) This eliminates two sites, such as A' and SEq 3, from being occupied at the same time in one full lepton sphere (unless there is link orbit expansion, as in quarks), but the main effect can be on disturbing the priority of balances for which the elimination of one orbit, but not both, from each interference may take effect on either one. Note that all of these 4 conflicts involve an SEq orbit. All but one of those has at least one potential conflict unless that orbit is expanded. If SEq orbits are not to be occupied, there are no conflicts in single spheres. But when all 3 SEq orbits are occupied to link and synchronize the 3 quarks of a baryon (2, 4, or 6 quarks in a meson), they must be at least pseudo-elliptically expanded in the right phase angles to clear each quark link orbit's potential conflicts. The linking attraction of the other two quarks makes this orbit expansion possible. It does not expand of itself.

The Up quark has just one more pair of component vortices than the electron in Table 6-4. The Down quark has just one additional pair over the up. This does not change with the heavier masses for each of these two only. The increased mass arises under the Charge-Mass Power Law by replacing two neutral + - and + - pairs in each quark with two ++ and - - charged pairs that neutralize each other, but with 3 times their previous mass. Both are then summed and multiplied by the Law's exponential term in each quark. Because of the exponential increase in interactions, the single added quantal pair in the down approximately triples that quark's mass increase to the heavier down compared to the up's increase to its heavier mass. (This difference has strong effects when it appears later in causing the otherwise puzzling sequence of mass steps in many baron and meson families of particles. This size of steps is in addition to causing with the dual quark masses themselves the large number of most of those steps between the numerous family members in the heavily proliferated families of baryons and mesons. Within each step group there are also isomer steps due to excited states of quanta that are not in a minimum energy ground state orbit.) The QM names themselves of up and down are arbitrary identifiers with no significant meaning.

It is important to recognize that the lighter mass of each quark is the one with the greater occurrence in Nature, and is the listed mass in Table 6-4. Either Up quark is always charged to +2/3 of the electron charge quantity, though that charge amount is arbitrarily defined as -1, while either Down quark is charged to -1/3. Here these charges are each the sums of the 1/6 charge forces of component vortices in each case from the turbulent spiral wave interactions off the bases of uniform vortex drive cones. In total, since light ring forms are stable, since the simplest closed baryon ring is a triplet, and since attractive opposite charge forces hold particles together, quark pair charges must balance in thirds. Four pairs with two charged make the smallest Up quark at even 2/3. The Down five pairs have the opposite 1/3 charge. Any others do not function. Polarity of charge force was assigned by human researchers 100 years ago. (Fractional charge is usually stated without referring to electrons. The heavy forms of the quarks are taken up with the numerous baryon families that make them important, though quite rare.)

(The definite dual quark masses are thoroughly developed in this MQP paradigm and are not recognized in QM at the time of writing. It is predicted from its usefulness later, in reorganizing the many proliferated families of the rare baryons made possible by the quark linking mechanism, that this feature of the paradigm will be adopted and adapted by QM in due course. The first reorganization occurs here with the complete, and simplifying, separation of the two smallest baryons from each other and from the two families of proliferated baryons of which these baryons are the separated baseline prototypes. In QM these two baryons and their families are merged in a single family called Nucleons, because they can and do occur principally in atomic nuclei. The QM classification has its point, but it obscures the differences in the structures. It is a major point of the paradigm that these structures can now be separately identified and clarified, and that they should be treated in that way for clarity of consequences, even though it may require later inconveniences of highly skilled labor to re-identify a century's old references in future use.)

Full discussion of the Pre-quark Precursor form and process by which quarks and a baryon come into existence together is deferred until the full range of similar processes arises with Expanded Neutrinos. The general nature of the typical pre-quark process is covered previously with the SEq linking orbits in Figure 2-6 on page 56. This is not a case of preformed bricks being brought together to shape a building. Each pre-quark is formed with the natural interaction forces of two other pre-quarks, all three of which form each other as they interlink into emerging as a baryon. It is a main thesis of the MQP Paradigm that this is the only way in which such a structure can be built up.

The structural necessities of a quark sphere triplet form of the simplest baryon, a proton from Fig. 2-7, are clearer after the electron. In Figure 7-1a the 3 link orbits lock their planes in a right angle corner with 3 S0s pointed to sum in a new baryon corner axis that brings 3D order throughout 3 quark structures with synchronization of all orbits. (It is significant that only a least minimum of such constrictions on structure of particles occur with the least CLs, electrons and positrons, or with neutrinos up through nu mu.)

157

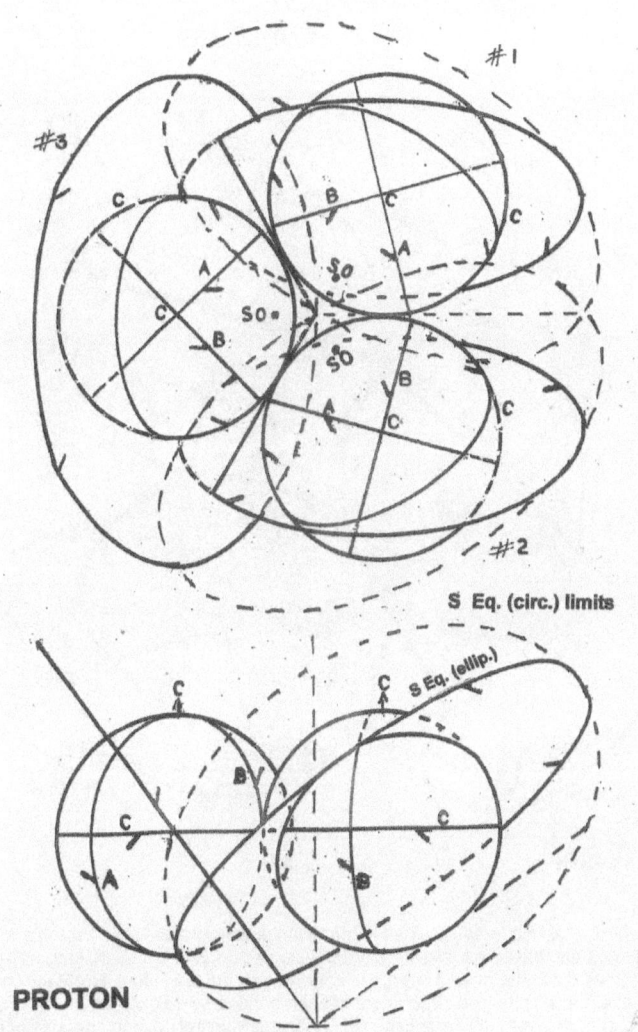

Fig. 7-1a Proton Plan and Elevation. The right angle corner of S Eq planes brings the two Up quarks #1 & #2 and the #3 Down quark in contact on their C orbits with their C axes vertical and with their S0 sum axes pointed together on the new central proton sum axis just out of sight behind the #2 sphere in the lower elevation view. The odd single negative Down has its link orbit expanded to the full circle, while the two positive Ups repel each other somewhat and are only expanded to ellipsoids (not true ellipses due to their constant angular rates of orbit.) For clarity of contrast with the Down their S Eqs are shown tangent to the sphere, but that reduced minor axis should be about 1.5 x radius of the sphere. The charge is +1. Obviously there will be many sphere-to-sphere orbit-only 15° clearance conflicts here & in the Neutron.

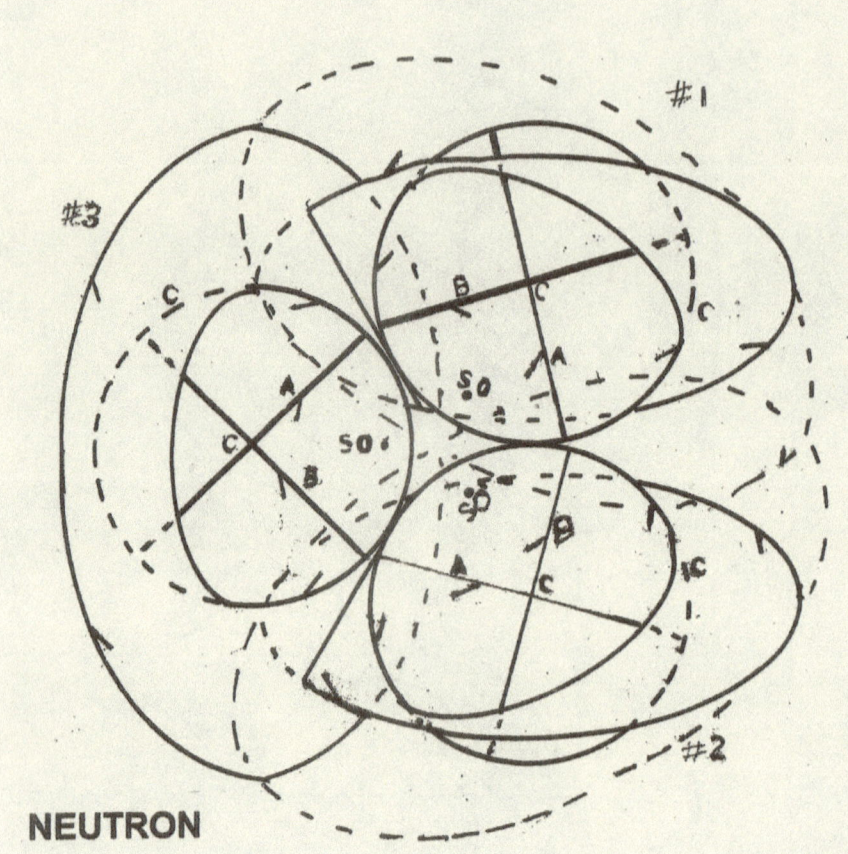

NEUTRON

Figure 7-1b Neutron Plan Schematic. The Neutron plan is very similar to the Proton's with the quark positions exchanged and the orbits reversed. With a single odd +2/3 Up quark in the #3 position and now two -1/3 Downs in #1 & #2, the overall charge is not reversed but is neutral. However, the individual quarks have their standard charges. The angle of each orthogonal box corner wall from the axis is equal at 90° − arcos $1/\sqrt{3}$ = 35.2644°. Differences in figure size are entirely due to changes of scale to fit the spaces available on the page since the Neutron elevation view is so similar to the Proton's that there is no need to show it here. Otherwise the comments on Fig. 7-1a are largely interchangeable with those here.

Despite the many similarities in overall plans, the reversal of orbits (with the change of charges and their predominant rotations) between the +1 charge Proton and the neutral Neutron makes a few changes in the sphere-to-sphere conflicts of orbits that occur because the quark spheres in the baryons are in direct contact on the C orbits. This also affects all other baryons that have these charge totals and so must follow the separate prototypes of these two main baryons. These conflicts are shown in Fig. 7-1c.

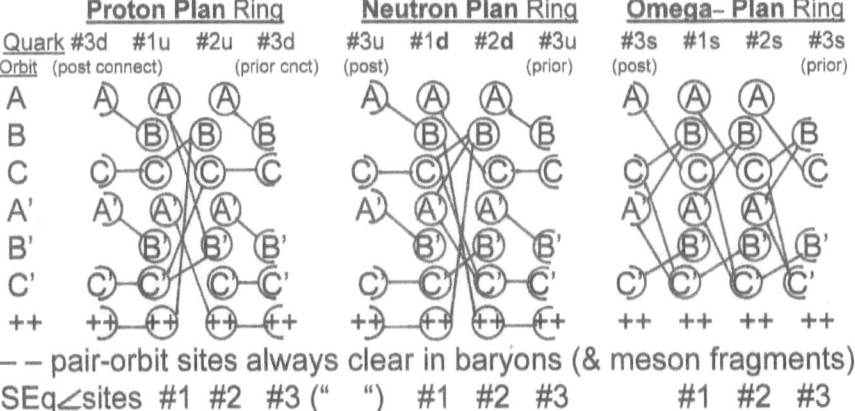

-- pair-orbit sites always clear in baryons (& meson fragments)
SEq∠sites #1 #2 #3 (" ")   #1 #2 #3   #1 #2 #3

Wherever a line connects two pair-orbit sites, that baryon can have only one of the two. Lines do not connect through a site to a third site. Half of each vertical #3 quark is left of its plan and half to the right where needed to show connections, or as in A & B orbits #3 quark orbits are shown either to left or to right for connections logic. NEUTRON C CONFLICTS force d to ABC', limited balance, stability, & mean life.
PAIR-ORBIT SYNC CONFLICTS INHERENT IN BARYON 3-QUARK RINGS

& INSIDE NON–QUARKS SPHERIC SEq #1 & #2 conflict with A', #3 with C' & ++

Rarity Increases Rapidly with Pairs No. in Balance of Force & Mass   61

Fig. 7-1c Quark-Sphere-to-Quark-Sphere Conflicts in the Three Baryon Prototypes. [The Omega Minus plan adds the very few baryons in which all three quarks have exactly the same charges, either + 2/3 (giving a baryon charge of +2 and requiring a mirror image of the conflict plan shown) or -1/3 (giving a baryon charge of -1.)   The Omega Minus baryon has no true small quarks, but 3 of the lighter version of order-of-magnitude heavier Strange quarks. There are small quark baryons that have been observed and fit this plan, but are rare and not so accreditably determined in charge, etc., as to make class prototypes.

    The quark-to-quark conflicts of orbits limit, first, the smaller baryons shown in Figure 7-1c, then, the medium-size baryons made by natural mixing in of a strange or charm quark with the small quarks, next, a baryon with two of those medium-sized quarks, and, ultimately, one with a heavy bottom or top quark. This progressively cuts workable mass numbers in the nine orders of magnitude (OM) of conceivable baryon masses beyond the previously noted barriers. The six sets of quark gyre and pair numbers effectively screen the potential particle mass numbers again and again for functioning combinations so that the few hundred baryons exist. Small quarks enable the finest natural steps of screening for valid hadron masses, and they build hundreds of baryons and mesons, while top quarks enable singles at most. In Fig. 7-1c the lighter Strange quark is only one OM in mass above the small range. Three of them make up the Omega Minus baryon orbit spheres in Figure 7-1d to complete the three prototypes for all baryon structures. Baryon triplet rings of three quarks in the proton plan are charged +1. Those in the neutron plan are neutral at 0. The few baryons in the omega minus plan are charged either -1 or +2. All baryons follow one of these three plans.

Table 7-1 of three prototype baryons defines the orbit and spin sites of quantal pairs of the lighter versions of quarks, u_A, d_A, s_A, in the PDG measured baryon masses under the Mass-Charge Power Law in the most nearly balanced, simplest structures.

Orbits & Spin Sites for Quarks & Quantal Pairs in Baryon Prototypes

Quarks Order Pair Orbits	Proton Type u_A1	u_A2	d_A3	Neutron Type d_A1	d_A2	u_A3	Omega Minus Type s_A1	s_A2	s_A3	
A			+−	+−	+−	+−	++	++	++	
B			+−	+−	+−	+−	++	++	++	
C		+−								Dual/Triple
A'										ABC conflicts
B'										forced to C'
C'				+−	+−		−−	−−	−−	in imbalance.
++	+−	+−				+−	++	++	++	Good-
−−	+−	+−				+−	−−	−−	−−	balance.
S0 Spin Site	++	++	+−	+−	+−	++	−−	−−	−−	Auto-balance.
SEq 1	++e			−−e			−−e			"
SEq 2		++e			−−e			−−e		"
SEq 3			−−c			++c			−−e	"

Table 7-1 Prototype Orbits and Single Spin Site for the Basic Baryons. e= ellipse-like orbits; c= circular expanded orbits; +−= neutral, ++ & −−= charged pairs. Baryons have a few avoidable conflicts between two quark spheres in 2 orbits of which one, but not both, can be occupied. In the neutron and Ω^- the necessary quantal shifts from balanced A-B-C orbits to less well balanced A-B-C' orbits are sufficient cause of decay in their measured PDG short mean lifetimes. <u>The single down quark in the proton has it perfectly balanced!!!</u> The 3 strange Ω^- imbalances do not cancel, and they are heavy and self-repellent. <u>Up quarks have 4 pairs of quanta; downs have 5; stranges have 7 or 8. There is no quark with 6 pairs.</u>

It is significant here that in each quark the single SEq orbit site must always have a charged pair of gyre quanta for the binding forces of these expanded links from each quark to the two other quarks to be sufficiently strong to hold even the lightest baryon together against the centrifugal inertia of all its orbiting quanta. Also, under the C-M Power Law the linking pair's three times higher charged mass than a neutral pair in its expanded orbit is the major contributor to the un-moving quark's spin, which might otherwise be inexplicable (even with the aid of the virtual "gluon" concept.) (Bass, 2007)

It is not obligatory in quarks that the S0 spin site always be occupied by a pair as it is in each of the prototypes. Since the S0 axis is the axis around which the entire quark sums its balance effects, the site is a perfect location for a single pair that would otherwise have no balanced site to go to under the self-sorting forces between the quanta during the quark/baryon Precursor state that must necessarily last less than 1 orbit cycle or fly apart. Without the lone S0 site, the lighter A form of the up quark would have no way to balance its fly-wheel effects while having one of its two charged pairs attracted into an SEq site. The rapidity with which a clump of randomly gathered pairs of uniform gyre quanta can sort themselves into a particle geometry and expel the pairs that do not fit can be estimated with the aid of the force equations of Chapter 5. The gyres quickly circle a center of interactivity or scatter. (Exact calculations on this require data that are not yet available nor feasible within available funds.) Omega minus (Fig. 7-1d) even turns its major ellipsoid axes 45° further in the box planes about its spheres.

161

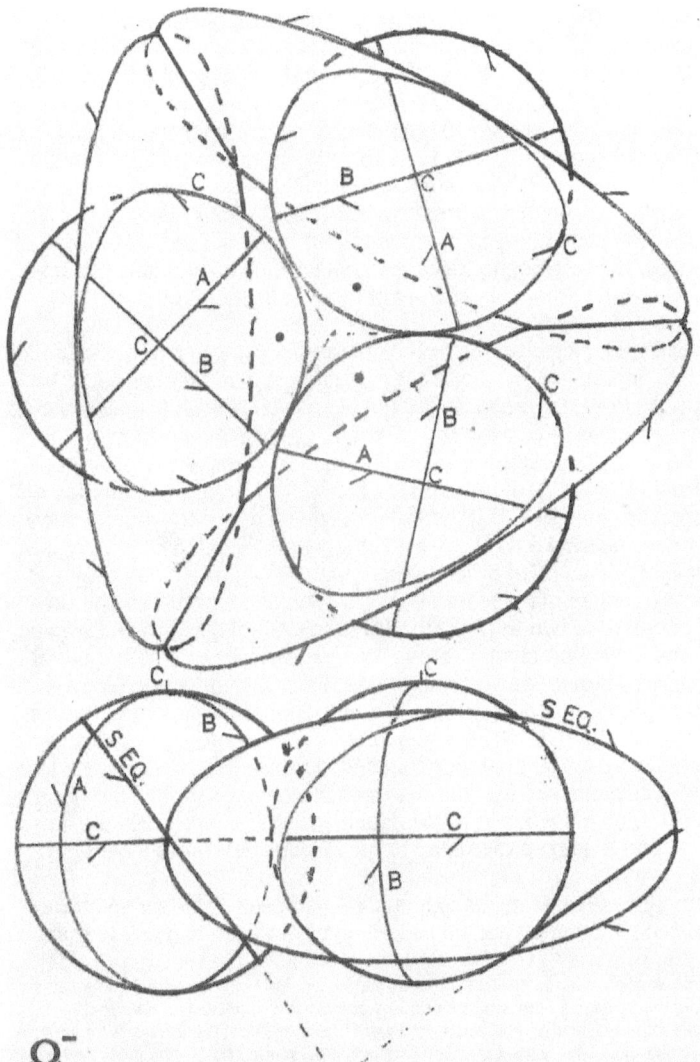

Ω⁻

Figure 7-1d The Omega Minus Baryon Prototype for 3 Like-Charge Quarks. Here SEq orbits are CCW, Counter-Clockwise around each SO dot (vs Clockwise, CW.) They again follow the practice of showing the elliptical orbits as tangent to the sphere for clarity. The typical available magnetic moment indicates that the minor axis is actually about 1.5 times the radius of the sphere in at least some cases. In any case none of the quasi-elliptical orbits are true ellipses about two foci, but proceed instead at a constant angular rate around the single center of symmetry. - - Obviously, carefully scaled models of quark spheres (with and without expanded links) from about 6 cm to 30.5 cm (2 3/8 to 12 inch) diameters can be very useful for step-by-step orbit clearances with tape orbits, movable colored tabs, and large calipers. (Well made earth globes with clearly drawn latitude and longitude angles are ideal for orbit visualizations.)

Local & Lifetime Effects of Three Prototype Structures of Baryons

The schematic sketches of Figs. 5-1-4, 6-1-2, 7-1a-d, necessitated by the LQ Charge-Mass Power Law and Exponential Law of Hadron Masses (Equns. 1-1-2) as in Figs. 2-6, 2-7, plus 2-10, though limited by the conflicting orbit restrictions and quark-to-quark constraints of Table 7-1 and Figure 7-1c, resolve basic inner structural systems of subatomic to subnuclear particles. These working functionals also embody the unified particle forces and masses scaled from laboratory force data to the prototypic numeric electron of Chapter 5. That builds baryon prototypes. Refinements and extensions of these simplest essentials will continue to be worked out with increasing detail in other particles in later chapters (and potentially much further over decades.)

[The growing MQP paradigm brings increasing auxiliary support and clarification to remaining mysteries, generalities, and gaps in particle Quantum Mechanics. QM continues to advance science very productively, but with much difficult complexity in its Standard Model of the particles. For over half a century physicists have known with great precision and certainty the QM probabilities for many nuclear end effects that quarks bring about from within nuclear particles, including the baryons in everyday earth and air (or in weapons and reactors.) But the six deductively identified types of quarks (and their symmetrical antiquarks) have been still largely unresolved in QM beyond a few very general qualities. These QM properties are usually thought of as heaviness or lightness in six uncertain ranges of mass (recently much reduced for the up and down), positive or negative one-third or two-thirds of the electron's quantity of forceful electric charge, both strong and weak non-electric forces, bound connections through virtual "gluons", detectable spins, "strangeness" or other odd "flavors" of difference from each other, "color" tag labels, and the quarks' apparent non-existence as separate particles freely available for close study. The MQP resolution of empiricly required structures for quarks and their subquanta simplify that uncertain status of quarks and steadily build their basic numeric quantification, as the rest of this chapter demonstrates in organizing, explaining, and re-classifying the ways in which dual masses of quarks build the large structures and families of the many proliferated baryons identified by the PDG.]

Table 7-2 summarizes prior Key Concepts herein that give auxiliary support to QM in general and extend it, but may not yet be fully familiar in use to many readers:

- The mass of any QM nuclear/subatomic particle measures the contained energy of multiple inner and outer interactions between the particle's components and any component quarks' subcomponents.
- New natural laws of this relation demand that each particle's mass consist of the interactive energy of many conserved coaxial pairs of smaller vortex microquanta of mass charged by direction of gyre spin.
- Such microquanta interact with all the forces of nature through turbulently energetic vortical spinning and synchronous orbiting in spheric quark structures that embody the mass law and expand single linking orbits to interlock closed triplets of quarks in baryons (that may decay in meson fragments with quarklets.)
- Under those laws the long previously uncertain mass of each of the six kinds of quarks resolves into two definite masses. (Recent 5-year-anticipated QM up & down masses approach the lighter dual mass.)
- Systematic combinations of these six dual quark masses build the proliferated series of hundreds of PDG observed particles and their isomers seen in cosmic ray showers and particle collision machines.
- This deterministic paradigm exhibits structural causes for varied stability and other features of quarks and of the particles they build and interact with in all visible nature, as is to be more fully demonstrated.

<u>Table 7-2 Key Concepts of Quark and Baryon Structures of Linked Orbiting Spheres of Vortex Quanta.</u>

An additional significant MQP paradigm result of these particle features as they occur in quarks is that the paradigm's usual small quark spheres, also the extended SEq linking orbits, and the final linked quark triplets in small PDG baryons are all scaled within the range of size variation of the highly accelerated electron sphere (Chap. 5.) This provides that the size of a tightly bound small baryon, a proton, is approximately consistent with the PDG empirically determined radius of the charge of a proton (in its simplest common terms, about 0.875 tenths of a millionth of a millionth of a centimeter.)

In one type of QM uncertainty, up and down quark masses are accredited by the PDG as theoretical current masses, not as running masses in the light quarks (only in heavy ones), and never as the altogether different constituent masses that some QM specialists use for quarks, especially at low energy baryon velocities. Herein the MQP paradigm's empirical power law masses are derived from the overall PDG accredited Summary Data tables, and are thus consistent across its orders-of-magnitude widely spread range with both current and running masses in their appropriate regions, as well as being consistent with the PDG in not correlating with "constituent quark" masses.

Under the simple power law derived (in two published analyses, Howard, 2005 and 2006) from the PDG data, the paradigm resolves the mass uncertainty for each kind of quark (as noted earlier) by finding two precisely quantized masses near the prior long-continued PDG uncertainties, one mass usually toward each PDG upper limit and one usually toward each PDG lower limit. In this respect the five year-MQP-anticipated 2010 QM paper (Davies and Lepage, et al., 2010) with its low uncertainty for an up and a down quark is much less accurate than the earlier wide uncertainties that did enable matching the mass steps of the baryon families (as does the paradigm's dual masses. The PDG 2014 up quark uncertainty also covers this.) Here it is quite notable that the PDG accredited 2014 masses of the up and down quarks (Olive et al., 2014) have been adjusted from the values of the 2010 paper such that the PDG down as well as its up mass still confirms the MQP 2005 smaller quark values of Table 1-1 herein within the stated PDG/QM uncertainties. (in the cited 2010 QM paper only the value of the up mass was in that close a confirmation of these two MQP anticipations of QM progress.)

Table of Small Quark Mass Calculations & Data Over Ten Year Period

Quark	Table 1-1 Herein (Howard, 2005)	Yao, et al.(&prior) (PDG, 2006-10)	Davies & Lepage et al. (2010)	Olive, et al., (PDG, 2011-14)
up smaller	1.914...MeV	(Single masses)	(Singles)	(Singles)
	(Two masses)	1.5—3-3.3+	2.01 ± 0.14	2.3, +0.7,-0.5
up larger	2.871...MeV	MeV	MeV	MeV
	(Both within PDG uncertainty)			
down smaller	5.11.... MeV			
	(Two masses.)	3-3.5—6-7	4.79 ± 0.16	4.8, +0.5,-0.3
down larger	8.032...MeV	MeV	MeV	MeV
	(Smaller within PDG uncertainty)			

Table 7-3 MQP Paradigm Anticipated QM Small Quark Masses At Least Five Years. Mass and force equations per Chapters 1 to 5 and references thereto herein, as correlated in Chapter 6 & herewith.

Natural combinations of the MQP two mass values for each kind of quark account for most of the complexities and further uncertainties of the many PDG series of baryons and mesons, and also emphasize the uniqueness of the very long life and nuclear stability of the proton among the numerous unstable baryons. The balanced orbital structure that provides excellent flywheel balance in the proton is clearly defined in Table 7-1 above, in conjunction with the definite cause of imbalance in the neutron. The imbalance of the neutron is caused by the orbital conflict between quarks (from Figure 7-1c) which forces the C orbit in the neutron to delay into the C' orbit site. That delay in two orbits is just enough to make the neutron only marginally stable, even with its fairly light inertial mass increase over the proton and its definitely higher coherent or "strong" forces than the proton (due to one more micro-quantal gyre. This shows that QM strong force is marginal.) The same type of delay imbalance in three orbits shown in the same table for the omega minus baryon (with its exponentially increased inertial mass momentum to be held together by only a linear additive increase in coherent forces) is sufficient to cut the PDG omega minus mean life to about 80 micro-micro-seconds. The PDG measured life of protons (from low decay rates) is greater than 10^{29} years, many millions of millions times longer than the presently computed 13.7 Gigayear duration of the QM Big Bang universe. The equivalent PDG mean lives of other baryons and mesons made of quarks range from far less than a millionth of a millionth of a second to an orders-of-magnitude jump to <15 minutes for a neutron, unless it is stabilized to cosmic lifetimes (as in this planet) by binding its quarks to a proton's in the common nuclei of the elements of moderate mass. Consequently, the lighter quarks in the lightest baryons cause most of the stars and planets to continue to exist over very long time periods. The proton, with its stability margin (which can stabilize a neutron when they are bound together by attractive strong plus quark-to-quark charge forces) is by far the most consequential and abundant baryon of all. Because it can be stabilized by the proton, the neutron is next in baryon abundance. All others are fairly rare.

Overall then. the proton (p) sets the model for the quark structures of those PDG baryons that are unit positively charged (+1, compared to the electron's electric charge of −1.) This charge is the simple sum of conserved charges of each baryon's three quarks' from their ±1/6 charged quantal gyres. In the same way neutrons are neutral, 0. The other rare baryons are charged +1, 0, -1, or +2, as the quarks and gyres add up.

In the proton prototype baryon structure of Figure 7-1a protons are the lightest +1 baryons, with 2 small up quarks (1.914…rounded MeV, lightest of all quarks, vs. 2.871… MeV for larger up quarks) at +2/3 charge each, & with 1 small down quark (5.11… MeV, next lightest, vs. 8.032… MeV for larger downs) at −1/3 charge. This gives the highest attainable baryon ratio of calculated attractive-forces-to-mass ratio. Linkage by charged pairs of quanta in 3 SEq orbits locks the orbit planes (dashed or solid outer circles) in a right-angled box corner. The 3 reference planes & tilted S0 axes meet on the corner's center axis for full particle summing. The major axes of the 2 ellipse-like SEq. orbits of the proton's up quarks are parallel in 3D and link with the circular SEq orbit of the down quark. The quarks (uud) and the phases of their balanced SEq orbits are each separately numbered CW in the + SEq orbit order. A lighter up quark's ++ &-- orbits exactly balance internally; and when 2 light up quarks each lock 1 light down quark firmly in place, its orbits also balance each other internally in light proton perfection. No other baryon can quite do this, though all are in the box corner form. The 3 quarks need not all balance together in a proton as they are not also spinning about a common axis.

The micro-quantal orbits, spins, and links of quarks within protons preserve the exact structural balance of each kind of quark, bind the 3 quarks with a peak ratio of forces-to-mass, synchronize the orbits of all three quarks, and have many open orbit and spin sites for larger quarks to build heavier, less perfect baryons.

Essentially the gyre-to-gyre forces of the neutral neutron baryon ring of three quarks in a rectangular box-corner of extended SEq planes and the -1 charged Omega minus ring similarly attempt to do the same thing as the proton ring, but cannot achieve the same balance because of the quark-to-quark conflicts of orbits that force each C orbit to delay 45° to C'. Departing from the proton (p) type of +1 charged particle with one negative 1/3 and two positive 2/3 quarks, there are a reversed neutron (n) type of neutral plan with one positive 2/3 and two negative 1/3 quarks, and an Omega minus type of plan with the same −1/3 (or +2/3) charge in all three quarks. These will spin in suitable directions with the preponderance of charge spin. (No other basic types of configuration are possible except mirror antiparticles.) Consequently, in Fig.7-1b, in reverse of protons, a neutron's neutral box corner plan (shown alone) has 1 CW circular SEq. orbit for a light up quark at +2/3 charge & 2 CCW quasi-elliptical SEq. orbits (in ddu order) for 2 light down quarks at −1/3 charge each, with 0 charge total.

In plan and elevation, the Omega minus prototype for the few baryons with 3 equally charged quarks has 3 sss lighter strange quarks with −1/3 charges. All 3 have quasi-elliptic SEq orbits in the suitable rotation sense with major axes each at 90° to its orbit reference plane & box corner links at their ends. The number order may start at any link. It is not necessary to show separate classes of non-identical -1/3 quarks or of 3 +2/3 quarks with reverse identical rotations at +2 charge. The defining feature is that all 3 quarks have the same charge. Neither of these two alternates has stable balance. - - Larger quarks on the 3 models may or may not have improved balance, but as with the charged leptons, the pseudo-centrifugal force effects of the exponentially increased masses build up so much faster than the linearly additive cohesive forces that the larger quarks and their baryons are all very unstable in net result. They have very short mean lives (leaving them rare in abundance) until they "decay" with mixed internal and external causes (to be established in Chapter 9, after covering the empirical types of meson and other fragments they decay into in Chapter 8.) Every definite action has causal consequences (or vice versa) in the MQP paradigm, as distinct from the more uncertain purely probabilistic event chains of QM and the Standard Model (which were brilliantly developed with very few observed sequential actions that could demonstrate causes.) The MQP paradigm is now able to fill many QM gaps as results of the Charge-Mass Power Law of LQs and Exponential Mass Law of Hadrons. This expanded series of causes has resulted only from the accumulated store of PDG empirical data and advancing observation. (This source in data, as well as Occam's importance of the simplest and most direct connections, or assumptions, should always be kept in mind.)

Tables 7-4a-e summarize the small quarks and their most immediate effects.

THE SMALL QUARKS (Balanced, <u>as If</u> quarks could exist alone)
(In expanded TABLE format with small type instructions for general use hereafter)

QUARKS	u_A	u_B	d_A	d_B	(or in dual d_B)
Masses MeV	1.914...	2.871...	5.11...	8.032...	
ORBITS (clear 15°)	2++, 2+-	3++, 1--	1--, 4+-	2--, 1++, 2+-	
A			+-		+-
B			+-		+- (<u>not</u>
C			+- (balanced		balanced)
A'			if single d)		
B'			<u>(Apparent simple changes require Fig. 7-1c checks)</u>		
C'					++
++	+-	++		++	(<u>not</u>
--	+-	--		--	-- balanced)

(Octant Centroid Spin Axis Sites & 180° opposites, 1-2-3 balance around S0 pole site)

S0 (as below)	++	++	+-	+- (balanced if single d)	
S0' ('= **in sync cylinders** per Fig. 6-2)				+- (high energy only)	
S1 (at centroid or ")	(if no data in this section, it may be omitted)				
S1' "	(all high-energy low-priority, balanced if all 3 symmetric)				
S2 "					NC (no change
S2' "					↓ in <u>column</u>
S3 "					below.)
S3' "					

(SEq pair Quark link orbit plane up to +100% larger radius [or on Lepton Sphere Surface?])

SEq1 (at 7.5° phase)   ++**E**llipsoid (in proton)   (Interchangeable only in initial formation in triplets, but
SEq2 (at 67.5°)   ++**E** (in positive baryons)   only 1 per orbit in any particle.)
SEq3 (at 127.5°)   --**C**ircle (in p)   --**C** (in pos. heavy baryons)
(+100% radius outer shell orbits of <u>c, b, & t quarks only;</u> angles as in ABC; min clear 7.5°.)
A100
B100
C100   (Explanatory notes above do not appear in later use of this table format.)

Table 7-4a. Larger B & Smaller A SMALL QUARK Components, including general notes to Tables 7-4b-c.
NC↓ = no change in Isomer quark column/below from Isoton to left. (This reduces confusing +- clutter.)

As noted, QM PDG recognizes only single mass values for quarks, but the MQP Paradigm has dual quark masses for baryon mass steps. Dual masses are needed in QM to account for partially explained (but brilliantly built up) complex QM lists of baryon families. Here, Tables 7-4a, -b, -c begin to introduce a necessary re-classification of the smaller dual quarks and their baryons (possibly not in the ultimate or final structures which may eventually appear by refinement of these initial structures.) The PDG/QM has for years classified small baryons in big families by secondary spin relations that always appear and by absence of strange quarks, more than by primary structure such as charge. Since Protons and Neutrons build charged nuclei, the PDG classifies them as Nucleons. This is accurate, but does not separate their functions or structures in nuclei. Next QM Delta Baryons have all possible charge functions, and Σs, etc., almost do so. But QM Lambda Baryons have single uds and charge formation in a definite family, which the MQP paradigm finds more informative. Accordingly, taking the <u>possible</u> + Deltas may start a distinctive Proton family which any + Ns may build on:

PROTON FAMILY SERIES (Baryon Series Gyre Pair Orbits/Sites. MQP Paradigm p+ Plan)

LQ/B	p+ 938.	Δ(1232)+			Δ(1600)+			Δ(1620)+			Δ(1700)+			Δ(1905)+			Δ(1910)+			Δ(1920)+ (CONT.)		
Type	Grp.0 Base	Grp.1 Isoton Set			Grp.2 Isoton Set			Isomer Set			Isomer Set			Grp.3 Isoton Set			Isomer Set			Isomer Set		
P+r/Q	Quark Set																					
Quark	u_A u_A d_A	u_A	u_B	d_A	u_A	u_B	d_A	u_B	u_B	d_A	u_B	u_B	d_A	u_A	u_A	d_B	u_A	u_A	d_B	u_A	u_A	d_B
Part/# orbit/Site	1 2 3	1	2	3	1	2	3	1	2	3	1	2	3	1	2	3	1	2	3	1	2	3
A		+-			+-			NC	NC	+-	+-	NC NC	+-				NC	NC		NC	NC	
B		+-			+-			→	→	+-	+-	→	+-				→			→		
C		+-			+-			→	→	+-	+-	→	+-				→			→		
A'																						
B'																						
C'																						++
++	+- +- +-	+-	++		++	++					+-			+-	+-	--			++		--	
--	+- +- +-	+-	--	--		--					+-		+	+-	+-	--						
S0	++ ++ +-	++++	+-	++	++	+-																
S1														+-								
S2														+-								
S3														++								
SEq1	++E	++E			++E									++E								
SEq2	++E	++E			++E									++E								
SEq3	--C	--C			--C			-C			-C			-C			-C			-C		

GRPS. 2,3,4,5 MAY NOT BE FULLY PDG ACCREDITED IN POSITIVE CHARGE & HAVE WIDE MASS UNCERTAINTIES.
Table 7-4b Larger B and Smaller A SMALL Quark Components in PROTON FAMILY SERIES (Continued on next page.)

This Table 7-4b is the first of the complete baryon family series, primarily on the PDG accredited baryons, with reclassifications carried out or recommended under the MQP Paradigm. There are two types of these tables for each baryon family. This form emphasizes the full introduction of the orbits and spin sites of the individual quarks in their smaller or larger structures required for the dual mass prototypes under the MQP Paradigm Mass Laws derived in Chapter 1 from the Particle Data Group accredited Summary Data Tables.

PROTON FAMILY SERIES CONT. (Baryon Series Gyre Pair Orbits/Sites. MQP Paradigm p+ Plan.)

LQ/B	$\Delta(1950)^+$				$\Delta(1930)^+$				$\Delta(2420)^+$				MQP Paradigm p+ Plan.) NO FURTHER PDG ACCREDITED Δ
Type	Grp.3 Isomer				Isomer				Grp.4 Isoton				No Grp. 5
P'r/Q	Quark Set				Set				Set				
Quarks	u_A	u_A	d_B	d_B	u_A	u_A	d_B	d_B	u_A	u_A	u_B	d_B	
Part/#	1	2	3		1	2	3		1	2	3		
Orbit/Site													
A					NC	NC						- -	
B					→	→						++	
C												+-	
A'													
B'													
C'													
++	+-	+-							+-	++			
- -	+-	+-					+-	+-					
S0	++	++	+-				+-	- -	- -	++	++	+-	
S1			+-				++						
S2			++				+-						
S3			- -										
SEq1	++E								++E				
SEq2	++E								++E				
SEq3	- -C								- -C				

Continued Table 7-4b Larger B & Smaller A SMALL Quark Components in PROTON FAMILY SERIES from prior page.

(SERIES FOLLOWING THIS FORMAT LARGELY 2010 PDG ACCREDITED PARTICLE STATUS WITH MQP PARADIGM ORBIT GROUPS.)

The second type of table shown next in Table 7-4c emphasizes the particle mass names versus the empirical measured masses of the PDG lists and the computed MQP Paradigm masses. Original baryon mass identification names from very early measurements have been preserved by the PDG to permit maintaining contact with early reports and data tables listing each baryon by its first identification even when later measurements have corrected the mass or the order of baryon masses. The PDG mass names of baryons have been followed herein, even when their later mass measurements have led to recalculation of matching paradigm structure and mass.

Here the proton is seen as the leading baryon of its own family of similarly charged baryons taken from the indefinitely identified PDG family of Delta baryons. (This means that in PDG terms these baryons have not yet been observed well enough to be certain that they are each individually +1 in charge, but that they are candidates for more definitive observation as such or at another possible level of charge. Later any PDG N Baryons that measure positive in charge must also be added to this table.)

The first part of Table 7-4b above minimizes the excessive repetition of the + or – charge sign <u>pairs</u> by using the NC↓ indicator of No Change with a down arrow for one quark's column in every case in which an isomer of the Group leading isoton particle to the left would exactly repeat the isoton orbit listings for the quark in the same position in that isomer all the way down to the bottom of the list. (With ten fewer similar columns this should be easier to read. But it does require looking 3 or 6 columns to the left to see what has been omitted under each NC↓.) Note particularly that each + and each – sign indicates a vortex spinning to the right or left for the proper electric charge force from that specific vortex, and that the predominant gyre spin in each quark will set its rotation directions. The second table above is the completion of the very long Table 7-4b organizing the large number of extra isomers listed by the PDG in Delta groups.

The next type of table does not attempt to show the detailed orbit structure of the coaxial <u>pairs</u> of gyres in order to emphasize the classification of the entire family of closely related baryons with different masses in steps. As mentioned elsewhere herein, it takes just 6 group isoton leaders, 0, 1, 2, 3, 4, & 5, to show the groupings of the A and B mass steps of 2 kinds of quarks with different charges in one baryon family. Some families of this sort will have blanks where no group member has been PDG certifiably observed as yet. Others, as with the Proton Family, have multiple observed isomers of isotons in several groups. Wherever there are 3 different types of quarks in one family, it will require 8 different groups with 8 different isotons to organize the family. A family with only 1 kind or charge of quark in the baryon triplet, as in the Omega Minus prototype family, will need only 4 groups to demonstrate the mass step process of the family (though there may be numbers of isomers in any one group with the same gyres in different orbits, usually of poorer balance, lower stability, and shorter mean life with usually slightly higher mass due to inclusion of an orbit that requires more energy and is less like a ground state.) It is also quite possible that a baryon which has never been certifiably observed before may be advanced to the accredited Summary Data lists by the PDG at any time. Here some marginal baryons have been included in families to fill out its groups, as in the Omega Minus prototype family. [In other such cases the MQP orbit data may have been worked out and recorded only in an early appendix to a web paper. If kept up, the web site for the MQP paradigm, electron-particlephysics.org, or its master site, particlephysics.info, carry the original papers, appendices, and tables, or they appear in Vol.2. Early equations and discussions that were not frequently used in later research (or have been amended) are listed at web sites, but are dropped here.]

To demonstrate in the next table the amendment process found necessary to set up well organized baryon family mass step structure among excessively large numbers of some PDG listings, tables are shown at least twice, in initial and in final estimate.

MQP PARADIGM <u>INITIAL</u> CLASSIFICATION ESTIMATE OF PROTON FAMILY SERIES (BY MASSES)
Proton⁺ **Prototype** Series Trials (in Masses MeV)

Group	PDG Name	Mass	Quarks	y	Σ component mass summation
0. Base	p⁺ 938	938.27	$u_A u_A d_A$	4.23590	8.939 Fixed by the group quarks.

(Mid Δ⁺ masses fill proton series)

1. Isoton	Δ(1232)⁺	1232±	$u_A u_B d_A$	4.39123389	9.896 **(SEE FIG. 7-2a next)**

<u>Since y is not a smooth</u>
(PDG did not separate or show <u>curve</u>, it looks like final
mixed charges beyond Δ(1232).) real Grp.1 or 2 is empty,
two members go really up

2. Isoton	Δ(1600)⁺	1600±	$u_B u_B d_A$	4.545113	10.853 a Grp. with reduced y, or
Isomers	Δ(1620)⁺	1630±	"	4.56202	" PDG will invalidate listings.
3. Isoton	Δ(1700)⁺	1700±	$u_A u_A d_B$	4.51953085	11.860
4. Isoton	Δ(1905)⁺	1890±	$u_A u_B d_B$	4.5453339	12.817
Isomers	Δ(1910)⁺	1910±	"	4.55491547	" **<u>This group has excessive</u>**
	Δ(1920)⁺	1920±	"	4.559668685	" **<u>numbers of isomers</u>**. At
	Δ(1950)⁺	1930±	"	4.564397	" least two look like PDG will
	Δ(1930)⁺	1960±	"	4.578437	" re-list not positive in charge

(PDG inverted mass order) (or merges listings), but can
<u>occur in low priority static sites</u>
usually of high energy.

5. Isoton	Δ(2420)⁺	2420±	$u_B u_B d_B$	4.704789957	13.774

<u>All MQP groups are logically filled.</u>
(in 6 Groups of 2 sets of A/B masses from Table 1-1 in single steps)
Where y = Log (Mass / Σ)/Log 3 and Σ = Mass Sum ($u_A + u_A + d_A$) (or any baryon three of
u_{AB}, d_{AB}, s_{AB}, c_{AB}, b_{AB}, t_{AB}) with plots of y vs M as set in Quark Groups for smoothest curves.
(Note PDG names <u>only fully</u> accredited particles <u>without</u> parentheses, but all partly accredited with them.)
Table 7-4c Proton Family Series Quarks $u_A u_A d_A$ to $u_B u_B d_B$ Mass Steps (<u>Preliminary Estimates</u>.)
From Equation 1-1 where $m_p = N_c^y \Sigma m_c$, M/Σ = 3^y for Baryons, and thus Log (M/Σ) = y Log 3, etc., above.
(Such Equn. 1-1 calculations run to more significant figures than are strictly valid, but do show suitable
relative values that continue to apply as mass measures are refined.) These initial groups were selected
entirely on the basis of clusters of the PDG particle masses. In these Groups above, the particles of
Group 2 can at once be listed in Group 3 where the larger Σ of 11.860 will smooth the y of Mass 1600
MeV to 4.4643479 and the y of 1630 MeV to 4.4812569. However, delaying that step permits a
demonstration of the internal logic of this re-classification in action in the figure here. In this empirical
paradigm system based on PDG accredited data, re-classification can only be tentative until the PDG
actually accredits measurement of each mass and charge combined in at least one (or more)
<u>measurement as a Delta+ or as an assigned positive nucleon N in the PDG classification of listings.</u>

From the beginning, since the perception in Figures 1-2 and 1-3 of the relatively continuous curve of the exponent y in Law Equations 1-1 and 1-2, it has been indicated that the orderliness of Nature is such that the exponent of Equation 1-1 usually has a smooth continuity. In Table 7-4c the y exponents for Group 2 versus particle masses are well above the otherwise fairly continuous curve (Fig. 7-2a) until changes indicated with this Table are tried. These particle masses taken from the PDG tables of baryon mass measurements that the PDG committee accredits as well measured may also merge with other PDG series. (All of these effects depend on the adaptability of the linking orbits in quarks. Fully simulating the quark links is another task yet to be done.)

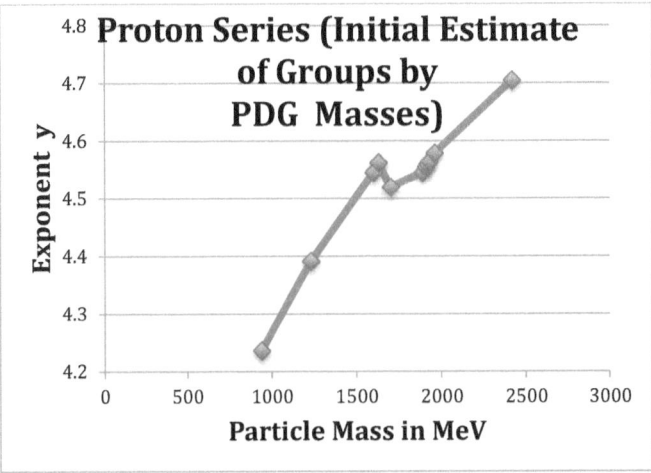

Fig. 7-2a Preliminary Initial Plot of Proton Series Groups Estimated by Particle Masses as in Table 7-4c. The required number of groups are each suggested by the PDG Delta Series listing of grouped numbers.

This figure makes it very clear that the initial preliminary estimate of group numbers, with their obligatory Σ sums of component quark masses and solution for y in each particle, definitely does not yield a smooth curve for the series. This confirmation, that this grouping is not consistent with natural grouping of mass steps by systematicly sequenced substitutions of heavier (B) mass levels of each quark for the lighter (A) mass levels in the base proton's lightest mass of the series, leads at once to trying the group amendments available without waiting for PDG mass refinements. The first trial leads to further trials until Table 7-4d, Fig. 7-2b next, appears suitable.

Proton $^+$ Series Amendment Trials

Group	PDG Name	Mass MeV	Quarks	y	Σ quarks mass sum MeV
0. Base	p 938	938.27	$u_A u_A d_A$	4.23590	8.939
2. Isoton	Δ(1232)	1232	$u_B u_B d_A$	4.307209	10.853 **No Grp. 1; 2 steps**
3. Isoton	Δ(1600)	1600	$u_A u_A d_B$	4.464348	11.86
Isomers	Δ(1620)	1630	"	4.481257	11.86 **(SEE FIG. 7-2b)**
	Δ(1700)	1700	"	4.51953085	11.86
4. Isoton	Δ(1905)	1890	$u_A u_B d_B$	4.5453339	12.817 **(Excessive**
Isomers	Δ(1910)	1910	"	4.5549155	12.817 **Isomers)**
	Δ(1920)	1920	"	4.5596686	12.817
	Δ(1950)	1930	"	4.564397	12.817
	Δ(1930)	1960	"	4.578437	12.817
5. Isoton	Δ(2420)	2420	$u_B u_B d_B$	4.70478996	13.774

Table 7-4d A Somewhat Improved Proton Series Structure Appears Suitable for a Time. Again groups are logically filled, but in 5 Groups of 2 sets of A/B masses in single steps. See Figure 7-2b for curve.

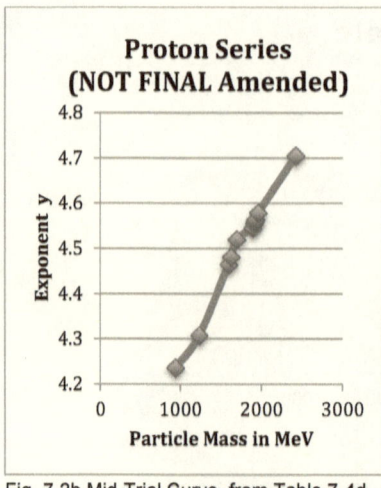
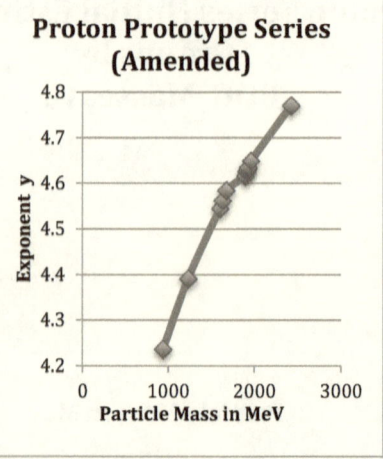

Fig. 7-2b Mid-Trial Curve, from Table 7-4d   Fig. 7-2c **Smoothest Curve**, from Table 7-4e Next.

PROTON PROTOTYPE SERIES

Group	PDG Name	Mass MeV	Quarks	y exponent	Σ quarks mass sum MeV	
0. Base	p⁺ 938	938.27	$u_A u_A d_A$	4.2359	8.939	
1. Isoton	Δ(1232)⁺	1232	$u_A u_B d_A$	4.391234	9.896	**See Fig. 7-2c**
2. Isoton	Δ(1600)⁺	1600	$u_B u_B d_A$	4.545113	10.853	above
Isomers	Δ(1620)⁺	1630	"	4.56202	10.853	
"	Δ(1700)⁺	1670	"	4.5840898	10.853	mass a bit high
3. Isoton	Δ(1905)⁺	1890	$u_A u_A d_B$	4.615969	11.86	mass a bit low
Isomers	Δ(1910)⁺	1910	"	4.625551	11.86	
"	Δ(1920)⁺	1920	"	4.630304	11.86	**Excess**
"	Δ(1950)⁺	1930	"	4.635033	11.86	**Isomers**
"	Δ(1930)⁺	1960	"	4.649073	11.86	
4. Isoton	Δ(2420)⁺	2420	$u_A u_B d_B$	4.7703366	12.817	**No Grp. 5**

Table 7-4e Proton Prototype Series Smoothest Curve Data (with PDG Delta Family) Final Amended. This final curve of the basic minimal proton family series was eventually discovered after the lenthy and tediously instructive experience of completing all the other baryon family series curves with mass steps. On coming back to the proton after that, it was obvious that the most important baryon's family was at that point not as well demonstrated in its still clearly re-curved graph (Fig. 7-2b above) as many others. Whether from the wider experience, or from stumbling again on the key structure as at the start of this venture into empirical support of QM, the improvement in predicting recognition of a fully systematic regularity in the stepped masses of baryon family series is a significant result that should lead further.

Added PDG improvements are predictable. It is difficult to find in Grp. 3 four non-conflicted sites for stable isomers. Until better force data at scale make the Table 7-4b isomers fully stable, expect that at least two will lose the PDG option of + charge. PDG may accredit a Grp. 5 isoton. Positive PDG charges may be found in the N series.

CONCEPTUAL PROTON SERIES EXPANDED FROM N & Δ SERIES

	PDG Name	Mass MeV	Quarks	y Am	Σ Quarks Mass Sum MeV
0. Base	p⁺ 938	938.27	$u_A u_A d_A$	4.236	8.939
1. Isoton	Δ(1232)	1232	$u_A u_B d_A$	4.391	9.896
2. Isoton	N (1440)	1450	$u_B u_B d_A$	4.45	10.853
Isomers	N (1520)	1490	"	4.48	10.853
"	N (1535)	1500	"	4.48	10.853
"	Δ(1600)	1550	"	4.516	10.853
"	Δ(1620)	1600	"	4.545	10.853
"	N (1650)	1640	"	4.567	10.853
"	Δ(1700)	1650	"	4.573	10.853
"	N (1675)	1660	"	4.578	10.853
"	N (1680)	1650	"	4.573	10.853
"	N (1700)	1680	"	4.589	10.853
"	N (1710)	1680	"	4.589	10.853
"	N (1720)	1700	"	4.600	10.853
3. Isoton	Δ(1905)	1870	$u_A u_A d_B$	4.606	11.86
Isomers	Δ(1910)	1890	"	4.616	11.86
"	N (1875)	1920	"	4.630	11.86
"	Δ(1920)	1920	"	4.630	11.86
"	N (1900)	1930	"	4.635	11.86
"	Δ(1950)	1930	"	4.635	11.86
"	Δ(1930)	1960	"	4.649	11.86
4. Isoton	N (2190)	2200	$u_A u_B d_B$	4.684	12.817
Isomers	N (2220)	2210	"	4.688	12.817
"	N (2250)	2220	"	4.692	12.817
"	Δ(2420)	2420	"	4.770	12.817
"	N (2600)	2500	"	4.799	12.817

NO GRP. 5

Table 7-4f Over-Expanded N & Delta Based Proton Series Data (Fig.7-2d.) Again this list of possible future members of the ultimate proton series far exceeds both the 6 unconflicted ground state isoton sites needed to fill the series and also the additional sites for higher energy isomer sites in the 5 groups (above the base group) in which one to three may usually occur, All these excess isomer candidates cannot actually be present in the series. Some must predictably be dropped from the PDG listings of accredited particle observations and measurements. Others will continue being listed, not with PDG accredited +1 charges, but only with eventual neutral 0, dual positive ++2, or -1 charges not in the proton series.

An alternative Proton Series concept is that it is made up of members from the PDG N series other than the proton base member , or better, from an undesignated combination of PDG recognized particles from both the PDG N series and its Δ series as shown above and next. (Tables & curves of this section are auto-plotted by Excel.)

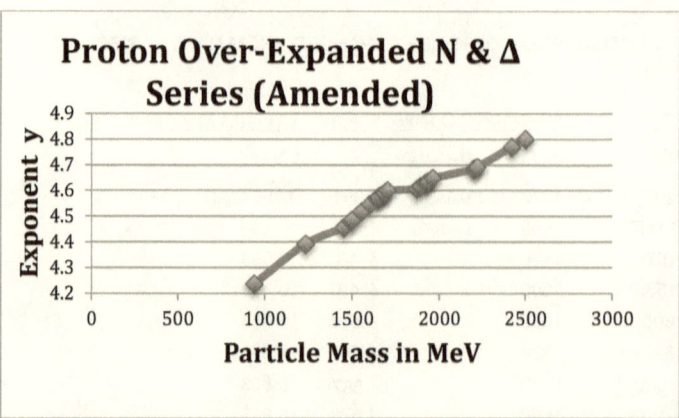

Figure 7-2d Over-Expanded N & Delta Based Smoothest Series Curve. It is predictable from this curve that two changes will occur in the current PDG N and Delta listings, that the particles responsible for the most irregular departures from a smooth cure here will become ineligible for this expanded Proton family for the reasons noted with Table 7-4f, or that more precise mass measurements will be accredited that move the particle into a smooth position on this graph. This curve is not at present smooth enough to qualify under the smooth curve criterion. (Since all the current PDG Delta listing do qualify and produce an adequately smooth and complete curve in Fig. 7-2c, it is possible that all the PDG N family listings will be disqualified for the Proton family and that there will be no expanded proton family such as the conceptually possible family noted here as perhaps within PDG future limits.)

In their condition of indefinite charge for the PDG small baryons other than the proton and neutron, it is necessary to list here all members of both these PDG series until later measurements are accredited by that international group to separate the legitimate positives from the neutrals and other charges such as the negative charge or the double positives. This necessity leaves the conceptual proton family series unrealistically over-expanded in the initial number of seeming isomers of the 5 group isotons for the base member, the proton, until the elimination of the excessive numbers by the PDG.- -There has not been any serious indication of an isomer of the proton, and none is expected in the MQP paradigm at the time of writing.- - (The isoton group leaders produced by a substitution of a heavier model of a single quark than in the next lower mass isoton are not considered isomers. Isomers are considered to have the same constituents in a different form, such as a different orbit for at least one differently excited pair in one of two otherwise identical particles. The pairs are different in their quantal charges in at least two micro-quanta between the lighter (A) or the heavier (B) up or down quark. In the four much heavier quarks from the strange to the top quark, there is also an increase in the number of quanta in going from the lighter to the heavier model of each type of quark. To put it another way, there are no isomers between two separate groups, only between two particles within a single group. But an eventual +I charged particle that the PDG now lists in the N series, may then be found to have isomers in one group of what the PDG now lists in a Δ series, because the quarks may then be known as the same in quanta pairs, though one quark may have one or more pairs in a more excited site. Isomers are simple, black or white in similarity, at the quanta pair number and charge level for their two gross A or B mass distinctions in quarks, no matter what orbit or static sites occur at smaller differences in energy mass levels.

Recent PDG reports give only very limited or no support in the Delta Baryon Summary Tables to an MQP re-classification as the primary proton family series with the Delta listings alone, as shown in Table 7-4e above. The heading for Delta Baryons does include, "$\Delta^+ = uud$," as one of four potential options for the particles listed by mass names, and the entry for the first particle alone for nominal mass ≈1232 in MeV (million electron-Volts of mass energy) does parenthetically state, "(mixed charges)." No reference is made there to the proton, which is clearly entered in the N Baryon family series as its first or smallest mass member with, "p, $N^+ = uud$," in the family heading. (Olive, et al., 2014) However, in the MQP Paradigm these identical quark components are the exact and only necessary reason (cause) for this obligatory re-classification for all Delta Baryons that do prove to be composed of these quarks. It is clearly predictable that some number of the family will be so proven in due course, according to the thoroughly accredited PDG entries. Whether those family particles will eventually be the smallest mass isotons in their stepped groups of the series or isomers of particles similarly re-classified from the N family (or newly accredited) is less predictable. [Since PDG presently shows 17 Ns to be divided in 2 parts between protons and neutrons, and 10 Deltas to be divided in 4 parts between 4 types of Deltas, the PDG membership of several hundred leading experts may elect on a probability basis of at least 3.4 to 1 to maintain the present classification indefinitely rather than recognizing the significance of the proton distinctively. It is empirically true that beyond the nucleus of potassium the nuclear protons begin to need increasingly the shielding from the mutual repulsion of their own kind provided by excess neutron attachment greater than 1 to 1. This enables protons to build all the stable and almost stable heavier elements. (Firestone, 2000 Chart) The neutron particle is therefore important in building the heavier atomic nuclei, but still not as important as the proton which must stabilize each neutron in each case.]

If an attempt is made to find a suitable N^+ Proton Family Series in the PDG N series as it stands (Olive, et al., PDG, 2014), using all accredited N data as below, the group quark amendments for smoothest y curve produce Fig. 7-2e, per Table 7-4g:

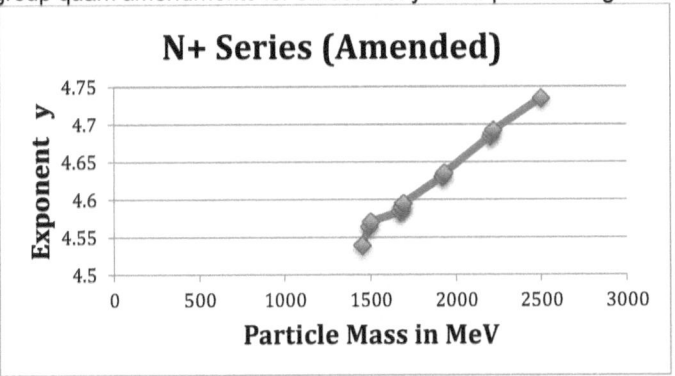

Fig. 7-2e An N+ PDG Series Smoothest Curve for a (proton) Family (2014) Without Proton (Table 7-4g.)

N+ Series (Amended)

Group	Name	Mass MeV	Quarks	y amended	Σ quarks mass sum MeV	
1. Isoton	N (1440)	1450	$u_A u_B d_A$	4.539534331	9.896	**NO GRP 0** $u_A u_A d_A$
Isomers	N (1520)	1490	"	4.564304273	9.896	**ELIMINATING PROTON**
"	N (1535)	1500	"	4.570392853	9.896	(NOT
2. Isoton	N (1650)	1670	$u_B u_B d_A$	4.584089781	10.853	A
Isomers	N (1675)	1680	"	4.589524061	10.853	PROTON
"	N (1680)	1680	"	4.589524061	10.853	FAMILY
"	N (1700)	1680	"	4.589524061	10.853	SERIES)
"	N (1710)	1680	"	4.589524061	10.853	
"	N (1720)	1690	"	4.59492609	10.853	
3. Isoton	N (1875)	1920	$u_A u_A d_B$	4.630304179	11.86	
Isomer	N (1900)	1930	"	4.635032705	11.86	
4. Isoton	N (2190)	2200	$u_A u_B d_B$	4.683581531	12.817	
Isomers	N (2220)	2210	"	4.687709607	12.817	
"	N (2250)	2220	"	4.691819046	12.817	
5. Isoton	N (2600)	2500	$u_B u_B d_B$	4.734393842	13.774	

Table 7-4g. N+ PDG Series with Quark Steps Amended for Smoothest y Curve. This table of y data becomes a nearly smooth curve only by cutting base Group 0 for the proton at y = 4.239, far out of line.

Other Family Series of the Small Quark Baryons

The rather detailed discussions above of the MQP Paradigm re-classification of a most significant, uniquely stable, and lowest stable mass Proton, as the base of the Prototype Baryon Family Series, with the necessary table and figure data displays, make it possible to study and understand the structures of the remaining MQP family series in the small quark baryons, and later the larger quark baryons, with much more condensed procedures. The principles and methods in working out stepped series masses in groups from dual quark masses are the same. The applications to supporting and resolving Quantum Mechanics difficulties, with re-classifications where necessary, are the same. Many of these family series are much closer to the PDG tables, though most are of far less importance to the every day structure and functions of planetary Nature, especially those beyond the Neutron family series taken up next.

As with the Delta family series earlier, the PDG data in its Nuclear family series provides accredited particle charge and quark information only on the smallest base particle, the neutron (as well as on the proton re-classified in the MQP Paradigm.) Consequently, all of the PDG N series table must initially be assigned to the Neutron series, with a prediction that future PDG accreditations of reported improved measurement data will at some time clarify which of the N series particles are actually neutral or change some of their accredited masses or the sequential order of their masses. That may re-determine the MQP table and graph for this N series. A number of such PDG changes are predictable from the MQP organization (or classification) of the PDG data in Tables 7-4h & i next.

Group	Neutron PDG Name	Prototype Mass MeV	Series (initial estimate) Quarks	y initial est.	Σquark mass MeV	y amended
Base	n	939.56	$d_A d_A u_A$	3.9588493	12.136	3.958844702
Isoton	N (1440)	1440	$d_A d_A u_B$	4.27842	13.093	4.278416239
Isomers	N (1520)	1520	"	4.3276	13.093	4.327630345
	N (1535)	1535	"	4.3366	13.093	4.336568936
Isoton	N (1650)	1655	$d_A d_B u_A$	4.27786	15.057	4.277863389
Isomer	N (1675)	1675	"	4.2885	15.057	4.288797325
	N (1680)	1685	"	4.29313	15.057	4.294215432
Isoton	N (1700)	1700	$d_A d_B u_B$	4.24619	16.014	4.246193
Isomers	N (1710)	1710	"	4.25153	16.014	4.251532008
	N (1720)	1720	"	4.25684	16.014	4.256839541
Isoton	N (2190)	2190	$d_B d_B u_A$	4.37143	17.978	4.371431199
Isomers	N (2220)	2250	"	4.396	17.978	4.396033757
	N (2250)	2275	"	4.4061	17.978	4.406091752
Isoton	N (2600)	2600	$d_B d_B u_B$	4.480285	18.938	4.480285073

Table 7-4h Neutron Family Initial Estimate of Series Groups by Particle Masses

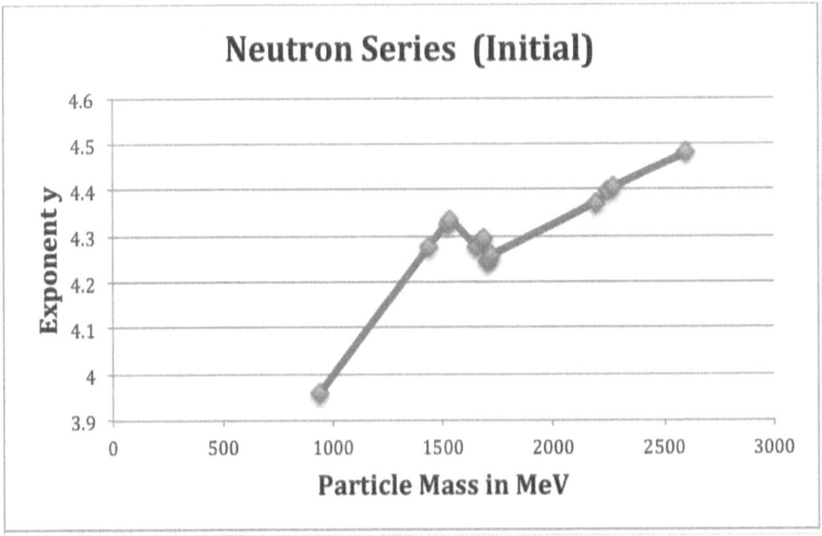

Neutron Series (Initial)

[Graph: Exponent y vs Particle Mass in MeV, ranging from ~3.95 at 940 MeV to ~4.48 at 2600 MeV]

Fig. 7-2f Initial Graph of Neutron Family Series. Obviously not an appropriate series group structure.

It is immediately necessary to begin amending the group structure of this series to find a workable family series for a prototype structure based on the smallest neutral baryon. It develops very quickly that this can be done only with a gap from that base to Group 2 with no accredited PDG neutral particle in Group 1 within range of 1200 MeV.

Group	Neutron PDG Name	Prototype Series Mass MeV	Quarks	(Amended) y initial est.	Σ quarks mass MeV	y amended
0. Base	n	939.565	$d_A d_A u_A$	3.9588493	12.136	3.95885
2. isoton	N (1440)	1430	$d_A d_B u_A$	4.27842	15.057 **NO GRP.1**	4.14485325
Isomers	N (1520)	1515	"	4.3276	15.057	4.1974114
"	N (1535)	1535	"	4.3366	15.057	4.20934911
3. Isoton	N (1650)	1655	$d_A d_B u_B$	4.27786	16.014	4.2217741
Isomers	N (1675)	1675	"	4.2885	16.014	4.2327081
"	N (1680)	1685	"	4.29313	16.014	4.2381262
"	N (1700)	1700	"	4.24619	16.014 Excess	4.2461933
"	N (1710)	1710	"	4.25153	16.014 Isomers	4.2515320
"	N (1720)	1720	"	4.25684	16.014	4.2568395
"	N (1875)	1875	"		16.014	4.3353789
"	N (1900)	1900	"		16.014	4.3474352
4. Isoton	N (2190)	2190	$d_B d_B u_A$	4.37143	17.978	4.3714312
Isomers	N (2220)	2250	"	4.396	17.978	4.3960338
"	N (2250)	2275	"	4.4061	17.978	4.4060917
5. Isoton	N (2600)	2600	$d_B d_B u_B$	4.48043	18.938	4.4802851

Table 7-4i Neutron Family Prototype Series Structure (Amended & Up-Dated with 2 PDG added, 2014.)

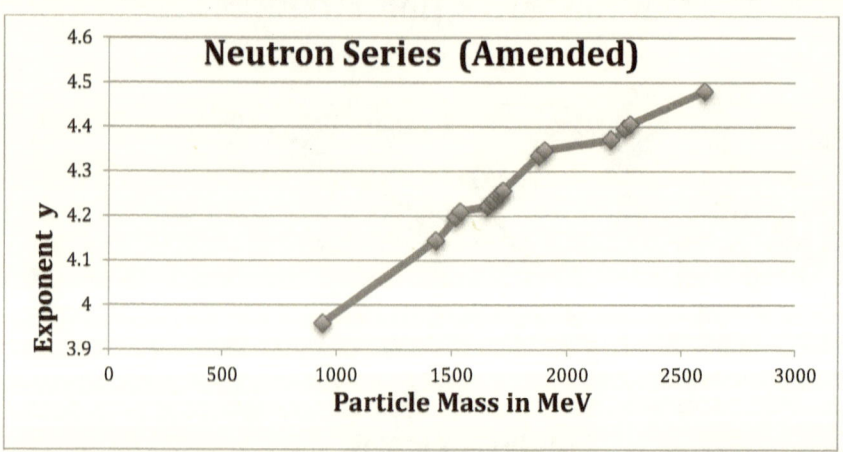

Fig. 7-2g Final Amended and Up-Dated (2014) Smoothest Neutron Family Series Curve. This only approaches a suitable smoothness. Some of the lower and higher PGD particle masses for the MQP Group 3, with its excessive number of isomers, are also a bit high in y for a smooth family series.

It is clear in Table 7-4i and Figure 7-2g that the Neutron Family Series can be predicted to be further amended by eventual PDG accreditations of charges that remove at least four members of Group 3 from this series, perhaps the curve's peaks. Until such improvements in the PDG accredited data occur, the irregularities in this curve are almost as broken and numerous as those in Fig. 7-2d. (For accurate comparison the two figures must be at exactly the same scale on both axes, which they are not in this reproduction, or must be carefully corrected by plotting on one large plot.) The contrasts with the much smoother re-classified proton curve in Fig. 7-2c are clear.

Note that in Table 7-4i if the missing Group 1 were present, it would have the first heavier B substitution of a quark mass in the lighter up quark rather than in the down quark shown for Group 2. Because of this, the first step increase in the quarks mass sum is about 3 MeV rather than about 1 MeV for u_B as Group 1 would have the step (if there were a lighter Group 1 particle of about 1200 MeV in this series. But the PDG N Series does not show a particle of that smaller mass range here. 1440 MeV is the mass range of the lightest particle above n.) Note also that in the n (or N) series, with only one u quark, the second step to the B mass of the down quark has the up quark returning to its lighter A mass, so that the steps are always in the smallest possible change to a higher mass of the quark mass sum Σ. In the p and Δ series earlier with two u quarks, this drop back of a B mass to an A mass for the u quarks did not occur until a later step after both u quarks had been stepped up. It is the sum Σ which must take the lowest step up in each case of a group change. [This requires careful checking, especially later in a series with three different quarks, such as uds in the PDG Lambda (Λ) series, for instance, or usc in the multiple PDG Charmed Series.]

The general name of Small Quarks applied here to the u and d quarks and their low mass series does not imply that these small quarks will not occur in series with the heavier quarks later. The small quarks will continue to appear in the series of the medium mass quarks and the most massive bottom and top quarks. There are only three PDG accredited series out of about a hundred for which this is not true. Those three are the PDG Omega (Ω) Series in which the strange quark, at low medium mass, borders on being a member of the Small Quarks' set of only two quarks. All three of these small and low medium mass quarks appear frequently, largely because they raise the forces-to-mass ratio to a more stable level for particle relative survival under impacts in use, as often noted earlier.

Further Importances of the Proton Series (and the Neutron Series)

Now that the structural basis of separate proton and neutron series has been made certain and those two classes of series have been thoroughly separated, as well as before taking up the family mass step classifications of the other proliferated baryon families, it will be clearer in the long run to summarize again the long-term effects of the overall orbital structure of baryons and quarks in this chapter stemming from the proton as the prototype of all baryons, plus the further significance (beyond its introduction in pages 160 to 163) of the outstanding importance of the proton itself, the leading baryon.

As to the orbital structure of baryons in general, note that SEq orbits in their necessary linkages create herein the special quark-to-quark bonds assigned in QM to hypothetical "gluon" particles which are separate from the continuing quarks. The QM gluons no longer appear to be necessary. [Alternatively, it may be considered from the

QM point of view that link orbits only provide structure for gluons, which have become deeply enmeshed in QM (QED) theory. It may require a long and significant discussion to transition QM from dependence on "gluons" to much simpler and more explanatory empirical link orbits as demonstrated in protons.]

The three SEq orbit numbers and their three angular start locations in each quark sphere in any baryon also inherently show a simpler structural cause for the previously mysterious QM chromodynamic (QCD) assignments of three "colors" and "color" quantum numbers to quarks in hadrons. "Color" rules were needed in QM to limit quarks in freedom of exchange in particle reactions. In MQP once SEq pairs are in place in a structure they are each 60° out of place (or out of phase in orbit) for the other quarks present and are internally blocked by other orbits from relocating except in the most destructive collisions. (A direct connection between QM color rules and SEq orbit phase angles may be much simpler to establish among particle physicists generally than a similar MQP supporting connection between QM gluons and link orbits.)

It is difficult to over-emphasize for both QM and the MQP paradigm that the proton's uniquely stable mean lifetime, PDG-listed at greater than 2.1×10^{29} years (much more than the QM computed life of the Universe squared), is the main cause for the cosmic accumulation of atomic nuclei in stars, sun, planets, earth, and moons. The proton's long lifetime is causally explained in the MQP paradigm. The ultimate cause of the proton stability here comes from the nearly perfect balances of its individual quarks on and along their S0 axes, like spinning machinery which does not waste energy in off-balance vibrations. [That cause is conservative. It can be estimated from the limited present non-symmetric gyre force data, that it is not entirely certain that the present proton configuration with the odd single down quark's link orbit in full circle and the two up quarks' links in quasi-ellipses is not actually reversed in Nature to reach even more perfect balancing of the quarks on the proton's combined axis as well. The heavier down quark's single negative charge on the link orbit and its total body may actually reduce the comparative size of the down enough that it comes in closer to the center of gravity (CG) of the proton, and the double positive charge of each up quark with its asymmetric link orbit may balance at a somewhat larger radius from the proton CG. Such a further perfected total balance over the entire proton could further contribute to the proton's very long mean life and might affect the current QM uncertainty in accounting for the exact proton spin cited earlier.] At the proton's low mass that balance is combined (in a high charge-to-mass ratio) with high <u>average</u> attraction between unlike charges of the micro-quanta in the SEq. orbits and the spheres (as in Chapters 4 and 5.) This adds a critical topping off of other QM strong force attractions between the quarks and their microbit quanta in protons. [These so-called QM "strong" forces (due here to spinning in either like or unlike sense along the sides of the main gyres) are unable to stabilize single neutrons to more than 15 minutes mean life at only slightly higher mass energy (939.6 MeV for neutrons compared to 938.3 MeV for protons), but with some definite neutron imbalance of the C' orbits of the downs as shown earlier. An MQP inward point-thrust force of turbulent vortices orbiting in particles (Fig. 4-3, page 86) definitely aids the marginal "strong" force from vortex side flow in forcefully holding the total structure of the proton together stably.]

When a neutron is bound to a proton and they become a deuterium (hydrogen two) nucleus, a new balance of charge attractions arises between the resulting matched set of 3++ and 3– – S Eq. linking pairs of microbits, which binds the neutron in long-life stability by topping off fully the QM marginally attractive strong forces of both sets of quarks. (Since these link forces include those between asymmetric gyres in charge force hemispheres, the exact required amount of the well-grounded topping-off force estimates should be refined in much-expanded lab measurements on asymmetricly located gyres, including vortex triplets and force band-frequency filters. There is no doubt that these measured charge forces are in the right range for this function.) The deuteron internal bond then is an increased combination of charge and other forces, added to the tiny but inherent force of interactive gravity between the eddy gyres of turbulent vortices especially in heavier particles (which use up available deuterons.)

As stable partners under the extended mass law in the deuteron combination, the two smallest baryons bond together in the positive nuclei of other typical atoms heavier than hydrogen. In this the closely spaced absorption of opposite proton uud charge force by nominally neutral, but slightly separate, ddu neutron charges clearly reduces the mutual repulsion of protons in nuclei and makes possible the build-up with extra neutrons of the nuclei of heavy elements, especially those near iron and heavier. Consequently, the MQP quark paradigm becomes a scale of atomic nature as one continuous process from the micro-quanta in the quarks to the variety of atomic nuclei (Fig. 1-5). Every step in making the energetic structural masses of nature is defined by essentially the same mass power law at a fixed power in the micro-quantal beginning of the MQP process to a law with positive and negative variable exponents respectively in the baryon and atomic nuclear phases of structure building up to and around protons.

In the same way that three quark spheres in a proton link the three quarks of a neutron to build the stable deuterium nucleus, the basis for the distinctive even-numbered atomic nuclei, the enlarged SEq. orbits of a deuteron also must aid linking an added proton and neutron to stabilize the helium 4 nucleus (alpha particle) prominent in the relative abundances (cosmic percentages) of the more stable atomic elements listed in the orders of their masses where there are repeated cyclic sequences of abundance peaks 4 mass units apart. (However, no one has established a feasible way in fusion for earthly nuclear energy to fuse two deuterons directly into helium 4. The process goes an indirect way in two steps, with a stopover at helium 3. The additional proton and neutron must be separately locked on in that process.)

All the accredited baryons have been gathered by the PDG and QM generally into family series that depend in part on the sources and circumstances from which the baryons have come under observation and in part on the six deduced types of quarks from which they appear to be composed. Both influences strongly affect the functions the individual baryons can perform. In the MQP paradigm the most stringent effects on classification into family series as well as the steps of mass within a series arise from identical accredited quarks that compose each baryon. The first and most important such family series is also that of the proton above. Its most significant members would initially be those that appear to lead each group (in the series) as its lightest member resulting from a substitution of the heavier B form of each quark for its lighter A form. This must begin with the lightest Base member composed only of the A configurations with greatest stability and cosmic abundance due essentially to having the highest ratio of coherent forces to the mass to be held together over ranges of conditions from near absolute zero degrees Kelvin in nearly empty space to the very highly energetic and hot conditions in outer regions of ordinary stars such as the sun or in cosmic ray collisions. -- [The usual Kelvin scale goes from zero at -273.16° Centigrade in ideal space to 0° C (273.16° K) at the triple point of water (freezing-melting-evaporating) and to 100° C (373.16° K) at the nominal boiling point of pure water at exactly mean sea level at the earth's equator (on a calm day with no atmospheric or tidal activity.)] -- Since the apparent lightest member of each particle group would be the one nearest a ground state of minimal energy with no excited quanta in orbits requiring more than the least amount of continuing energy, these apparent group leading isotons are the family members to examine first with every type of quark to confirm actual group transitions rather than the possible groups. This would have a criterion of smoothly varying ratio of the exponent y (in Equ. 1-1 for the particle) to (preferably) the PDG particle mass (since the sum Σ of the best values of quark masses as seen in Equation 1-1 can only yield heavily jointed curves within any one full family series.) Because the groups are initially only apparent by PDG mass variation, the possible sequences of PDG mass changes in natural groupings must be tried until a smoothest y curve for the family series results. Comparisons of such curves beginning with a closely related family series based on the proton gives the most informatively systematic classifications of all the baryons.

Return to the Three Remaining Small Quark Series

After re-working the PDG N series and completing the important significance of the positive Proton-based series with their closely related, but secondary, neutral N series, the remaining small quark series are only the Delta sub-series with other than positive charge. The remaining entirely small quark particles the PDG does not at present accredit by charge, which is determined explicitly by the ringed triplet of quarks that compose a baryon. They are the Δ^-, the Δ^{++}, and the Δ^0 Series.

The Δ^{++} Series may have only one group, but there are multiple members, and its curve is smooth. Being made up only of uuu, the lightest quarks, it creates and marks a perceptible trend in baryons, that positive baryons in mixed series tend to have masses near the low end of the range of PDG accredited uncertainties. (That taken alone is only a trend guide for selecting where to start in investigations. It has shortened some of the repetitious comparisons in the MQP paradigm.) Neutral baryons may appear to trend toward masses near the middle of uncertainties. Negative baryons tend to have more component quanta in their quarks on average in mixed-charge PDG series (especially since positive top quarks are very rare, and since strange quarks with lower medium mass abundances accentuate the negative charge influence of down quarks against the greater abundances of light protons with positive up quark advantages), and thus negative baryons tend toward the heavier range of PDG mass uncertainties in mixed charge PDG series. Then, it can be systematicly relevant that MQP generally follows this trend initially in the prior Proton Δ^+ Series and later.

183

Group	Δ Series (Paradigm Name)	Mass MeV	Ω⁻plan) Quarks	y initial	Start Low end PDG Δ mass range Σ quarks sum MeV	y Re-amend
3. Isoton	Δ(1232)⁺⁺	1230	$u_B u_B u_B$	4.885959	8.613 **NO GRPS 0,1,2**	4.51614897
Isomers	Δ(1600)⁺⁺	1500	"	4.9553896	8.613	4.6967868
"	Δ(1620)⁺⁺	1600	"	4.98428	8.613	4.75553229
"	Δ(1700)⁺⁺	1670	"	4.9017194	8.613 (Not a prototype)	4.794508736
"	Δ(1905)⁺⁺	1855	"	4.8950332	8.613 **Excess Isomers**	4.890139443
"	Δ(1910)⁺⁺	1860	"	4.89747026	8.613 **possible with**	4.892589617
"	Δ(1930)⁺⁺	1900	"	4.91195	8.613 **smallest quark**	4.91195714
"	Δ(1950)⁺⁺	1915	"	4.9191115	8.613	4.919115021
("	Δ(1920)⁺⁺	1920	"	4.92148853	8.613	4.921488528
	Δ(2420)⁺⁺	2200		Omit, beyond limit 5.0	(Over-excited ?)	

Table 7-4j Amended Δ++ Series Data. Excess isomers may have very brief and highly unstable lives.

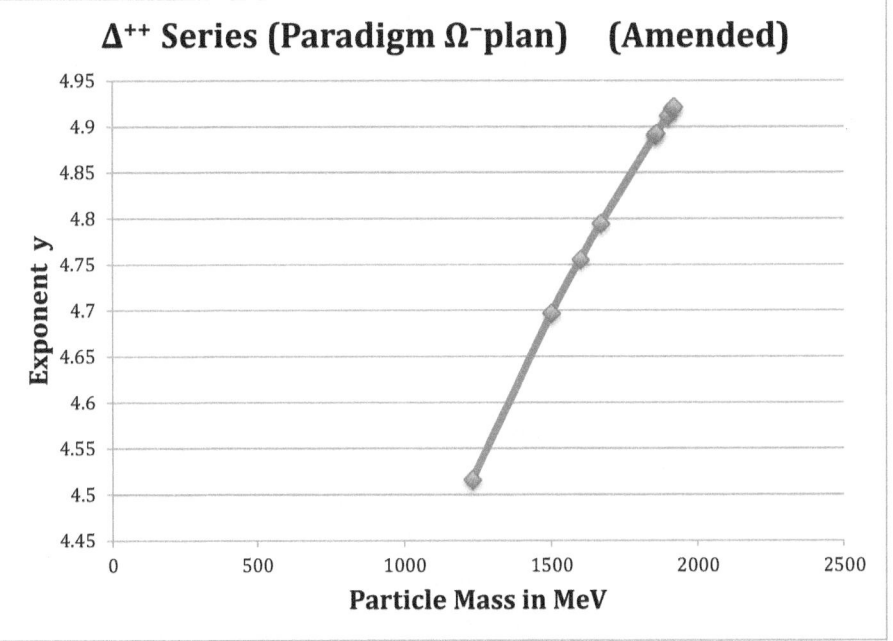

Fig. 7-2h Amended Δ++ Series y Versus Particle Mass. This curve of y is exceptionally smooth.

In spite of the great irregularity of having only one data group, where 4 are needed, and no base group from the PDG data in this series structure, this curve is so smooth and regular as to appear impossibly ideal. The initial estimate curve (not shown) was extremely irregular. Aside from the smooth Proton Series curve amended earlier from the same PDG Delta Baryons initial data, the two other smoothest Delta

Zero (Neutral) and Delta Minus (Negative) curves (required by the PDG) look useful in their data tables, but definitely irregular in curves. Better accredited data are needed.

	Δ⁻ Series		(Negative Plan)		Start high PDG mass range.
Group	Name	Mass MeV	Quarks	Σ quarks sum	y Amended
0. Base	Δ (1232)	1234	$d_A d_A d_A$	15.33	3.994315881
1. Isoton	Δ (1600)	1600	$d_A d_A d_B$	18.252	4.07194084
Isomers	Δ (1620)	1620	"	18.252	4.083248305
"	Δ (1700)	1640	"	18.252	4.094417024
2. Isoton	Δ (1905)	1905	$d_A d_B d_B$	21.174	4.09558797
Isomers	Δ (1910)	1910	"	21.174	4.09797392
"	Δ (1920)	1970	"	21.174	4.126127893
"	Δ (1930)	2000	"	21.174	4.139884919
"	Δ (1950)	1950	"	21.174	4.116839657
3. Isoton	Δ (2420)	2500	$d_B d_B d_B$	24.096	4.225330635

Table 7-4k Amended Δ⁻ Series Data. (4 Grps. are sufficient.)

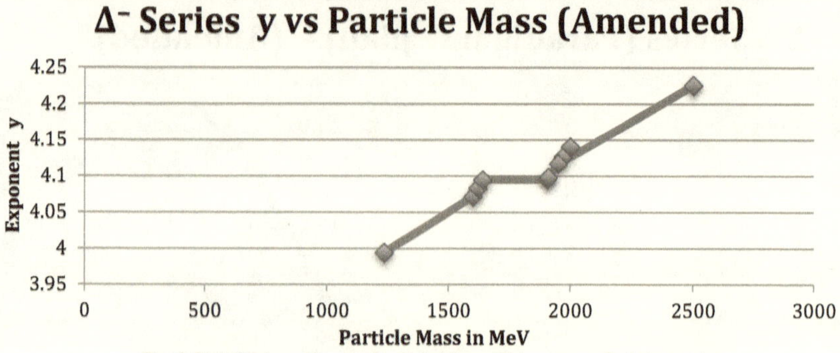

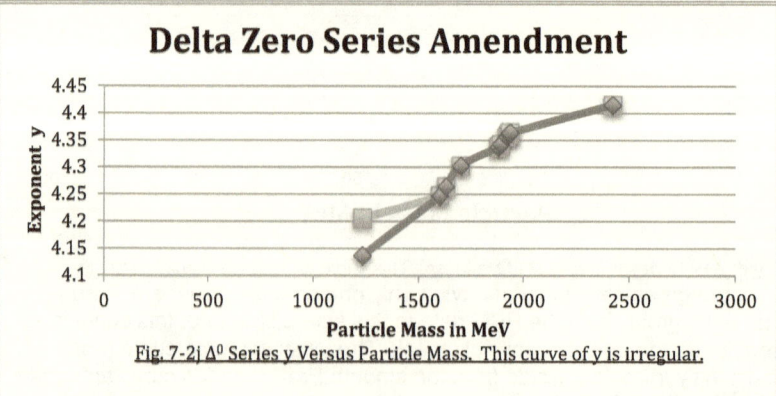

Δ⁻ Series y vs Particle Mass (Amended)

Fig. 7-2i Δ⁻ Series y Versus Particle Mass. This curve of y is irregular.

Delta Zero Series Amendment

Fig. 7-2j Δ⁰ Series y Versus Particle Mass. This curve of y is irregular.

Group	Name	Mass MeV	Quarks	y amended	Σ quarks sum		y Re-amended
	Δ^0 Series (Paradigm neutral plan)				Amend to mid PDG mass range.		
1. Isoton	$\Delta(1232)^0$	1232	$d_A d_A u_B$	4.20550357	13.093	No Grp 0	4.136415053
2. Isoton	$\Delta(1600)^0$	1600	$d_A d_B u_A$	4.24709969	15.057		4.247099688
Isomers	$\Delta(1620)^0$	1630	"	4.2640086	15.057		4.264008643
"	$\Delta(1700)^0$	1700	"	4.3022826	15.057		4.302282597
3. Isoton	$\Delta(1905)^0$	1880	$d_A d_B u_B$	4.33780303	16.014		4.337803031
Isomers	$\Delta(1910)^0$	1890	"	4.34263189	16.014		4.342631898
"	$\Delta(1920)^0$	1920	"	4.3569667	16.014		4.35696667
"	$\Delta(1950)^0$	1930	"	4.361695	16.014		4.361695196
"	$\Delta(1930)^0$	1935	"	4.3640503	16.014	No Grp 4	4.36405028
5. Isoton	$\Delta(2420)^0$	2420	$d_B d_B u_B$	4.41512516	18.935		4.415125161

Table 7-4I Δ^0 Series Data Amended for Smoothest Curve. Another series with no Group 0.

This set of three remaining related series (after the Proton Delta Series) is so unusual or irregular as to raise questions about the completeness or accuracy of its accredited PDG data. It is possible that these questions will be resolved in time by later PDG accreditations, or by PDG abandonment of some charge options, or by transfers of listings with the PDG N Series after further or more precise data measurements by some research organization. In the single page PDG Summary Table of all baryons for 2014 there are only these 10 Δ particles that are shown with 3 or 4 stars as established baryons, but there are 12 more with 1 or 2 stars as observed, but uncertain and with fragmentary data. It can be expected that some of these uncertain observations will be improved or removed in the future. Removals may include some PDG series charges.

The same expectation applies to the N baryons for which there are 30 listed in the single page summary, but only 17 with the 3 and 4 star qualifications for the truly accredited Summary Tables for 2014. Certainly some of these will change in status. This consideration will also continue for the more massive baryons composed at least in part from the heavy quarks. The total 2014 single page summary lists 151 baryons, of which only 89 are accredited with 3 or 4 star ratings of their data and level of certainty that they exist at the PDG criteria. Those with 1 star ratings have been described as observed by a published research paper, but the PDG is not sure from the paper's evidence that these particles ever even exist in a decay, cosmic ray shower, or reactor.

The Low-Medium Mass Strange Quark Baryon Family Series

The Strange Quark series are notable for the beginning of a succession of oddities in their regularities. It may be remembered that in Fig. 1-7 of Chapter 1 there was between the two closely related small quarks and the strange quark a reversal of the sequence of quark charges with increasing mass. In the small quarks the sequence is from + to – charges, while in the medium quarks the sequence is from – to +, which continues with the heaviest quarks. There is also in that figure at that point a reversal from the close logarithmic spacing of nearly vertical lines bearing quark sites to their wide logarithmic spacing, a reversal which continues with the heavy quarks. In addition, there is also a reversal from the positive slopes of these lines for small quarks to negative slopes for the heavier quarks. All of these changes between the small quarks and the strange quark (and heavier) are associated with the beginning of overlapping of these mass-charge graph lines at higher mass rather than standing clear of each other as among the small quarks. Also, the separations of masses between the lower and higher dual values of mass of individual quarks change from being on a single logarithmic line in small quarks to being on separate lines for strange and heavier quarks, which defines not having or having increases in the number of quantal pairs from the lower to the higher masses of each quark. Instabilities of the quarks and their baryons take a sharp turn into micro-micro-seconds of the longest stated mean lives at about this point, with further decreases by many orders-of-magnitude (OM) beyond the strange quarks (though bottom quarks, and thus their baryons, return to about the micro-micro-second range of longest stated mean lives.)

One other major change at that break-point between charges of small and medium quarks may disappear with further PDG data accumulation. That is the point now and much earlier at which possible PDG negative charges of triplet baryons (as in Delta baryons) begin to appear unequivocally in the PDG data. There are none of 3 or 4 star rating of certainty of existence and negative in the small quark baryons, nor in the PDG's first Strange quark's accredited Lambda Summary list. They appear initially at the time of writing only in two of the first four 4 star cases of the Strange quark's Sigma (Σ) dds baryons, similarly in the Strange Xi (Ξ) dss baryons, and finally one in the only two Accredited Strange Omega (Ω) sss baryons. Five of these baryons are negative 4 star, and the one Omega baryon is the only such single quark baryon. Unfortunately the Strange Omega baryons only have two accredited members, and this forces the use of the only two unaccredited Omegas to complete an initial trial Prototype series.

Omega⁻ Prototype Trial Series   (mid mass range)

Group	Name	Mass MeV	Quarks	y initial	Σ quarks mass sum MeV	
0. Base	Ω⁻1672.5	1672.45	$S_A S_A S_A$	1.73949847	247.401	PDG
1. Isoton	Ω(2250)⁻	2252	$S_A S_A S_B$	1.92362326	272.124	PDG
2. Isoton	Ω(2380)⁻	2380	$S_A S_B S_B$	1.894789	296.847	(unaccredited)
3. Isoton	Ω(2470)⁻	2474	$S_B S_B S_B$	1.819674	321.57	"

Table 7-4m Initial Trial Omega⁻ Series with the Necessary 4 Groups.

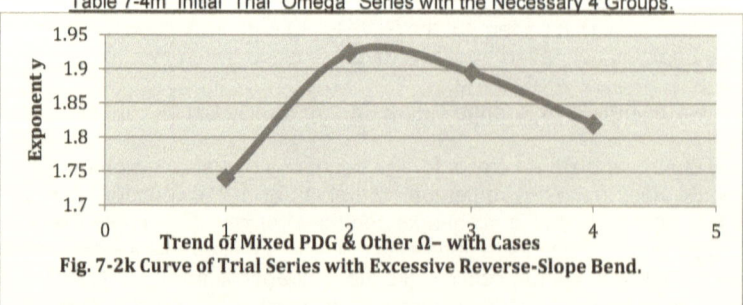

Fig. 7-2k Curve of Trial Series with Excessive Reverse-Slope Bend.

The initial trial curve is excessively bent in reversed slope to qualify as a smooth curve, though there is justification in the PDG footnotes for including at least the lighter of the two non-accredited particles. This situation is still marginal and more so for either of the other two possible particle sites for a single-charged prototype baryon. The two 3-4 star ratings of the accredited Omega Minus baryons are arbitrarily filled out here as the single-charge prototype without trying to forecast whether other particles might be more appropriate after another PDG century of data. To become fully appropriate an Ω^- or other series of the future must be accredited with curve smoothing equivalent to the following, which must be considered hypothetical as having two over-large deviations:

Group	Omega⁻ Prototype Series Amended (Hyp.)				(mid mass range)	
	Name	Mass MeV	Quarks	y Hypothetic	Σ quarks mass sum MeV	
0. Base	Ω^-1672.5	1672.45	$S_A S_A S_A$	1.73949847	247.401	PDG
1. Isoton	$\Omega(2250)^-$	2252	$S_A S_A S_B$	1.92362326	272.124	PDG
2. Isoton	$\Omega(2380)^-$	**2600 (?)**	$S_A S_B S_B$	**1.9752645**	296.847	required
3. Isoton	$\Omega(2470)^-$	**2900 (?)**	$S_B S_B S_B$	**2.0018443**	321.57	"

Table 7-4n Hypothetical Amended (Desired) Approx. Omega⁻ Prototype Series Data

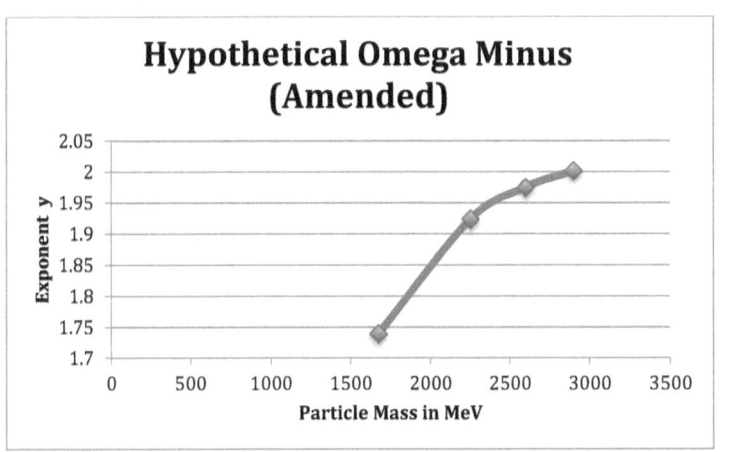

Fig. 7-2l Hypothetically Amended Approximate (or Desired) Omega⁻ Prototype Series

This hypothetical cannot be predicted to occur in future PDG data, but an equivalent smoothing of a data curve with the four obligatory groups may possibly appear. Though unlikely, it may not be ruled out of potential cosmic ray showers, etc.

The Lambda baryons (Λ) begin with the lightest of the Strange quark particles, the Λ^0 Series base, designated simply Λ, at 1115.683 MeV composed of uds. (It would appear that the Σ^+ of uus should be lighter, but it is evidently at an excited state.) This series has one of the longer runs of mostly 3 and 4 star PDG baryons with only one set of unconfused quarks and neutral charge, a definite family, but with odd structure. As is

	Λ^0 Series (Paradigm n plan) Amended				(Uses mid mass range)		
Group	PDG Name	Mass MeV	Quarks	y initial	Σ quarks mass sum MeV		y Amended
4. Isoton	Λ^0 ☐☐☐☐☐☐	1115.683	$u_A d_A s_B$	2.297	114.215	**N0 GRPS 0, 1, 2, 3**	2.0745621
Isomers	$\Lambda(1405)^0$	1405.1	"	2.497	114.215		2.2845012
"	$\Lambda(1520)^0$	1519.5	"	2.549	114.215		2.3557482
"	$\Lambda(1600)^0$	1600	"	2.586	114.215		2.4027368
"	$\Lambda(1670)^0$	1670	"	2.625	114.215		2.4417133
"	$\Lambda(1690)^0$	1690	"	2.635	114.215	**EXCESS ISOMERS**	2.4525496
"	$\Lambda(1800)^0$	1800	"	2.51	114.215		2.5099476
"	$\Lambda(1810)^0$	1810	"	2.515	114.215		2.5149905
"	$\Lambda(1820)^0$	1820	"	2.52	114.215		2.5200056
"	$\Lambda(1830)^0$	1830	"	2.525	114.215		2.5249932
5. Isoton	$\Lambda(1890)^0$	1890	$u_B d_A s_B$	2.547	115.172		2.5467633
6. Isoton	$\Lambda(2100)^0$	2100	$u_A d_B s_B$	2.627	117.136		2.6272753
Isomer	$\Lambda(2110)^0$	2110	"	2.63159	117.136		2.6315995
7. Isoton	$\Lambda.(2350)^0$	2350	$u_B d_B s_B$	2.722	118.093		2.7222508

Table 7-4o Heavily Amended Lambda Zero Series with Smooth Curve But Many Excess Isomers

shown in the y initial column, the sudden break back down from 2.635 to 2.51 indicates a very irregular initial y curve. This requires a long series of amendments of the quark groups and elimination of Groups 0, 1, 2, and 3 (quarks and mass Σ $u_A d_A s_A$ 89.492, $u_B d_A s_A$ 90.449, $u_A d_B s_A$ 92.413, and $u_B d_B s_A$ 93.42) to find a smooth y curve. The excess of isomers in Grp. 4 must predict that at least the 3 star particles at masses 1600, 1800, and 1810 MeV will eventually be merged with others by PDG, and/or the larger mass uncertainties will be modified to smooth the well distributed initial groups.

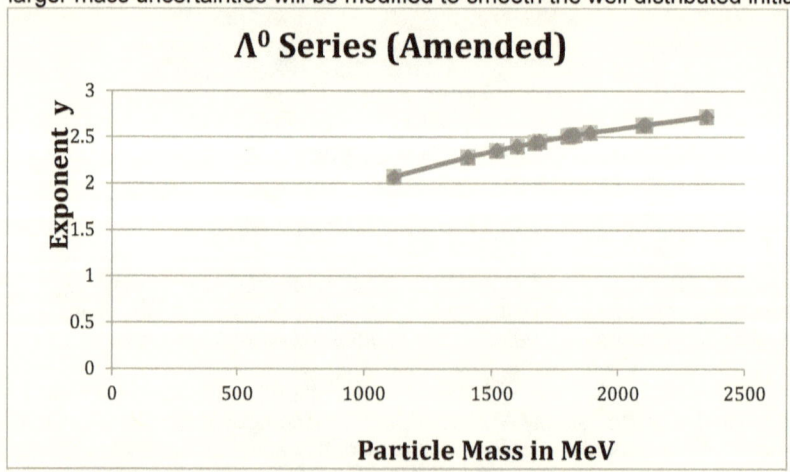

Fig. 7-2m Smooth Curve for Heavily Amended Baryon Series Lambda Zero

Next, the Σ Family of Strange quark baryon series appears at first to be similar to the PDG Small quark Δ series because of the similar way in which the PDG has classified the three differently charged Σ sub-families based on variations of the u, d, and s quarks, but that could be misleading. Again the MQP paradigm finds it necessary to re-classify the sub-families to emphasize the differences from the two-quark set-up with four variations of sub-family charges. (There is no ++ sub-family division this time with the obligatory negative strange quark dominating the quark triplets. Also, with that emphasis, it Is the neutral sub-family that must be shared with the alternate major family rather than the + sub-family as before.) While the re-classification of Strange baryons now also results in a new combined sub-family with an excess of isomers, the new sub-family otherwise fits together much more smoothly than the Proton's over-expanded mixed sub-family, though this time four groups are omitted from each of the combinants. In other words, main features of the two re-classifications are very different, not alike. But both carry out strictly the new MQP rule that exact sets of quarks and consequent particle charges are the precise and only criterion of baryon (& meson) family series.

	Σ^0 Series (n plan, PDG uds quarks) Amended (mid mass)				(as above)		
Group	PDG Name	Mass MeV	Quarks	y initial	Σ quarks mass sum MeV		y Amended
4. Isoton	Σ^0 1192	1192.642	$u_A d_A s_B$	2.357316666	114.215	NO GRPS 0, 1, 2, 3	2.135279
Isomers	$\Sigma(1385)^0$	1387.2	"	2.482887022	114.215		2.272831
"	$\Sigma(1660)$	1660	"	2.619183725	114.215		2.436246
"	$\Sigma(1670)$	1670	"	2.624650643	114.215	EXCESS ISOMERS	2.441713
"	$\Sigma(1750)$	1750	"	2.48430539	114.215		2.484305
"	$\Sigma(1775)$	1775	"	2.4972168	114.215		2.497217
5. Isoton	$\Sigma(1915)$	1915	$u_B d_A s_B$	2.358724554	115.172		2.558725
Isomer	$\Sigma(1940)$	1940	"	2.57053067	115.172		2.570531
6. Isoton	$\Sigma(2030)$	2030	$u_A d_B s_B$	2.596416815	117.136		2.596417
7. Isoton	$\Sigma(2250)$	2250	$u_B d_B s_B$	2.682668928	118.093		2.68 2669

Table 7-4p Σ^0 Series Not in Combination with Lambda Zero with Same Quarks from Table 7-4o above.

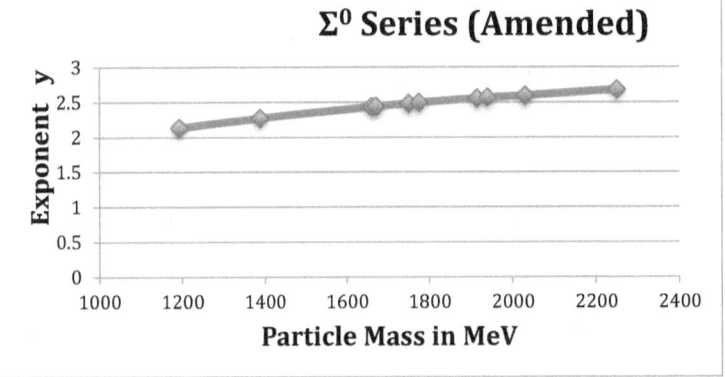

Fig. 7-2n Σ^0 Series Not in Combination with Lambda Zero with Same Quarks from Fig. 7-2m above. Here three separate quarks require eight groups to run out all the group steps involved.

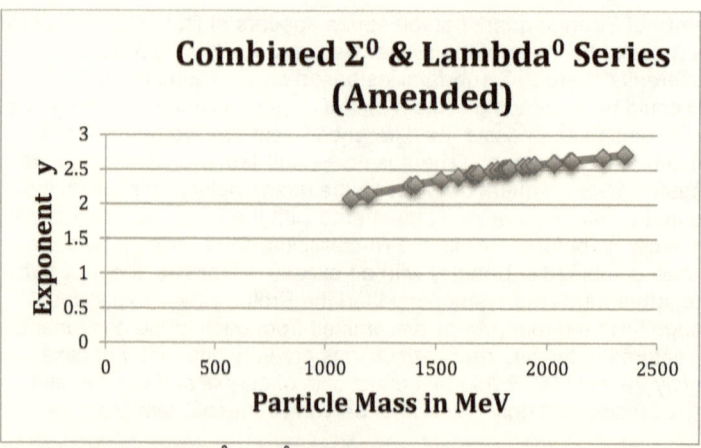

Fig. 7-2o Re-Classified Λ⁰ and Σ⁰ Series Combined in a Single Perfectly Smooth Amended Curve.

Again it must be predicted by the excesses of isomers in each of the two now separated PDG data sets in this Series, re-classified in the MQP paradigm as combined into a single series, that the PDG data itself will in the course of future measurements be re-accredited as having merged or eliminated data points to a less excessive isomer level, and/or will have found large adjustments in uncertainties that will produce a more even distribution and smooth curve than the irregular curve now obviously indicated in both initial y data sets. (Also again, precise forecast calculations of these estimates will require a larger investment in asymmetric scaled model measurements of lower uncertainties for quarks than have been feasible.) This chapter is prepared largely on the basis of the accredited empirical PDG Summary Tables, as that continues to demonstrate accounting generally for the step-by-step masses of Baryon Series members by systematic MQPP substitution of the heavier of dual quark masses for the lighter of those masses in the typical base structure of the lightest and usually most abundant particle member of each Baryon Series or the base member's closest approximation. This is a method capable of continuing improvement and more general valid application to established national and international research objectives, as will be further demonstrated chapter by chapter for every type of particle (other than virtual particles to be eliminated.) - - Still, overall, a type of irregularity does appear in these two series, in that neither of their very light baryons, the Σ^0 1192 & Λ^0 1115, can initiate a smooth curve structure as a series base Grp. 0, nor at less than Grp. 4. An ultimate set of PDG data should change this peculiar type of result for Strange quarks, or this result must be recognized as frequent in both small and strange quark baryons.

The other Strange Σ Baryon Series, the Σ^+ and the Σ^- Series, continue this seemingly odd structural pattern. Again their initial y curves are very erratic. They have to be put through the group change process to find smooth y curves. But there are no other series with the same quarks, with which they must combine in a completely new re-classification beyond the established PDG series. The changes are all amendments within the usual PDG series, as these MQP paradigm computations demonstrate.

Σ⁺ Series (P'digm p plan)   PDG Low Mass Range
Amended

Group	PDG Name	Mass MeV	Quarks	y initial	Σ quarks mass sum MeV		y Amended
3. Isoton	Σ⁺ 1189	1189.37	$u_A u_A s_B$	2.387928	111.018	NO GRP 0,1,2	2.15862
Isomers	Σ(1385)⁺	1382.8	"	2.510503	111.018		2.29578
"	Σ(1660)⁺	1630	"	2.654828	111.018		2.44549
"	Σ(1670)⁺	1665	"	2.674166	111.018		2.46483
"	Σ(1750)⁺	1730	"	2.499685	111.018	EXCESS ISOMERS	2.49968
"	Σ(1775)⁺	1770	"	2.520491	111.018		2.52049
4. Isoton	Σ(1915)⁺	1900	$u_B u_A s_B$	2.577191	111.975		2.57719
Isomers	Σ(1940)⁺	1915	"	2.584349	111.975		2.58435
"	Σ(2030)⁺	2025	"	2.635187	111.975		2.63519
5. Isoton	Σ(2250)⁺	2210	$u_B u_B s_B$	2.706614	112.932		2.70702

Table 7-4q Series Σ⁺ Data Amended Less Grps. 0, 1, 2 ($u_A u_A s_A$ 86.295; $u_B u_A s_A$ 87.252; $u_B u_B s_A$ 88.209)

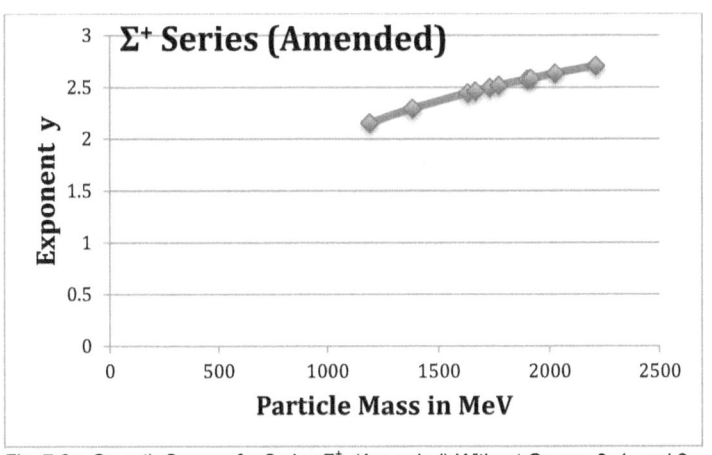

Fig. 7-2p Smooth Curve y for Series Σ⁺ (Amended) Without Groups 0, 1, and 2

As indicated in the last prior paragraph this curve is quite smooth, which shows good continuity of structure of the series of baryon particles made from the listed quarks, except for a possible shortage of unconflicted excited orbit (or static) sites for quark gyres of group three to be occupied simultaneously in some given impact of larger particles. This limitation might also inhibit reactions in the next series. (In any event the PDG observed mean lives of these particles run around 10^{-10} seconds.) This finding does account for the step-by-step masses of this Σ⁺ series. The Σ⁻ series curve next is not quite so smooth, in part because it is drawn at higher scale so that the small irregularity can be seen. If drawn at a much higher scale, two more slight offsets could show the very small reversals of particle mass order here. On the 2014 PDG 1 page Baryon Summary 61% of observed Σs are 1 or 2 star. This is about 50% high, showing a need for more good data. When that is available, the Fig. 7-2q Σ⁻ curve may improve.

Σ⁻ Series (P'digm Ω⁻ plan)

Group	Amended PDG Name	Mass MeV	Quarks	PDG High Mass Range y initial	Σ quarks mass sum MeV	(Rows in PDG Name order)	y Amended
3. Isoton	Σ⁻ 1197	1197.449	$d_A d_A s_B$	2.32903	117.412	**NO GRP 0,1,2**	2.11381
Isomers	Σ(1385)⁻	1387.2	"	2.43468	117.412		2.24770
"	Σ(1660)⁻	1690	"	2.58700	117.412		2.42742
"	Σ(1670)⁻	1685	"	2.58431	117.412		2.42472
"	Σ(1750)⁻	1800	"	2.48482	117.412	**EXCESS**	2.48482
"	Σ(1775)⁻	1780	"	2.47465	117.412	**ISOMERS**	2.47465
4. Isoton	Σ(1915)⁻	1935	$d_B d_A s_B$	2.52828	120.333		2.52828
Isomers	Σ(1940)⁻	1950	"	2.53532	120.333		2.53531
"	Σ(2030)⁻	2040	"	2.57638	120.333		2.57638
5. Isoton	Σ(2250)⁻	2280	$d_B d_B s_B$	2.65579	123.254		2.65579

Table 7-4r Series Σ⁻ Data Amended Less Grp. 0,1,2 ($d_A d_A s_A$, 92.689; $d_B d_A s_A$, 95.61; $d_B d_B s_A$, 98.53)

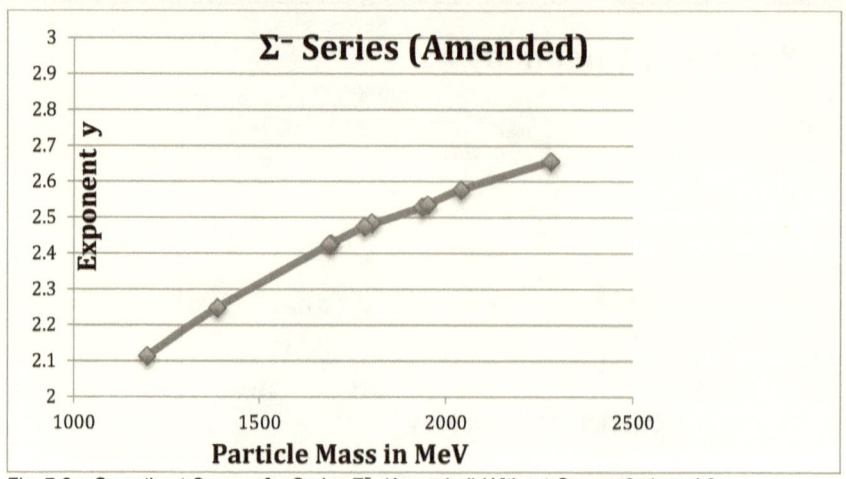

Fig. 7-2q Smoothest Curve y for Series Σ⁻ (Amended) Without Groups 0, 1, and 2

    The relatively brief PDG lists of Ξ (Xi) Strange Baryons are next in the sequence of increasing mass. It is interesting to note also that the PDG confirms in these tables the MQP observation (and consequent practice) described a few pages earlier that within several series with the same Greek Letter name the positive (+) charged series is lightest, the neutral (0) series is normally of a mean range mass, and the negative (−) charged series is most massive. It is also worth noting that the groupings of these two series are entirely in groups 0 and 1, exactly the opposite of the Σ series. However, these two series cannot be combined with prior Σ series, which have only one strange quark per baryon, since these Ξ baryons have two strange quarks in each baryon. That is their distinction in structure. The neutral Ξ⁰ combines the two singly charged negative 1/3 s quarks with a doubly charged positive 2/3 up quark. The negative Ξ⁻ has a single −1/3 down quark combined with its two strange quarks for its unit negative charge.

193

Ξ⁰ Series (P'digm n plan) (Amended) Starts mid mass

Group	PDG Name	Mass MeV	Quarks	y initial	Σ quarks mass sum MeV		y Amended
0. Base	Ξ⁰	1314.66	$u_A s_A s_A$	1.87908	166.848	NO GRP 2,3,4,5	1.87896
1. Isoton	Ξ(1530)⁰	1531.8	$u_B s_A s_A$	2.01290	167.805		2.01290
Isomers	Ξ(1690)	1680	"	1.97639	167.805		2.09696
"	Ξ(1820)	1818	"	2.04372	167.805	EXCESS	2.16882
"	Ξ(1950)	1935	"	1.99454	167.805	ISOMERS	2.22559
"	Ξ(2030)	2020	"	2.02965	167.805		2.26472

Table 7-4s Series Ξ⁰ w/o Grp. 2,3,4,5 ($u_A s_B s_A$, 191.57; $u_B s_B s_A$, 192.53; $u_A s_B s_B$, 216.29; $u_B s_B s_B$, 217.25

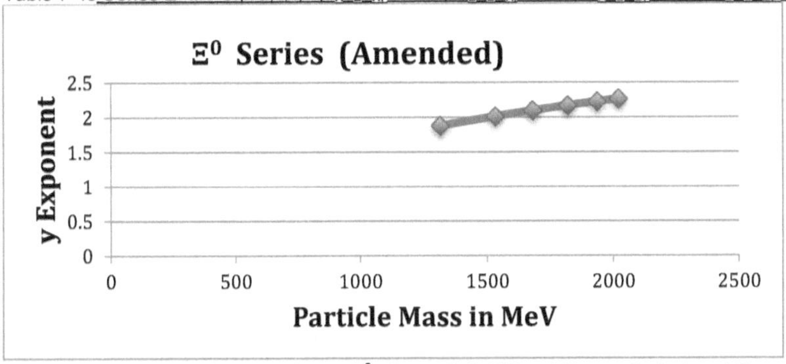

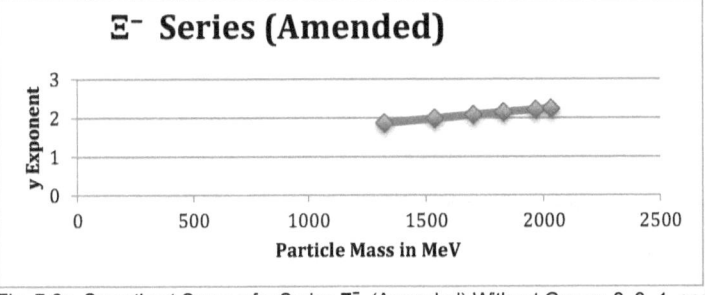

Fig. 7-2r Smoothest Curve y for Series Ξ⁰ (Amended) Without Groups 2, 3, 4, and 5

Ξ⁻ Series (P'digm Ω⁻ plan) (Amended) Starts High Mass Rng

Group	PDG Name	Mass MeV	Quarks	y initial	Σ quarks mass sum MeV		y Am'nd'd
0. Base	Ξ⁻	1321.71	$d_A s_A s_A$	1.86628	170.045	NO GRP 2,3,4,5	1.86655
1. Isoton	Ξ(1530)⁻	1535	$d_B s_A s_A$	1.98728	172.966	EXCESS ISOM'RS	1.98723
Isomers	Ξ(1690)	1700	"	1.97210	172.966		2.08016
"	Ξ(1820)	1828	"	2.02463	172.966		2.14624
"	Ξ(1950)	1965	"	1.99519	172.966		2.21202
"	Ξ(2030)	2030	"	2.01278	172.966		2.24164

Table 7-4t Series Ξ⁻ w/o Grp.2,3,4,5 ($d_A s_B s_A$, 194.77; $d_B s_B s_A$, 197.689; $d_A s_B s_B$, 219.491; $d_B s_B s_B$, 222.412

Ξ⁻ Series (Amended)

Fig. 7-2s Smoothest Curve y for Series Ξ⁻ (Amended) Without Groups 2, 3, 4, and 5

This completes the MQP Paradigm re-classification and step-by-step mass structuring of the Strange Baryons (since the Omega Minus Series, as a prototype, was done earlier above.) However, the strange quarks, like the small quarks, will continue to occur in the considerably more massive Charmed Baryon Series, in which the charm quark itself is an order of magnitude more massive than the strange quark (Table 1-1.)

The Charmed Baryon Series

The PDG names of the Charmed Baryons and of their Series are intended to indicate the close relations of these much heavier baryons and of their quark listings to the lighter Small Quark Baryons and Strange Baryons. That would be very confusing on the various kinds of differences involved if it were not for the PDG practice of adding a "c" subscript to the name of each lighter Baryon to obtain names for these less frequently observed (less abundant), much more massive, and typically more unstable particles. The use of such reasonable names does require added alertness in reading, writing, listening, and speaking about these unavoidably more complex particles. For a Lambda $_c$ Plus, or a Xi $_c$ Zero, the extra "c" is often not noticed until a misunderstanding arises. But the simple subscript is much easier and quicker to write or say than any other fast designation. (For the same reason of writing ease the subscript and a charge superscript sign are sometimes left on the one line. All of this will also apply later to "b"ottom quarks and their baryons, not to mention, beyond that, to all the scattered parts of broken or decayed baryons in Mesons {which brings the over-line marks, such as Ū or d¯ in the real, but rarer, or mysteriously missing worlds of anti-quarks.})

Group	Σ_c Series (P'r'digm Ω⁻ plan) Amend'd PDG Name	Mass MeV	Quarks	y initial	(Stated mid range masses) Σ quarks mass sum MeV	y Amended
0. Base	$\Sigma_c(2455)$	2453.98	$u_A u_A c_A$	0.67398	1170.328 **NO GRPS 3,4,5**	0.673966
1.Isoton	$\Sigma_c(2520)$	2517.9	$u_A u_B c_A$	0.69681	1171.285	0.696628
2.Isoton	$\Sigma_c(2800)$	2800	$u_B u_B c_A$	0.63077	1172.242	0.792546

Table 7-4u Baryon Series Σ_c w/o Grp3,4,5 ($u_A u_A c_B$,1398.828; $u_A u_B c_B$,1399.785; $u_B u_B c_B$,1400.742)

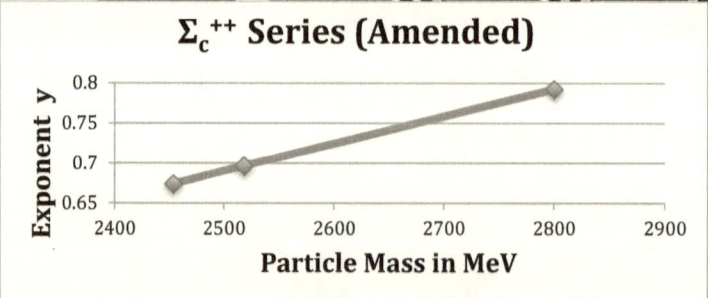

Fig. 7-2t Smoothest Curve y for Series Σ_c (Amended) Without Groups 3, 4, and 5.

This is the only remaining series charged ++. With only three PDG particles, it is too small for certainty of little change with more data. It does show stepped structure. Its nearest PDG relative, Σ_c, udc, is best re-classified combined with Λ_c^+, also udc.

	Combined Λ_c^+ & Σ_c^+ Series			(Each series fills the other series gaps.) Amended Mid range masses separately stated			
Baryon	(P'r'dm p plan)						
Group	PDG Name	Mass MeV	Quarks	y initial	Σ quarks mass sum MeV		y Amended
0. Base	Λ_c^+	2286.46	$u_A d_A c_A$	0.607123	1173.525		0.607123
1.Isoton	$\Sigma_c(2455)^+$	2452.9	$u_B d_A c_A$	0.670340	1174.482		0.670340
2.Isoton	$\Sigma_c(2520)^+$	2517.5	$u_A d_B c_A$	0.692481	1176.446	**NO GRP 4,5,6,7**	0.692481
3.Isoton	$\Lambda_c(2595)^+$	2592.25	$u_B d_B c_A$	0.719480	1177.403		0.718375
Isomers	$\Lambda_c(2625)^+$	2628.11	"	0.571943	1177.403		0.730880
"	$\Sigma_c(2800)^+$	2792	"	0.626589	1177.403		0.785943
"	$\Lambda_c(2880)^+$	2881.53	"	0.625115	1177.403	**EXCESS**	0.814673
"	$\Lambda_c(2940)^+$	2939.3	"	0.652742	1177.403	**ISOMERS**	0.832742

Table 7-4v Combined Baryon Series Λ_c^+ & Σ_c^+ w/o Grps. 4,5,6,7 ($u_A d_A c_B$, 1402.025; $u_B d_A c_B$, 1402.982; $u_A d_B c_B$, 1404.946; $u_B d_B c_B$, 1405.903) Note three separate quarks need eight groups to take all the steps.)

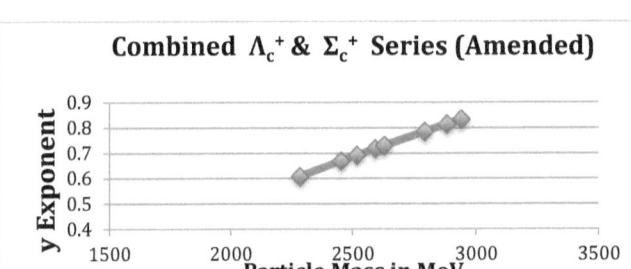

Fig. 7-2u Smoothest Curve y for Combined Baryon Series Λ_c^+ & Σ_c^+ (Amended) Without Grps 4,5,6,& 7.

	Σ_c^0 Series (Paradigm n plan)			Amended	(mid range masses, separately stated)	
Group	PDG Name	Mass MeV	Quarks	y initial	Σ quarks mass sum MeV	y Amended
0. Base	$\Sigma_c(2455)^0$	2453.74	$d_A d_A c_A$	0.668925	1176.722 **No Grps.3,4,5**	0.668917
1.Isoton	$\Sigma_c(2520)^0$	2518	$d_A d_B c_A$	0.690192	1179.643	0.690192
2.Isoton	$\Sigma_c(2800)^0$	2806	$d_B d_B c_A$	0.624414	1182.564	0.786515

Table 7-4w Series Σ_c^0 without Groups 3, 4, 5 ($d_A d_A c_B$, 1405.22; $d_B d_A c_B$, 1408.142; $d_B d_B c_B$, 1411.064)

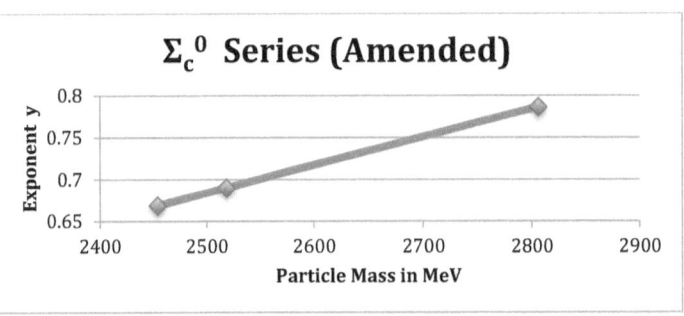

Fig. 7-2v Smoothest Curve y for Baryon Series Σ_c^0 (Amended) Without Grps 3,4,5.

The series in the Charmed Baryons, including those yet to be discussed, are consistent in regularity within the MQP Paradigm about a feature that is (strangely) lacking in the MQP Strange Series compared to the Small Quark Series. The Charm Series repeats the Small Quarks' functionally occupying the 0, 1, and 2 group sequences in most Series in the PGD 2014 data. Wide failures to do this are primarily in the Strange Family Series, where the emphasis is on not having these lighter mass groups present in the Families. This is actually an additional <u>unexpected validation of the MQP Paradigm (MQPP)</u> as demonstrating within its causal system a true and real systematics of the subatomic and sub-nuclear particles rather than their being superficially random in operation. Table 7-4w looks ahead in the much heavier remaining Charm and Bottom Families of Series to show the regular and systematic **p**resence, **a**bsence, or **i** incomplete inclusion of a baryon in each Family Series group that is needed to cycle through the number of step-by-step mass increase combinations for each set of three quarks in each baryon. None actually uses 8 groups currently.

REGULARITY of PDG DATA on PARTICLE **P**RESENCE in BARYON QUARK GROUPS

Grp	Small Quark families						Strange Quark fam's						Charm Quark families						Bottom Quark families					
	p	N/Δ	N⁺	N⁻	Δ⁺⁺	Δ⁻	Δ⁰	Ω⁻	H	Λ⁰	Σ⁰	Σ⁺	Σ⁻	Ξ⁰	Ξ⁻	Σ_c⁺⁺	Λ_c⁺/Σ_c⁺	Σ_c⁰	Ξ_c⁺	Ξ_c⁰	Ω_c⁰	Σ_b⁺	Σ*_b⁺	Λ_b⁰	Σ_b⁻	Σ*_b⁻	Ξ_b⁰	Ξ_b⁻	Ω_b⁻
0	P	P		P		P		P	P					P	P	P	P	P	P	P	P	i		P	i		i	i	i
1	P	P	P		P	P		P	P					P	P	P	P	P	P	P	P	i		P		i	i		
2	P	P	P	P				P	P	P	P					P		P	P	P	P								
3	P	P	P	P	P	P	P	P	P			P	P					P		P	P								
4	P	P	P	P						P	P	P	P																
5			P	P		P				P	P	P	P																
6										P	P																		
7										P	P																		

Table 7-4w<u>Regularity of **P**resence of Baryon (3 Quark) Particles in Quark Family Series Groups (/= 2 in 1)</u>

This table displays the very systematic regularity of the Presence of working baryon particles at the Series Zero Base and other group levels in Family Series generally, except in Strange Quark families. Group 0 to 2 baryons are missing in the Small Quark families exactly 6 times In 21 possibilities. But in the Strange Quark families those baryons are missing 14 times in 24 possibilities with 4 more gaps in Group 3 sites. Charmed families lack only 1. Bottoms lack 6 again in 24. Actually the Charm and Bottom Quark families instabilities due to the very large masses of these quarks (or mass-to-force ratios) may prevent their primary groups from ever being completely filled out. The Strange Quark families are opposite in nature generally, possibly due to the reversal of systematics of these quarks themselves noted earlier. (The Strange disparity would be 15/21 if corrected for two factors: The Hypothetical Omega Minus family is only a conjectural duplicate, and the Group 2 and 3 Omega Minus particles of the Prototype Omega Minus are not PDG accredited like the rest.)

197

This systematic variation of functions applies to the remaining baryons confirming coverage through some Charm and rare Bottom series in their lower number groups.

Ξ_c^+ Series (P'digm p plan) (mid range PDG masses)

Group	PDG Name	Mass MeV	Quarks	y initial	Σ quarks mass sum MeV	y Am'ndd
0. Base	Ξ_c^+	2467.8	$u_A s_A c_A$	0.618525289	1250.881	0.61848
1. Isoton	Ξ'^+_c	2575.6	$u_B s_A c_A$	0.656745312	1251.838	0.65671
Isomer	$\Xi_c(2645)^+$	2645.9	"	0.664343566	1251.838 NoGrp 4,5,6,7	0.68122
2. Isoton	$\Xi_c(2790)^+$	2789.1	$u_A s_B c_A$	0.577210965	1275.604	0.71208
Isomer	$\Xi_c(2815)^+$	2816.6	"	0.585548821	1275.604	0.72101
3. Isoton	$\Xi_c(2980)^+$	2971.4	$u_B s_B c_A$	0.6197235	1276.561	0.76903
Isomer	$\Xi_c(3080)^+$	3077	"	0.6509318	1276.561	0.80082

Table 7-4x Series Ξ_c^+ w/o Grp 4,5,6,7($u_A s_A c_B$,1479.38; $u_B s_A c_B$,1480.34; $u_A s_B c_B$, 1504.10; $u_B s_B c_B$, 1505.06)

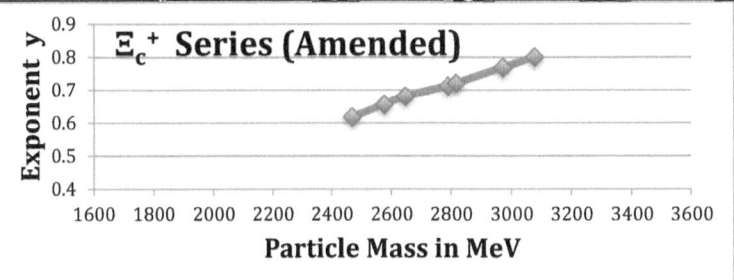

Fig. 7-2w Smoothest Curve y for Baryon Series Ξ_c^+ (Amended) Without Grps 4,5,6,7.

Ξ_c^0 Series (Paradigm n plan) (mid range PDG masses, separately stated)

Group	PDG Name	Mass MeV	Quarks	y initial	Σ quarks sum MeV	y Amended
0. Base	Ξ_c^0	2470.88	$d_A s_A c_A$	0.61517	1254.078	0.617300
1. Isoton	Ξ'^0_c	2577.6	$d_B s_A c_A$	0.65381	1256.999	0.653672
Isomer	$\Xi_c(2645)^0$	2645.9	"	0.66189	1256.999	0.67748
2. Isoton	$\Xi_c(2790)^0$	2791.8	$d_A s_B c_A$	0.57613	1278.801 No Grp	0.710682
Isomer	$\Xi_c(2815)^0$	2819.6	"	0.58103	1278.801 4,5,6,7	0.719701
3. Isoton	$\Xi_c(2980)^0$	2968	$d_B s_B c_A$	0.61675	1281.722	0.764313
Isomer	$\Xi_c(3080)^0$	3079.9	"	0.65044	1281.722	0.798000

Table 7-4y Series Ξ_c^0 w/o Grp 4,5,6,7($d_A s_A c_B$,1482.58; $d_B s_A c_B$,1488.50; $d_A s_B c_B$, 1507.30; $d_B s_B c_B$, 1510.22)

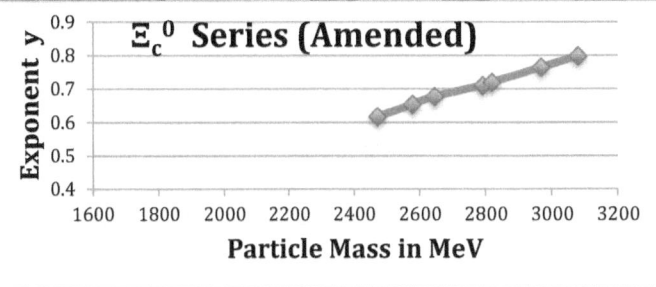

Fig. 7-2x Smoothest Curve y for Baryon Series Ξ_c^0 (Amended) Without Grps 4,5,6,7.

The last remaining Charmed Baryon Series is the Ω_c^0. This PDG series is very sparsely represented by two 3 star baryons. Their masses are close enough together that they are probably in adjacent groups. Though there were indications that y should be expected to decrease at the start of the series (from the start of the prior series), and it did not (but increased instead), the groups were left at Groups 0 and 1 until later PDG data is accredited. (This leads to an expectation, or prediction, that when PDG data accredits four to five or more baryons in this series, the two present baryons will be forced on average into higher numbered groups or at least into higher isomer positions with lower y in order to provide the smoothest curve of y.) Until more baryons than two are accredited in a series, there is no purpose in presenting such an arbitrarily assigned straight line as the smoothest series curve of y, not even for curve slope, though the vertical value of the points does roughly approximate where some part of an eventual smoothest curve based on adequate data will probably lie relative to other curves.

Ω_c^0 Series (P'd'm n plan) (mid range PDG masses, separately stated) INCOMPLETE

Group	PDG Name	Mass MeV	Quarks	y initial	Σ quarks sum MeV	y Amended
0. Base	Ω_c^0	2695.2	$s_A s_A c_A$	0.64269	1331.43 **No Grps**	0.6419179
1. Isoton	$\Omega_c(2770)^0$	2765.9	$s_A s_B c_A$	0.64874	1356.157 **2,3,4,5**	0.6487376

Table 7-4z Series Ω_c^0 w/o Grp 2,3,4,5.($s_B s_B c_A$ 1380.88; $s_A s_A c_B$ 1559.974; $s_A s_B c_B$ 1584.697; $s_B s_B c_B$ 1609.42)

That completes the Charmed Baryon Series at the time of writing. The PDG may issue a limited off-year update at any time, or it may be a full year before a new edition.

The Bottom Baryon Series

About half a century after the small quark baryons were established structures, baryon series containing a Bottom Quark are still very scantly accredited by the PDG. About eight are empiricly well known with one or two particles in a prospective family series, but only one, the Λ_b^0 has at the time of writing as many as three members on which a preliminary curve of possible y exponents can be usefully based. Even more limiting, where in the two-particle structure of the Charm Quark's Baryon Ω_c^0 just above there was a full assembly of 6 other Charm Series on which to base estimates of whether it is reasonable that such particles might possibly be in the 0 and 1 groups, in the Bottom Families there is no basis for such estimates beyond the Λ_b^0 series itself. Consequently, there will be only one Series curve and table in the Bottom Baryons until more PDG accredited data in this area is available. For the Top Quark there are none.

Λ_b^0 Series (Prdm n plan) (Mid range PDG Masses, separately stated.)

Group	PDG Name	Mass MeV	Quarks	y initial	Σ quarks sum MeV	y Amended
0. Base	Λ_b^0	5619.5	$u_A d_A b_A$	0.303627	4028.825 **No Grp 2,3,4,5,**	0.302898
1. Isoton	$\Lambda_b(5912)^0$	5912.1	$u_B d_A b_A$		4029.781 **6,7**	0.348885
Isomer	$\Lambda_b(5920)^0$	5919.73	"		4029.781	0.350587

Table 7-4za Series Λ_b^0 w/o Grp 2,3,4,5,6,7($u_A d_B b_A$ 4031.75; $u_B d_B b_A$ 4032.70; $u_A d_A b_B$ 4417.624; $u_B d_A b_B$ 4418.581; $u_A d_B b_B$ 4420.546; $u_B d_B b_B$ 4421.503)

When more data are accredited, they will probably displace these Group 1s to a higher group on a truer curve than the one shown next, which is really only two points.

Λ b0 Series (Amended)
Exponent y
0.36
0.34
0.32
0.3
5600 5650 5700 5750 5800 5850 5900 5950 6000
Particle Mass in MeV

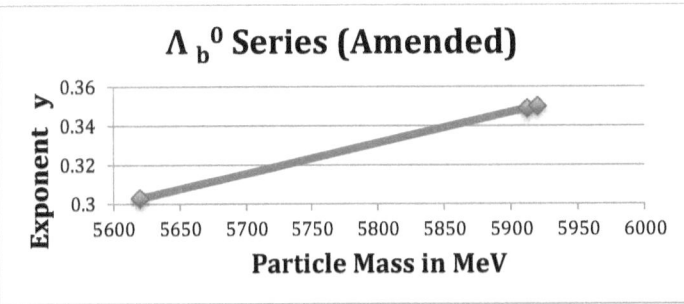

Fig. 7-2y Smooth Preliminary Curve y for Baryon Series $\Lambda_b{}^0$ (Slightly Amended) W/o Grps 2,3,4,5,6,7.

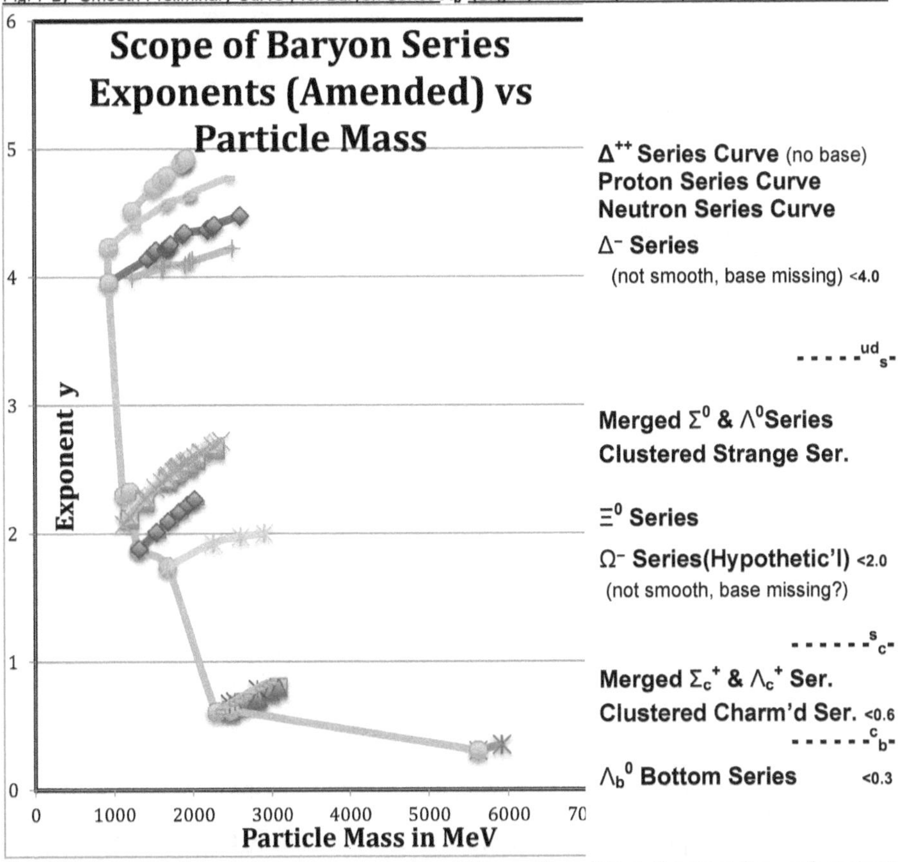

Scope of Baryon Series Exponents (Amended) vs Particle Mass
Exponent y
Particle Mass in MeV
0 1000 2000 3000 4000 5000 6000 70
0 1 2 3 4 5 6
Δ++ Series Curve (no base)
Proton Series Curve
Neutron Series Curve
Δ- Series
(not smooth, base missing) <4.0
- - - - -ud s-
Merged Σ0 & Λ0Series
Clustered Strange Ser.
Ξ0 Series
Ω- Series(Hypothetic'l) <2.0
(not smooth, base missing?)
- - - - - -s c-
Merged Σc+ & Λc+ Ser.
Clustered Charm'd Ser. <0.6
- - - - - -c b-
Λb0 Bottom Series <0.3

Fig 7-2z Sampled Baryon Series Curves on Arc of Linear Plot of Their Smallest Bases' Exponents.

It is interesting to confirm by comparison with the next logarithmic rather than linear plot, and also with the log-scaled plots of Figs. 1-1 to -3, that the Exponential Law and Power Law on which the MQP paradigm is based would not have been discovered if the initial plots had not been limited to the base line smallest baryons and mesons of each type. The confusion from members of all the series branch curves would certainly have obscured the simplicity of the baseline curve, and the law equations would not have been found. Also as noted earlier, a common equation for these branch curves would be useful. But the baseline curve with systematic branching is informative.

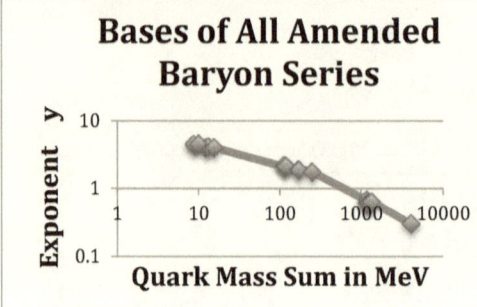

Fig. 7-2za A Logarithmic Plot of the y Exponents of the Smallest Base Members of the Baryon Series vs the Sum (Σ) of the Masses of the Quarks in Each Baryon. With the same y the curvature of the base log plot is downward, not upward as in Fig. 7-2z. (Mesons may be found later with this type of curve from which Series exponent curves could branch in the same way.) Compare with Figs. 1-1 to -3 in Chapt. 1.

     This concludes re-classifying the PDG present accredited baryons, including, as far as they can be taken at the time of writing, the demonstration of their series steps of masses derived from the dual masses of each type of quark by stepwise substitution of their heavier masses for the lighter masses. The series should each begin with the lightest mass of that baryon, wherever the clearly lightest possible mass has been accredited. That lightest possible mass for any given baryon series is definitively the one that is composed only with the lighter form of every quark in its composition. It may be that some of these ideally lightest baryons will never be found because of their rarity. (It is not intended to be implied that Nature, in its precursor process of forming quarks and baryons simultaneously, must begin at the lightest form and go through all the quark mass substitutions to arrive at the measured baryon mass. The precursor must go where the available components in each one take it. The stepped process is merely the analytic recapture of where the possible components can take each final baryon mass in the unexcited ground state, as these MQP paradigm calculations clearly show. Any residual excitement energies will raise the final structure and the exact final mass energy up an added substep isomer (to be taken up after a significant amount of more refined and precise lab measurements with further scaled equations.) To that baryons need only to add the magnetic "force" effects long known by Maxwell's Equations to arise by relative motion in time with distortions of static electric charge fields.

     MAJOR REFERENCES for Chapter 7

Olive, K. A., et al., (PDG) (2014), Chin. Phys., **C38**, 090001, (www.pdg.lbl.gov), Quarks Summary List & Baryons Summary List, and prior biennial editions.
Howard , (2005)
Davies, C., and Lepage, G. P., et al., (2010), Phys. Rev. Lett, **104**, 2 April issue
Bass, S. D., (2008), Science, The Spin Structure of the Proton
Bass, S. D., (2007), How Does the Proton Spin, Science 315, 1672, 23 March 2007
Firestone, R. B., Experimental Chart of Nuclides (2000), LBNL publication.
Lorentz, H. A, (1909 & -15) The Theory of the Electron, Dover Publications, 1952

Chapter 8

The Mesons as Broken Baryons (with Quarklet Debris)

SUMMARY: An MQP exploratory analysis has been made of the PDG Light Unflavored Mesons, about 40% of all Mesons. This stands for now as a prototype initial approach toward total quark structures of the Mesons, with radical antiquark differences from structures of Baryons. In contrast with the brief mean lives of LU mesons, their source is often in impact breakage of most stable baryons. Further marked differences may arise in the heavier classes of mesons. It is perhaps not impossible that some mesons may originate in some fraction through the same Precursor initiation process required for simultaneous generation of baryons and their quark components, since quarks also build mesons. However, the total lack of balanced and meta-stable meson outcomes with lowest light quark masses of diquarks shows an unsuitability of the form from its start, leaving only a secondary Recursor process restricted to beginning from baryons that have themselves originated in a preselection for stabilization at a critical balance which has been broken by energetic impacts. That is a bias of mesons that it is not necessary to change. This is like the sharp edges of a broken perfect crystal that had no second fluid precursor opportunity to adjust themselves under their own component forces. Something is missing from the smoothly closed rings of the baryons, a perfection that is always missing in the mesons, with their unsatisfied projecting ends that are only half-satisfied in their middles. From that obligatory fragility which always "decays" with observed extremely short mean lives under presumed random impacts there is no necessity to depart. - - Complete calculation of all observed meson masses under the simplified hadron's Exponential Law of Masses also requires that quarks have projected broken ends of quarklets that are missing various levels of what could have been torn away from quarks as whole neutrinos. - - Further, those calculations of all observed larger masses can only be done with specific sequences of joining single pairs of masses by twos, a process necessity for joinders of two, four, or six quarks/quarklets, or some cosmic chains of them, a tedious process of randomly impacted variations under the quarks' own forces.

Explorations of the structures of mesons started smoothly with the very smallest base mesons of the families of the Light Unflavored Mesons (LUM). These are the most easily and frequently observed mesons for which masses can be readily computed under the hadrons' Exponential Law of Masses with whole quark components identical and anti-identical to those in baryons. This is especially true for those light mesons composed from only two quarks/anti-quarks. As soon as LU meson families with PDG members of larger mass were considered, the smooth progression ran into difficulties. The deviations of computed masses were frequent and excessive. These deviations were evidence of necessary requirements for fractured quarklets and for building only in arbitrary sequences of steps by pairs of quarks/quarklets/combinations (with conserved coaxial pairs of micro-quanta in orbiting or static sphere and link sites), having two, four, or six quark-like "quarks" and "anti-quarks" (or similar quarklets) in final better matches of mass. These factors are all necessarily indicative of mesons as fractured baryons.

The Meson (and Quark) Fragments of Baryons
The many series of very unstable PDG meson particles arise mainly from impact destruction or "decay" of baryons, resulting in scrambled reorganization of conserved coaxial pairs of quanta into both quarks and antiquarks (and quarklet-antiquarklet debris, as well as mixtures of whole/fractured dual-mass quarks.) More and smaller baryons in their usual quark triplets or leptons may be among the results. It is the essential characteristic of mesons that they occur as simple balancing antiquarks and quarks in even diquark units, often with new link orbit phases. These scrambled sets of coaxial pairs of gyre quanta are forced together in the moment of break-up. They sort themselves by their own vortex forces into re-organizations with sufficient stability to be

observed for mean lives in the sub-microsecond to micro-microsecond range typically. There is a natural emphasis on decay or breakage under collision impacts of the weaker links of the less stable, heavier baryons, which compensates in the process for their lower percentage abundance. Here, though the very lightest mesons appear to be most abundant (as with the baryons), listed varieties of empirical mesons constructed only by lighter versions of quarks are much less predominant than in the baryons for the MQP paradigm. (This is true also in full consistency with, and within, the PDG meson series.)

The mesons are treated as one large class of particles at three (or more) points in the 2006 and prior, as well as later, biennial PDG reports. In the one-page 2006 Meson Summary Table there are 105 accredited and many non-accredited particles, to which a column about two-thirds of a page long has since been added, largely in the heavier charm and bottom meson area. From 2006 to the 2014-15 present, on the first page of detailed tables on the Light Unflavored (LU) Mesons (whose 45 accredited members in 2006 are taken as the paradigm example here), an equation states that about half have zero contribution from S (strange) quarks. On the line below that equation there is another one which states that for many accredited LU mesons an additive contribution from strange quarks does exist. Further, the 2006 (etc.) PDG explanatory note on The Quark Model cites mounting experimental evidence for a strange quark component of one of the accredited LU mesons, and there is also discussion of four-quark/anti-quark and six-quark/anti-quark mesons as well as of interchangeability of up, down, and strange quarks. This is changed slightly in the 2014-15 PDG Notes, but only to reduce somewhat the mention of six-quark LU forms and quark interchangeability. Still, the actual component structure of most individual mesons continues to be left much more uncertain than is the case in the baryons. In addition only two PDG accredited LUM are added to the 45 over the eight intervening years; the recent activity has been mainly in the heaviest quark mesons. The 2006-8 MQP analysis of LU mesons is essentially still valid and should remain valid. (There are always minor updates consistent with the prior structure in most particle studies.)

In this MQP paradigm, many LU meson uncertainties are resolved. Under the mass/charge power law and its exponential variant in their proper phases, there is a structural necessity that the great majority of the 45 accredited members of the 2006 PDG "Light Unflavored" (LU) mesons require not only a strange quark component, but also two or even three pairs of the three up, down, and strange quarks and anti-quarks, rather than a fixed single pair, to have the PDG measured masses. It is the measured mass values that require it. Like the baryons, the mesons also often must include (under the power/exponential law) series combinations of the lighter and heavier mass versions of each type of quark. However, this is not regularly added step-by-step, but at irregular random within the PDG series. In addition, systematicly derived quarklets with other exact power law masses (reduced by loss in high energy collisions of a neutral one or two pairs of quanta to largely unobservable neutrinos) are also required to construct many of the experimentally observed mesons to the PDG accredited masses. This quarklet factor of mass variation necessarily makes any meson series system far more complex than the baryon series because of the near doubling of the quark variables (in the next table), and the PDG acknowledged possibility of having one, two or three pairs of quarks in observed mesons rather than the baryons' fixed three quarks at two masses each (plus orbitally varied isomers of quarks in either case.) Two of the uncertain PDG LU Meson series are displayed here in a table as MQP clarified examples of all meson series, with the necessary microquantal assemblies of the three light quarks and their ablated quarklets, for which an introductory table is also computed first by the power law.

Quantal MQP Structures of Quarks/Quarklets in Light Unflavored PDG Mesons

Quarks	u_{A1}	u_{B1}	d_{A1}		d_{B1}		s_{A1}		s_{B1}		
Quarklets				d_{A2}		d_{B2}		s_{A4}		s_{B2}	s_{B4}
Quanta	2++	3++	1--	1--	2--	2--	4--	3--	2--	2--	2--
	2+-	1--	4+-	3+-	1++	1++	3++	2++	1++	1++	1++
					2+-	1+-			5+-	4+-	3+-
Net Charges	+2/3	+2/3	-1/3	-1/3	-1/3	-1/3	-1/3	-1/3	-1/3	-1/3	-1/3
Mass MeV (by	1.914	2.871	5.11	1.436	8.032	2.393	82.47	10.95	107.2	51.05	21.803
Power Law)											

(and their symmetric, oppositely charged anti-quarks and anti-quarklets)
Each quark/quarklet is a sphere of pair orbits with a main-charge pair in an expanded SEq linking orbit.
Table 8-1a MQP Quantal Structures, Charges, & Masses of Quarks/Quarklets for the PDG LU Mesons.

In this microquantal table for the observed LU mesons, up quarklets are not required (nor available; 4 pairs of quanta are the quark minimum.) Quarklet subscript 2 indicates a collision loss of two 1/6 charged quanta in one <u>conserved</u> neutral pair of quanta, which is impossible in the s_A case where there are no neutral pairs to lose. The subscript 4 shows a loss of four quanta, either in two neutral pairs as in the s_{B4} set, or in a neutral combination of a ++ pair and a -- pair as with s_{A4}. (The lighter quarks have reached their quarklet limit here, but s_B might exist further beyond these LUM series, as s_{B6} and s_{B8}. The quarklet possibilities in mesons with charm, bottom, or top quarks are further compounded.) The subscript 1 indicates the quarks themselves with no losses of quanta to neutrinos during the initial collisions or decays. Quark and quarklet masses are computed under the power law. Anti-particles have mirror-image form and charge.

MQP Structures of Typical Light Unflavored Mesons In Two PDG Series (Mixed)
<u>Computed</u> under the Exponential Law from Quark/Quarklet Masses

Quarks	u_{A1}	u_{B1}	d_{A1}	d_{B1}	s_{A1}	s_{B1}	(and their anti-quarks)
Quarklets			d_{A2}	d_{B2}	s_{A4}	s_{B2} s_{B4}	
Mesons (Charges Assigned)							MQP Structure
Mass Names (PDG MeV, rarely exact, often widely uncertain)							
π^+ 139.57	•		•				$u_{A1} \bar{d}_{A1}$
π^0 134.977	••						$u_{B1} \bar{u}_{B1}$
ρ (770)⁰						••	$s_{B1} \bar{s}_{B1}$
π (1300)⁰	•	•					$u_{A1} \bar{u}_{B1}$
$π_1$ (1400)⁰				•	••		$s_{A1} \bar{d}_{B1}$ $s_{A4} \bar{s}_{A4}$
ρ (1450)⁰				• ••		•	$s_{B1} \bar{s}_{A1}$ $d_{B2} \bar{s}_{A1}$
$π_1$ (1600)⁰				• •		••	$s_{B2} \bar{s}_{B1}$ $s_{B2} \bar{s}_{A4}$
$π_2$ (1670)⁰				• ••		•	$s_{B2} \bar{s}_{A1}$ $d_{B2} \bar{s}_{A1}$
$ρ_3$ (1690)⁰				••		• •	$s_{B2} \bar{s}_{A1}$ $s_{B4} \bar{s}_{A1}$
ρ (1700)⁰				•	•••		$s_{A1} \bar{s}_{A1}$ $s_{A1} \bar{d}_{B1}$
π (1800)⁻¹	•••••			•			$u_{A1} \bar{u}_{A1}$ $u_{A1} \bar{u}_{A1}$ $s_{A1} \bar{u}_{A1}$

[A bar over ū,d̄,& s̄ indicates a symmetric anti-quark/(-let) of opposite charge.]
It is obviously <u>difficult to find a clear distinction</u> between these PDG quark series.
<u>Either other options must exist, or there is no such direct similarity to baryons.</u>
Table 8-1b Two PDG LU Meson Series of MQP Computed Structures in PDG Measured Masses

This table demonstrates by the MQP structure that the present PDG meson series may be only apparently related, with random structural variations in the particle quarks which are not directly observable. The possible variations here of structure and exterior effects are not so simply mass, charge, and quark related as in the baryons, which have only one basic structural form for three-quark particles capable of existence (the box triple corner) with only three prototypical variations of net charge distribution and internal motion plus small isomeric variations in the mass energy of two specific quarks. These limits are removed in the unstable meson fragments of destroyed baryons. (The masses in these tables were calculated in 2006 to 2007, and they are still largely within the 2014 PDG stated uncertainties. This does not change the fact that only two of the structures found under MQPP could be considered to come within any stated PDG structural equations. Note that it is necessary in MQP mass calculations to assign and to track component charges or neutrality under the Mass-Charge Power Law for quarks in Table 8-1a, though not in Table 8-1b under the Exponential Mass Law of hadrons, but that charged particle masses are identical to those of their anti-charged anti-particles in both the PDG/QM and the MQP systems. To date, because of the added structural dimension of quarklets in addition to dual quark masses, full MQP reclassifications of meson series would be 3D configurations excessively elaborate for ready use, and the PDG classification titles are retained herein for continued identification of the prior data on individual particles. Study on further simplification of this unavoidable feature must continue as PDG meson mass and charge accreditations become less uncertain. Due to the necessities for their strong abundance, quarklets are not expected to vanish.)

This limited table shows especially on the last line the three pairs of quarks needed to construct a charged pi (1800) meson at the accredited mass within PDG uncertainty tolerances. Many PDG mesons have several options in possible structure. (See table of calculations below.) This structure for this meson combines typical (though not all possible) kinds of quark linkages so that the various parts of the structure demonstrate baseline structural connections in mesons generally (though not necessarily the optimum structural isomer for this meson.) The linkages between the six quarks/anti-quarks are variants of the basic rectangular (orthogonal) two-quark corners between S Eq. orbit planes of the baryons displayed earlier. Also in this formation, a single pair, a dual pair, or the entire triple pair of quarks can be a prototype of possible meson or meson series assemblies that might occur.

An even broader set of possible variations from the minimal structural change due to least baryon breakage in mesons is shown in the next figure repeated from Fig. 2-9 (page 59.) This is within estimates of the variation permitted by forces from individual quarks/quarklets with the general internal linking structure of Fig. 2-6 (page 56.) Note in this connection that, where baryon internal orbits are fully synchronized around the baryon triplet ring of quarks, the meson variations of this figure are available under those forces because it has not yet been found feasible to close a fully synchronized meson ring of quarks/quarklets at one, two, or three pairs in baryon-derived rectangular pairs, even with the freedoms shown here between pairs. (Possibly this expanded ring closure will be found with computerized improvement of the limited physical simulations used to date.) In any case, this configuration enables the potential of indefinite extension of cosmic meson chains such as would be required to provide the packing fractions of the astrophysicists' requirement of ultimate condensation of neutron stars into pure quark packs (sometimes improperly called quark soup.) [Lattimer & Prakash, The Physics of Neutron Stars, Science, (23 April 2004). 304, #5670, pp 536-542, and references, plus its citations in later updates on pulsars and magnetars up to this time of writing.]

Meson, 6 Quark Schematic

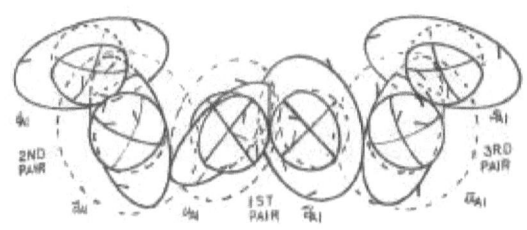

RANDOM MESON A-B-C AND S EQ. ORBITS

ILLUSTRATES OPEN ENDS OF ANY MESON CHAIN

– ENDS FAIL SEQUENTIALLY – <u>STRONG FORCE INEFFECTIVE</u>

Fig. 8-1a Structural Open Form of a Six-Quark Meson Ending with Diquarks on Ends of the Chain. One or both of the two links between diquarks may also join in a form of right angle link that turns them toward each other rather than extended. In either case, the 90° linkage angles are such that the two end quarks cannot close a ring by linking with each other and satisfying the end forms as they do in baryon triplets.
    Here, in nominal order of the pair masses, the first or #1 pair (a quark with an anti-quark) from the table is in the center of the plan figure with the second or #2 pair to the viewer's left. The C primary axes of the center pair are along the plan's line of view. Since this pair's two S Eq. planes are at 35.3... degrees from the usual C axis vertical, the right angle joints put the C axes of each of the two outer pairs symmetricly off that reference by about ten degrees. Also, the #1 central pair is shown (with an additional degree of less strongly linked meson freedom in the paradigm) near the middle of a quantally stepped, inward sliding range of interactive position offset which could occur in synchronizing marginal interferences of quantal orbits between the quark pairs of some putative mesons.
    The central #1 pair of quark spheres is coupled together internally by their S Eq. orbits in the same way as quarks 2 (up) and 3 (down) of the proton, and they retain the same number identification for their S Eq. orbits. Here a positive down antiquark takes the place of a proton's negative down quark 3 in the remaining pair of a baryon assembly which has lost its quark 1. The left #2 pair is mutually coupled

like quarks 3 and 1 of the proton, and they too retain the same numbering for their S Eq. orbits, but in this case with a positive down antiquark instead of an up quark/. Between these two pairs the linkage would nominally be at the tips of the elliptical S Eq. orbits in 1-2, 2-3, or 3-1 sequence like that between any two identical quarks of the omega minus prototype baryon, except for being shown in the available sliding offset between the two pairs along this new corner line (as just noted.)

The #3 pair to the viewer's right combines the lighter up antiquark and the lighter strange quark in an adaptation of the quark 2 to quark 3 linkage in the proton plan which can synchronize orbits of like-charge quarks here because the assembly is not required to close a ring of three quarks without interferences (as is necessary in any baryon.) The #3 pair can be linked to the #2 pair in the sliding adaptation of that same linkage for the same reason. In each of these interactively adaptive linkages the replacement negative up anti-quark may retain the same pattern of quantal orbit and spin sites as the positive up quarks in the proton or may adapt for orbit synchronization as closely as its quantal set permits to the basic site pattern of the negative down quark. Net charge of such a meson-like assembly would be neutral. (It is estimated that this layout for a meson chain in which three right angle pairs are turned away from each other pair would follow the same Exponential Law of Masses in building mesons from the quarks as would an only slightly closer linkage when the quarks turn toward each other at about the same distance spacing, but still cannot close a synchronized linkage. The mass-building interaction of currents between the gyres in separate quarks is limited in approximately the same ways, essentially by partial shielding from members of other closely knit quark/quarklet groups.)

Multi-Quark Meson Mass Calculations

To compute measured meson masses, quark/quarklet options are selected in pairs by trial and error from single quark mass energies (Table 8-1a above) under the exponential form of mass law. For single pairs and final steps of multiple pair summation Equn. 1-1b law is inverted for exact exponent y of N components (2 in all steps here) as: $y = \log(m_p / \Sigma m_c) / \log N$ Equn. 8-1
where m_p is the measured final particle mass, & Σm_c is the sum of the two input masses of particle components. For multiple pair steps the initial y exponent is taken from the latest prior curve for the available Σ (as in Chapter 1), starting with the separate #1 and #2 (lightest) pairs, and these are then summed as if for a two-pair combined mass with two components. The resulting mass is then summed with a separately calculated pair mass for any #3 heaviest pair to calculate the total meson mass as if it had only two components. That process accounts for distributed separations of the quanta (in up to 6 quark spheres) which reduces mass interactions from what would occur if all the quanta were in one quark sphere. (This should then be checked by running through the original Exponential Law, Equn 1-1b, with the final y to be sure that it does yield the required mass measurement which PDG accredited from a reported measurement.) This calculation process is shown here in a short table containing typical meson examples that are definitive in both parts put in and resulting outputs. (No other system works, though it is frustratingly very much like feeling under water for little eels in the dark.):

Typical Meson Calculation Records of Structures and y Exponents

|PDGName|AccrdtdMass |ΣCalculation |Chrg|yExpnt|Calc.Mass MeV|Structure
ρ (770)   775.5 ±0.4 s_{B1} 107.18899 –1/3   (or reverse anti-s with reversed charges)
              $\bar{s}_{B1}$ 107.18899 +1/3
              N=2,   Σ 214.37798   0   1.8548   775.46 $S_{B1}-\bar{S}_{B1}$
$\pi_1(1600)^0$ 1653 ±16 $s_{B2}-\bar{s}_{B1}$ 588.342 0   (AFTER LINKING EACH PAIR AS ABOVE)
              $s_{B2}-\bar{s}_{A4}$ 372.317 0
              N=2,   Σ 960.659   0   0.7830   1653.01 $S_{B2}-\bar{S}_{B1} - S_{B2}-\bar{S}_{A4}$

Table 8-1c Typical Meson Calculation Records Format for best calculations of LU Meson Mass

Table 8-1d **LIGHT UNFLAVORED MESON STRUCTURE CALCULATIONS (LAST STEP)** - with 8-year Meson PDG changes 2006-2014-15

PDG Name	MeV Prior Accrdtd Mass	Σ Calculation		Chrg	yExpnt	MeV Calc.Mass	Structure	2015 Change
$\pi^\pm$	139.57	u_{A1}	1.9141	+2/3				No Change (NC below)
		$\bar{d}_{A1}$	5.1117	+1/3				
		N=2, Σ	7.0257	+1	4.312	139.57	u_{A1}-$\bar{d}_{A1}$ (or reverse antimatter)	
π^0	134.9766	u_{B1}	2.871	+2/3				NC
		$\bar{u}_{B1}$	2.871	−2/3				
		N=2, Σ	5.742	0	4.5550	134.9766	u_{B1}-$\bar{u}_{B1}$	
η	547.51	d_{A1}	5.1117	−1/3				547.862 ±0.017
		$\bar{s}_{B1}$	107.18899	+1/3				over tolerance
		N=2, Σ	112.30016	0	2.2855	547.51	d_{A1}-$\bar{s}_{B1}$	
f_0(600)	400-1200	s_{A1}	82.4674	−1/3			f_0(500)	400-550
		$\bar{s}_{B2}$	51.05118	+1/3				over tolerance
		N=2, Σ	133.5186	0	2.16792	600 ±	s_{A1}-$\bar{s}_{B2}$	
ρ(770)	775.5 ±0.4	s_{B1}	107.18899	−1/3			775.26 ±0.25	
		$\bar{s}_{B1}$	107.18899	+1/3				in tolerance
		N=2, Σ	214.39798	0	1.85484	775.46	s_{B1}-$\bar{s}_{B1}$	
ω(782)	782.65 ±0.12	u_{A1}-$\bar{u}_{A1}$	108.128	0	(AFTER LINKING EACH PAIR AS ABOVE,) NC			
		u_{A1}-$\bar{u}_{A1}$	108.128	0				
		N=2, Σ	216.256	0	1.8556	782.64	u_{A1}-$\bar{u}_{A1}$ - u_{A1}-$\bar{u}_{A1}$	
η'(958)	957.78	u_{B1}-$\bar{d}_{B1}$	162.766	+1				NC
		$\bar{u}_{B1}$-d_{B1}	162.766	−1				
		N=2, Σ	325.532	0	1.5569	957.78	u_{B1}-$\bar{d}_{B1}$ - $\bar{u}_{B1}$-d_{B1}	
f_0(980)	980 ±10	d_{A1}-$\bar{d}_{B1}$	170.807	0			990 ±20	
		d_{B1}-$\bar{d}_{B1}$	181.847	0				in tolerance
		N=2, Σ	352.654	0	1.47453	980 ±	d_{A1}-$\bar{d}_{B1}$ - d_{B1}-$\bar{d}_{B1}$	
a^0(980)	984.7 ±1.2	d_{B1}-$\bar{d}_{B1}$	181.847	0			980 ±20	
		d_{B1}-$\bar{d}_{B1}$	181.847	0				in tolerance
		N=2, Σ	363.694	0	1.43695	984.7	d_{B1}-$\bar{d}_{B1}$ - d_{B1}-$\bar{d}_{B1}$	
φ(1020)	1019.46	s_{B4}-$\bar{d}_{A2}$	213.55	0				NC
		s_{A4}-$\bar{d}_{A2}$	166.674	0				
		N=2, Σ	380.224	0	1.422884	1019.46	s_{B4}-$\bar{d}_{A2}$ - s_{A4}-$\bar{d}_{A2}$	
h_1(1170)	1170 ±20	u_{B1}-$\bar{s}_{A4}$	179.651	+1				NC
if +1		s_{B4}-$\bar{s}_{B4}$	303.685	0				
		N=2, Σ	483.336	+1	1.27541	1170	u_{B1}-$\bar{s}_{A4}$ - s_{B4}-$\bar{s}_{B4}$	
if 0		d_{B1}-$\bar{d}_{B1}$	181.847	0				
		s_{B4}-$\bar{s}_{B4}$	303.685	0				
		N=2, Σ	485.532	0	1.26887	1170	d_{B1}-$\bar{d}_{B1}$ - s_{B4}-$\bar{s}_{B4}$	

PDG Name	MeV Accrdtd Mass	Σ Calculation	Chrg	yExpnt	MeV Calc.Mass	Structure	2015 Change
b₁(1235)	1229.5 ±3.2	u$_{A1}$-s̄$_{A1}$ 451.585	+1				NC
		d$_{A2}$-d̄$_{A2}$ <u>84.545</u>	0				
		N=2, Σ 536.130	+1	1.197444	1229.523	u$_{A1}$-s̄$_{A1}$ - d$_{A2}$-d̄$_{A2}$	
(alternate		s$_{A1}$-d̄$_{A1}$ 446.494	0				
or next item)		u$_{A1}$-d̄$_{A2}$ <u>96.627</u>	±1				
		N=2, Σ 543.121	+1	1.17873	1229.5	s$_{A1}$-d̄$_{A1}$ - u$_{A1}$-d̄$_{A2}$	
a₁(1260)	1230 ±40	s$_{B2}$-s̄$_{A4}$ 375.92	0				NC
(or alternates above		u$_{A1}$-s̄$_{A4}$ <u>167.215</u>	±1				
or isomers)		N=2, Σ 543.135	+1	1.178689	1229.5	s$_{B2}$-s̄$_{A4}$ - u$_{A1}$-s̄$_{A4}$	
f₂(1270)	1275.4 ±1	s$_{B2}$-d̄$_{B2}$ 367.080	0			1275.5 ±0.8	
(or 4 alternates/isomers		s$_{B4}$-d̄$_{A2}$ <u>213.550</u>	0				in <u>tolerance</u>
cluster next.)		N=2, Σ 580.630	0	1.135259	1275.4	s$_{B2}$-d̄$_{B2}$ - s$_{B4}$-d̄$_{A2}$	
f₁(1285)	1281.8 ±0.6	s$_{A1}$-d̄$_{A2}$ 442.845	0			1281.9 ±0.5	
(cluster)		u$_{B1}$-ū$_{B1}$ <u>139.252</u>	0				over <u>tolerance</u>
		N=2, Σ 582.097	0	1.13884	1281.8	s$_{A1}$-d̄$_{A2}$ - u$_{B1}$-ū$_{B1}$	
η(1295)	1294 ±4	s$_{B4}$-s̄$_{B2}$ 412.124	0				NC
(cluster)		d$_{A1}$-d̄$_{B1}$ <u>170.807</u>	0				
		N=2, Σ 582.931	0	1.1504	1294	s$_{B4}$-s̄$_{B2}$ - d$_{A1}$-d̄$_{B1}$	
π(1300)	1300 ±100	s$_{A1}$-d̄$_{B1}$ 463.688	0				NC
(cluster)		u$_{A1}$-ū$_{B1}$ <u>120.136</u>	0				
		N=2, Σ 583.824	0	1.154906	1300	s$_{A1}$-d̄$_{B1}$ - u$_{A1}$-ū$_{B1}$	
a₁(1320)	1318.3 ±0.6	s$_{B4}$-s̄$_{B4}$ 303.685	0			Name a₂(1320)	
(end 5 cluster)		s$_{B4}$-s̄$_{A4}$ <u>280.848</u>	0				NC
		N=2, Σ 584.534	0	1.17331987	1318.3	s$_{B4}$-s̄$_{B4}$ - s$_{B4}$-s̄$_{A4}$	
f₀(1370)	1200 to 1500	s$_{B1}$-s̄$_{A4}$ 506.484	0				NC
		d$_{A1}$-d̄$_{A1}$ <u>163.552</u>	0				
		N=2, Σ 670.036	0	1.0318654	1370	s$_{B1}$-s̄$_{A4}$ - d$_{A1}$-d̄$_{A1}$	
π₁(1400)	1376 ±17	s$_{A1}$-d̄$_{B1}$ 463.688	0			1354 ±25	
		s$_{A4}$-s̄$_{A4}$ <u>211.306</u>	0				in <u>tolerance</u> <u>barely</u>
		N=2, Σ 674.994	0	1 027534	1376	s$_{A1}$-d̄$_{B1}$ - s$_{A4}$-s̄$_{A4}$	
η(1405)	1409.8 ±2.5	s$_{B2}$-s̄$_{A1}$ 534.074	0			1408.8 ±1.8	
		d$_{B1}$-s̄$_{A4}$ <u>186.979</u>	0				in <u>tolerance</u>
		N=2, Σ 721.053	0	0.96731133	1409.8	s$_{B2}$-s̄$_{A1}$ - d$_{B1}$-s̄$_{A4}$	
f₁(1420)	1426.3 ±0.9	s$_{A1}$-s̄$_{A1}$ 582.354	0			1426.4 ±0.9	
		u$_{B1}$-ū$_{B1}$ <u>139.252</u>	0				in <u>tolerance</u>
		N=2, Σ 721.606	0	0.982994	1426.3	s$_{A1}$-s̄$_{A1}$ - u$_{B1}$-ū$_{B1}$	
ω(1420)	1400 to 1450	s$_{B1}$-d̄$_{B2}$ 511.856	0				NC
		s$_{B4}$-d̄$_{A2}$ <u>213.550</u>	0				
		N=2, Σ 725.406	0	0.9690303	1420	s$_{B1}$-d̄$_{B2}$ - s$_{B4}$-d̄$_{A2}$	

PDG Name	MeV Accrdtd Mass	LU MESONS Table 8-1d (Cont.) Σ Calculation	Chrg	yExpnt	MeV Calc.Mass	Structure	2015 Change
$a_0(1450)$	1474 ±19	s_{B1}-$\bar{s}_{A1}$ 629.143	0				NC
		d_{B1}-$\bar{d}_{A2}$ 151.478	0				
		N=2, Σ 780.621	0	0.9170462	1474	s_{B1}-$\bar{s}_{A1}$- d_{B1}-$\bar{d}_{A2}$	
$\rho(1450)$	1459 ±11	s_{B4}-$\bar{s}_{B4}$ 303 683	0			1465 ±25	
		s_{B2}-$\bar{s}_{B2}$ 469.139	0			in tolerance	
		N=2, Σ 772.822	0	0.9167718	1459	s_{B4}-$\bar{s}_{B4}$ - s_{B2}-$\bar{s}_{B2}$	
$\eta(1475)$	1476 ±4	s_{A1}-$\bar{s}_{A4}$ 476.277	0				NC
		s_{B4}-$\bar{s}_{B4}$ 303.685	0				
		N=2, Σ 779.962	0	0.920217	1476	s_{A1}-$\bar{s}_{A4}$ - s_{B4}-$\bar{s}_{B4}$	
$f_0(1500)$	1507 ±5	s_{B4}-$\bar{s}_{A4}$ 280.849	0			1504 ±6	
		s_{B1}-$\bar{s}_{A4}$ 506.484	0			in tolerance	
		N=2, Σ 787.333	0	0.936634	1507	s_{B4}-$\bar{s}_{A4}$ - s_{B1}-$\bar{s}_{A4}$	
$f_2'(1525)$	1525 ±5	s_{A4}-$\bar{s}_{B1}$ 506.484	0				NC
		s_{B4}-$\bar{s}_{B4}$ 303.683	0				
		N=2, Σ 810.167	0	0.912518	1525	s_{A4}-$\bar{s}_{B1}$ - s_{B4}-$\bar{s}_{B4}$	
$\pi_1(1600)$	1653 +18,−15	s_{B2}-$\bar{s}_{B1}$ 588.342	0			1662 ±9	
		s_{B2}-$\bar{s}_{A4}$ 372.317	0			in tolerance barely	
		N=2, Σ 960.659	0	0.782991	1653.01	s_{B2}-$\bar{s}_{B1}$ - s_{B2}-$\bar{s}_{A4}$	
$\eta_2(1645)$	1617 ±5	s_{B2}-$\bar{s}_{A4}$ 375.920	0				NC
		s_{B2}-$\bar{s}_{A1}$ 534.074	0				
		N=2, Σ 909.994	0	0.829391	1617	s_{B2}-$\bar{s}_{A4}$ - s_{B2}-$\bar{s}_{A1}$	
$\omega(1650)$	1670 ±30	s_{B1}-$\bar{d}_{B2}$ 511.856	0				NC
		s_{B4}-$\bar{s}_{A1}$ 472.503	0				
		N=2, Σ 984.359	0	0.7625916	1670	s_{B1}-$\bar{d}_{B2}$ - s_{B4}-$\bar{s}_{A1}$	
$\omega_3(1670)$	1667 ±4	s_{B1}-$\bar{d}_{A1}$ 515.995	0				NC
		s_{A1}-$\bar{d}_{B1}$ 463.688	0				
		N=2, Σ 984.359	0	0.759523	1667	s_{B1}-$\bar{d}_{A1}$ - s_{A1}-$\bar{d}_{B1}$	
$\pi_2(1670)$	1672.4 ±3.2	s_{B2}-$\bar{s}_{A1}$ 534.074	0			1672.2 ±3.0	
		d_{B2}-$\bar{s}_{A1}$ 451.009	0			in tolerance	
		N=2, Σ 985.083	0	0.763603	1672.4	s_{B2}-$\bar{s}_{A1}$ - d_{B2}-$\bar{s}_{A1}$	
$\varphi(1680)$	1680 ±20	s_{A1}-$\bar{d}_{B1}$ 463.688	0				NC
		s_{B2}-$\bar{s}_{A1}$ 534.074	0				
		N=2, Σ 997.762	0	0.7516893	1680	s_{A1}-$\bar{d}_{B1}$ - s_{B2}-$\bar{s}_{A1}$	
$\rho_3(1690)$	1688.8 ±2.1	s_{B2}-$\bar{s}_{A1}$ 534.074	0				NC
		s_{B4}-$\bar{s}_{A1}$ 472.503	0				
		N=2, Σ 1006.577	0	0.746541	1688.8	s_{B2}-$\bar{s}_{A1}$ - s_{B4}-$\bar{s}_{A1}$	
$\rho(1700)$	1720 ±20	s_{A1}-$\bar{s}_{A1}$ 582.354	0				NC
		s_{A1}-$\bar{d}_{B1}$ 445.670	0				
		N=2, Σ 1028.024	0	0.742535	1720	s_{A1}-$\bar{s}_{A1}$ - s_{A1}-$\bar{d}_{B1}$	

LU MESONS Table 8-1d (Cont.)

PDG Name	MeV Accrdtd Mass	Σ Calculation	Chrg	yExpnt	MeV Calc.Mass	Structure	2015 Change
$f_0(1710)$	1718 ±6	$s_{B1}-\bar{d}_{A2}$ 516.703	0				1723 ±5
		$s_{B1}-\bar{s}_{A4}$ <u>506.484</u>	<u>0</u>				<u>in tolerance barely</u>
		N=2, Σ 1023.187	0	0.7476602	1718	$s_{B1}-\bar{d}_{A2} - s_{B1}-\bar{s}_{A4}$	
$\pi(1800)$	1812 ±14	$u_{A1}-\bar{u}_{A1}$ 108.124	0				1812 ±12
		$u_{A1}-\bar{u}_{A1}$ <u>108.124</u>	<u>0</u>				
		N=2, Σ 216.248	0	1.65	678.659		
		$u_{A1}-\bar{u}_{A1} - u_{A1}-\bar{u}_{A1}$ 678.659	0				
		$s_{A1}-\bar{u}_{A1}$ <u>470.760</u>	<u>−1</u>				<u>in tolerance</u>
		N=2, Σ 1149.419	−1	0.6566781	1812	$u_{A1}-\bar{u}_{A1} - u_{A1}-\bar{u}_{A1} - s_{A1}-\bar{u}_{A1}$	
$\phi_3(1850)$	1854 ±7	$s_{B1}-\bar{s}_{A1}$ 629.143	0				NC
		$s_{B1}-\bar{s}_{A1}$ <u>629.143</u>	<u>0</u>				
		N=2, Σ 1258.286	0	0.55918137	1854	$s_{B1}-\bar{s}_{A1} - s_{B1}-\bar{s}_{A1}$	
$\pi_2(1880)$	1895 ±16	ADDED ENTRY					NOT YET RESOLVED
$f_2(1950)$	1944 ±12	$u_{A1}-\bar{u}_{A1}$ 108.124	0				NC
		$u_{A1}-\bar{u}_{A1}$ <u>108.124</u>	<u>0</u>				
		N=2, Σ 216.248	0	1.65	678.659		
		$u_{A1}-\bar{u}_{A1} - u_{A1}-\bar{u}_{A1}$ 678.659	0				
		$s_{B1}-\bar{s}_{A1}$ <u>629.143</u>	<u>0</u>				
		N=2, Σ 1307.802	0	0.571884084	1944	$u_{A1}-\bar{u}_{A1} - u_{A1}-\bar{u}_{A1} - s_{B1}-\bar{s}_{A1}$	
$f_2(2010)$	2011 +60,−80	$u_{B1}-\bar{u}_{B1}$ 139.252	0				NC
		$u_{B1}-\bar{u}_{B1}$ <u>139.252</u>	<u>0</u>				
		N=2, Σ 278.504	0	1.47	771.517		
		$u_{B1}-\bar{u}_{B1} - u_{B1}-\bar{u}_{B1}$ 771.517	0				
		$s_{B1}-\bar{s}_{A1}$ <u>629.143</u>	<u>0</u>				
		N=2, Σ 1400.660	0	0.5218063	2011	$u_{B1}-\bar{u}_{B1} - u_{B1}-\bar{u}_{B1} - s_{B1}-\bar{s}_{A1}$	
$a_4(2040)$	2001 ±10	$u_{B1}-\bar{u}_{B1}$ 139.252	0				1995 +10, −8
		$u_{A1}-\bar{u}_{B1}$ <u>120.136</u>	<u>0</u>				
		N=2, Σ 259.388	0	1.53	759.535		
		$u_{A1}-\bar{u}_{B1} - u_{B1}-\bar{u}_{B1}$ 759.535	0				
		$s_{B1}-\bar{s}_{A1}$ <u>629.143</u>	<u>0</u>				<u>in tolerance</u>
		N=2, Σ 1388.678	0	0.527009054	2001	$u_{A1}-\bar{u}_{B1} - u_{B1}-\bar{u}_{B1} - s_{B1}-\bar{s}_{A1}$	
$f_4(2050)$	2025 ±10	$d_{A1}-\bar{d}_{B2}$ 139.838	0				2018 ±11
		$d_{A1}-\bar{d}_{B2}$ <u>139.838</u>	<u>0</u>				
		N=2, Σ 279.676	0	1.47	774.76		
		$d_{A1}-\bar{d}_{B2} - d_{A1}-\bar{d}_{B2}$ 774.76	0				
		$s_{B1}-\bar{s}_{A1}$ <u>629.143</u>	<u>0</u>				<u>in tolerance</u>
		N=2, Σ 1403.903	0	0.52847865	2025	$d_{A1}-\bar{d}_{B2} - d_{A1}-\bar{d}_{B2} - s_{B1}-\bar{s}_{A1}$	
$f_2(2300)$	2297 ±28	$s_{B4}-\bar{s}_{A4}$ 280.849	0				NC
		$d_{B1}-\bar{d}_{B1}$ <u>181.847</u>	<u>0</u>				
		N=2, Σ 462.696	0	1.147	1024.654		
		$s_{B4}-\bar{s}_{A4} - d_{B1}-\bar{d}_{B1}$ 1024.654	0				
		$s_{B1}-\bar{s}_{A1}$ <u>629.143</u>	<u>0</u>				
		N=2, Σ 1653.797	0	0.4739687	2297	$s_{B4}-\bar{s}_{A4} - d_{B1}-\bar{d}_{B1} - s_{B1}-\bar{s}_{A1}$	

210

PDG Name	MeV Accrdtd Mass	LU MESONS Table 8-1d (Cont.) Σ Calculation	Chrg	MeV yExpnt	Calc.Mass	Structure	2015 Change
φ(2170)	2175 ±15	ADDED ENTRY					NOT YET RESOLVED
f2(2340)	2339 ±60	s_{B4}-$\bar{s}_{A4}$ 280.349	0			2297 ±28	
		s_{B4}-$\bar{d}_{B2}$ 222.845	0				
		N=2, Σ 503.194	0	1.13	1101.284		
		s_{B4}-$\bar{s}_{A4}$ - s_{B4}-$\bar{d}_{B2}$ 1101.284	0				
		s_{B1}-$\bar{s}_{A1}$ 629.143	0				well over tolerance
		N=2, Σ 1730.427	0	0.434763781	2339	s_{B4}-$\bar{s}_{A4}$ - s_{B4}-$\bar{d}_{B2}$ - s_{B1}-$\bar{s}_{A1}$	

Symbols (from Table 8-1a, page 203)

Quarks	u_{A1}	u_{B1}	d_{A1}		d_{B1}		s_{A1}		s_{B1}		
Quarklets				d_{A2}		d_{B2}		s_{A4}		s_{B2}	s_{B4}
Mass in MeV	1.914	2.871	5.11	1.436	8.032	2.393	82.47	10.95	107.2	51.05	21.803

ADDED PDG ENTRIES	2006 to 2015		2	(not yet calculated)	
Changed beyond tolerance	«	«	4	(need recalculation)	
Changed barely within «	«	«	3	(« «)	
Changed within tolerance	«	«	12	(may « «)	
NO CHANGE (NC)	«	«	26		
Total			47		

Table 8-1d LU Meson Structure Calculations (Last Step) with 2006 -2014-15 PDG 8-Year Changes

    This table has 47 condensed entries, plus a very few alternate cases (which are not intended to be all-inclusive, but show that alternatives are possible or exist.) These entries have been closely cross-checked with each other to insure their consistency and elimination of errors, including review of items not shown but necessary for items that are listed. Reviewing these data has been very informative. The table was computed over a period of time in 2006 to 2008 in an exploratory review of mesons. In the meantime the analysis of the stepped mass formation of the baryon series was done. This put a new light on the meson data tables, expecting them to have a similar clarity with some similar re-organization. However, over the latter part of the century the PDG has not found in mesons so many consistently definite quark groups as in baryons. The presence of anti-particles and, now, of quarklets, interferes with that clarity, even though the basic group of two, rather than three, quark-like components is helpful.

    One of the background items often not listed, is the <u>initial</u> value of the exponent y for the number of components N in the Exponential Law Equation for mass energies of Hadrons (both Baryons and Mesons). As stated above just before the table, these values are to be read from the latest working curves (if available) of y versus Σ, the sum of mass energies of the components in MeV. This method of obtaining a value for y with two components depends on the combination's having been used in a successful computation of a PDG-measured meson particle mass at some point in time, or it may have been carried over from a baryon calculation. Presuming that it was in meson work, it may have been in a case of a two-quark meson, or a case of a four-quark (or six-quark) meson. With four-quarks two separate y values are used in sequence. If the first one is read off the logarithmic data curve wrong by about 5%, the second one from Equn. 8-1 compensates for the earlier error in the final combination of the four quarks. Thus, there will be records of two successful y values (for a combination of two specific quarks) differing legitimately by a few percent. In other cases the prior y reading may be found in notes, and the identical value will be used repeatedly, though it may not be too accurate, or calculated on the inverse law. So the consistency of the long table may legitimately be somewhat irregular, especially if the initial use was in an overview survey with a minimal number of significant figures, or minimal checking for errors. There are four such spot variations up to about 5% of diquark sum mass data now found in the table, including one significantly higher, out of 92 such determinations, about 20 of which are duplicates. The learning from this current review might include future re-computing of all old records for the consistency of y values as well as those required for new PDG accreditation of empirical measurements, as listed at the bottom of the table. (Great care is required also in reading y from curves for new values.) The review does indicate that there must be quark structures which have not been found and included, for some of the final measured particle masses in the table. These alternates should be found, essentially as shown in the long table, even though the search may be lengthy and frustrating, in order to be certain that a complete and appropriate re-classification of the LU mesons can or cannot be provided, and to finish demonstrating a definite method for providing it on the heavier mesons in the future.

As a result of the computation that was done, a printable linear slide rule of quark-pair mass sums was prepared that is particularly suitable for finding the diquark groups that can give a four-quark mass to match PDG measurements (or as combined with a selected quark pair mass to yield a six quark structure for a measured mass.) The problem is really to select from all the possible quarks and quarklets the combinations of 2 quark/quarklet masses that will add under the exponential law to a required Σ (sum of component masses) for estimating the structure of 4 quark/quarklet PDG LU Mesons in that range. The rule is printed on card stock or stiff transparency (best for repeatedly taping the rule at a setting and easily resetting.) Then cut the upper and lower halves apart and quasi-permanently tape the lower half to the upper as its extension to the right (be sure to match scale values carefully.) Next cut left to right on a very straight edge between the two sets of scale markers. This produces the upper and lower scales of a slide rule. To use it, set any 2 points on the 2 inner scales to give the required Σ at a reasonable y. Then any second scale marks that are opposite each other add to that value approximately, and the exact pairs and pair masses that provide it may be read off above the second scale markers. Small variations of the exponent make the exact mass estimate. This slide rule can be extended as necessary for the other heavier classes of mesons. It can also be extended, or programmed in a computer, to add the six-quark and two-quark mesons by the demonstrated method dependent on the MQP paradigm of particle structure. Eventually, a smooth curve for y in meson series can again be sought to determine whether there is a true step-by-step additive meson structuring similar to that of the baryons, but it can be multidimensional due to the necessity for quarklet steps as well the steps for dual masses of quarks. At some point in this further usage the six-quark meson version of exponential mass increase can extend to quasi-infinite chain lengths of cosmic mesons by repeating mid-pairs of quarks, and, depending on density, fractional exponents of the mass factor 1/3 for power law neutral pairs can affirm mass gains for such chains in collapsed "neutron" stars. (Lattimer & Prakash, 2004)

213

Fig. 8-1b Diquark to Tetra-Quark LU Meson Quark Pair Estimator Slide Rule

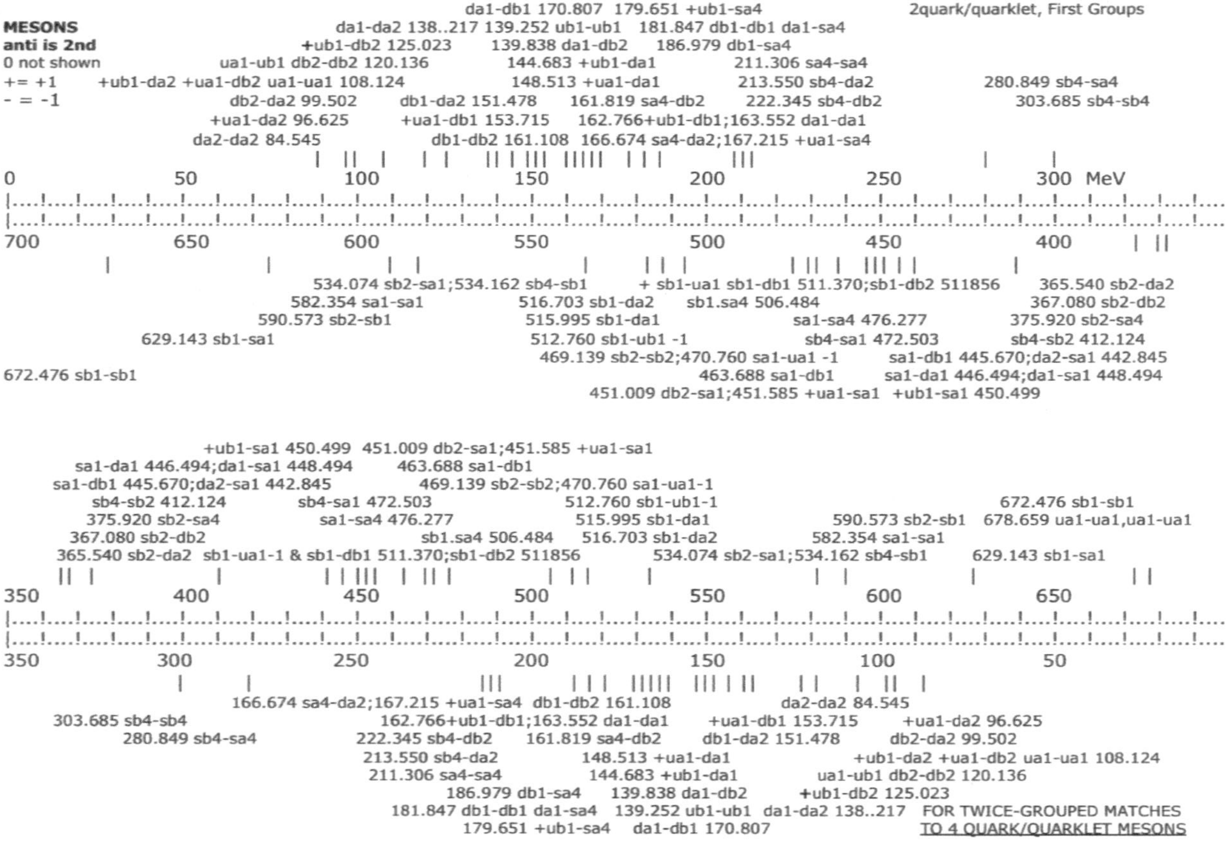

In Figure 8-1b note that in every quark or quarklet pair listed together on the scale the first listed is the ordinary particle component of the indicated meson, and the second or right hand component of the two is the meson's anti-particle, but is not marked with the usual bar (for anti-) to condense the crowded notation. Usually the quark or quarklet charge sign is the normal one for the particle, but occasionally a + sign or a -1 multiplier is shown to insure that the component has the proper charge. This permits the closest packing of information on the rule. Also occasionally two fairly similar mass pairs are shown with only a single mass number to fill in the long table for addition to obtain Σ, and in these cases the best y value will be slightly different for the two named pairs to obtain the same write-in value listed.

All these quarks and quarklets with four or more pairs of microquanta are scaled down permanently in size within their hadrons much like the simpler electron (with three pairs of microquanta in the A-B-C orbits) at its most reduced scale. When the electron is static or only orbiting a nucleus at a fraction of the speed of light, it was empiricly found above to have a particle radius near its classical large Compton radius, but when the electron has been compressed by the forces of acceleration to near the speed of light, it is found empiricly (PDG) to be at least several (four or more) orders of magnitude less than the Compton radius (Chapt. 5.) and thus compatible with the paradigm's quark spheres within the PDG empirical measurements of the size of the proton and pi meson.

At this small scale of separation between particles the conserved charge effect of repulsion between like charges (at quantal cone bases) is overridden by the greater PDG accredited attractions, called the strong force, between same or opposite charged quarks (and all aspects of the paradigm's quanta). This force (as in the PDG discussion) is then reduced in the paradigm by close approaches of the orbiting quanta to the 15 degree spacing limit with its very short range build up of resisting attraction. With the smallest three uud quarks in a proton, the net attractions (in the PDG strong force's intermediate range of close separation) also override the paradigm's stored momentum effects of the mass energy interaction between the rapidly spinning and orbiting component quanta. In the paradigm's neutron (ddu) the stored momentum effects in the slightly larger mass energy (coupled with the neutron's poorer structural balance) is estimated to be enough greater to destabilize the free neutron slightly to add to the account for its fifteen-minute mean life (as accredited by the PDG) rather than the extremely long (seemingly quasi-infinite) PDG mean life of the proton cited in the chapter on Baryons. In the charged pi meson the attractions of one lighter up quark cannot stabilize one lighter down antiquark (or the antiparticle's reverse) with its combination of same charge repellance reducing the strong near zone attraction and the stored momenta with the poorer balance. (It takes the attractive balance of two up quarks with a down quark to do that in the proton.) Also the two heavier mass u_B up quarks in the neutral pi meson cannot be stable, even when one is the antiquark to add the attraction of unlike charge in the SEq orbit. (The PDG accredited mean life of this meson is almost nine orders of magnitude shorter than that of the charged pi, due in the MQP paradigm to a sensitive transition to unstabilizable levels of momentum in the slightly larger orbiting mass energy, per Chapters 5 and 6.)

The MQP step-by-step meson calculation process takes account of the fact that in this paradigm's structure of mesons with two or three pairs of quarks the quarks are not all interacting closely in direct contact with each other, as occurs with the three quarks in a baryon box corner, but each pair is separated somewhat with linkages only from one member of the pair to one member of another pair. This partial separateness of pairs reduces the degree of overall mass energy accumulation by interaction of each quark's quanta with all others in the meson. Unlike the baryon box corners in a nucleus, four quark spheres cannot find a way of synchronizing and summing all their quantal orbits in a pyramid of direct contact between all four quarks for closer linkage and larger interaction mass of a two-pair or four-quark meson. Charges must be conserved in every case and be consistent with the rule that full hadron particles can have only plus or minus one electron level of net charge or be neutral (though there are also a few PDG empirical baryons with no negative quarks and a +2 net charge after the Ω prototype plan with all three quarks of the same charge.) In conserving charge, note that half the quarks (and quarklets) in every meson are antiquarks or antiquarklets which have the opposite charge from the same type of quark. Otherwise, quarklets retain the charge of the basic quark from which they are derived by loss of one or two neutrinos in destructive collision or decay of an original baryon or closely bound pair of baryons (or also of very heavy initial mesons from prior collisions and decays.)

    The meson mass calculation process may be easier to grasp if also summarized from a different point of view. Overall, where there are two or three pairs of quarks or quarklets in a meson, the exponential mass law must be applied in steps. The structures available to these mesons cannot bring all the quarks in the particle into the same close contact with all others with a directly synchronizable set of orbits as in the three quark box corners of the baryons. This can be done usually only for any two quarks that are left in a two-wall corner after a collision or decay. However, quarks in a meson pair may be each synchronizably adjacent to one of two other quarks in two other pairs, which together create the larger particle out of as many as three pairs with loose quark ends. The stepped calculations for this are generally consistent with empirical PDG masses as long as either the first connection step is between the two lowest mass quarks, as being first to assemble under their own forces, and then larger quark pairs with similar two pair steps in ascending order to the largest quarks. (The reverse order or mixed order may be used, but only if necessary in unusual structures, to obtain matches to measured accurate masses with exponents y below 5 in a suitable range for the type of particle mass involved.)

    The cause of the high level of irregularity in LU Mesons is also more readily understood by considering one of the major situations of their Recursor generation in everyday Nature. When ordinary atomic nuclei are destroyed in a collision a large number of rearrangements of the quarks and quantal pairs in mesons and other particles become possible. Typically, when a high energy cosmic ray particle enters earth's atmosphere the seven protons and seven neutrons in each nitrogen (N) nucleus (about 80% of atmospheric gases) provide a most likely collision site. In the disruption of these 14 baryons in an off-center impact (neglecting the impacting particle), it could occur that a number of their widely separated individual quarks are destroyed, though a third to a half of their up and down quarks are not broken up themselves but are very rapidly pushed into changed contacts with a fractured and re-arranged antiquark in sets of two. In that extremely short relative instant of action the quantal pairs of about half of the quark spheres within the volume of the original N nucleus might be pushed into combinations of new quark internal configurations such as strange and down antiquarks. By mutual attraction in such a highly condensed collision condition the quantal pairs and surviving quarks/antiquarks would re-assort themselves under their own forces without observable delay until all are exactly accounted for in some way. However, this might result in many leftover spherical balls of various neutral neutrinos without S Eq. orbits (since that requires a 1/3 or 2/3 charged particulate assembly, as well as another charged assembly of suitable matching charge to round the other out to a whole number charge while mutually attracting S Eq. orbiting quanta out into an enlarged orbit linkage.) Such uncharged isolated particles are rarely detectable in empirical measurements of a collision. (Without the more widely extended influence of a net particle electric charge, any single neutrino sphere that does not make an improbable random direct collision with another nearby particle in the relatively open space between particles will typically pass from its observable point of initiation by a prior collision well beyond the field of observation before its existence can be observed by another interaction. This condition is distinct from that of the net neutral neutron with three charged spheres and expanded S Eq. orbits of charged microquanta that are sufficiently isolated and spread for charged field potentials to be influential well beyond its much larger spatial dimensions, so that any neutron has a much larger capture cross section area than a neutrino, and is also more likely to be broken up before the action is completed. See Chapt. 10 Recursors for intermediate structures.)

For a more specific aspect of the overall possibilities, one nitrogen nucleus in the atmosphere contains the necessary structural quark and microquantum materials for: 6 pi+ mesons, 10 eta mesons, 1 antielectron, and 1 small mu neutrino; or 7 pi+, 9 eta, 1 electron, 1 antielectron, and 3 smaller mu neutrinos; or 8 pi+, 8 eta, 1 electron, 2 mu neutrinos, and also 6 electron neutrinos; or 8 "strange" K+ mesons (in a non-LU PDG series), 1 muon, 2 mu neutrinos, and 81 electron neutrinos (if they do not combine in some way such as tau neutrinos under the condensation of the impact); not to attempt to list all the other possible simple options in LU, "strange", "charmed", "strange-charmed", "bottom", etc., PDG accredited meson series, nor the numerous more complex cases that have only been observed a very few times or under questionable conditions and are only Listed by the PDG, not accredited in the PDG Summary Tables, much less the observed baryon residues that can occur. (Note, in this MQP paradigm, each neutral eta LU meson combines one heavier "strange" antiquark and a lighter down quark, $d_{A1}\bar{s}_{B1}$, to generate the measured PDG mass energy, 547.5 MeV.)

The patch-work irregularity of the PDG/QM classification of the LU Mesons as that classification has accumulated over the last half-century is most clearly visible if the nominal QM series are each graphed separately against a MeV mass scale with the MQP component structures, especially wherever the QM structures are indefinite, as shown next in continued Table 8-1e over a number of pages. The PDG 2006 (and present) names for particles include their much earlier mass measurements by which they are customarily labeled to keep old data traceable. The names are rarely still accurate in QM practice (Table 8-1d.) The most frequently seen mesons are the two lightest π (pi) particles. Their PDG related ρ (rho), a, and b labeled mesons follow in a continuing table with the QM consistently clear series format distinctions. (Superscripts used here with names indicate in most cases only the particle charge as structured here within the PDG indefinite accredited range per Table 8-1d.)

Table 8-1e PDG 2006 Accredited Light Unflavored Mesons by PDG Series and MQP Structures (per Table 8-1d)

Quarks Quarklets Mesons (MeV Mass Names)	u_{A1}	u_{B1}	d_{A1} d_{A2}	d_{B1} d_{B2}	s_{A1} s_{A4}	s_{B1} s_{B2} s_{B4}	(and their anti-quarks) Pair Structure	Mass Scale 0
π^0 134.977		••					$u_{B1}\ \bar{u}_{B1}$	100
π^+ 139.57	•		•				$u_{A1}\ \bar{d}_{A1}$ 200	200
								300
								400
								500
								600
								700
								800
								900
								1000
								1100
								1200
$\pi(1300)^0$	•	•		•	•		$u_{A1}\ \bar{u}_{B1}$ $s_{A1}\ \bar{d}_{B1}$	1300
$\pi_1(1400)^0$				•	•	••	$s_{A1}\ \bar{d}_{B1}$ $s_{A4}\ \bar{s}_{A4}$	1400
								1500
$\pi_1(1600)^0$					• •	••	$s_{B2}\ \bar{s}_{B1}$ $s_{B2}\ \bar{s}_{A4}$	1600
$\pi_2(1670)^0$				•	••	•	$s_{B2}\ \bar{s}_{A1}$ $d_{B2}\ \bar{s}_{A1}$	1700
$\pi(1800)^{-1}$	•••••				•		$u_{A1}\ \bar{u}_{A1}$ $u_{A1}\ \bar{u}_{A1}$ $s_{A1}\ \bar{u}_{A1}$	1800
								1900
								2000

Quarks Quarklets Mesons (MeV Mass Names) (rho)	u_{A1}	u_{B1}	d_{A1} d_{A2}	d_{B1} d_{B2}	s_{A1} s_{A4}	s_{B1} s_{B2} s_{B4}	(and their anti-quarks) Pair Structure	Mass Scale 0
								100
								200
								300
								400
								500
								600
$\rho(770)^0$						••	$s_{B1}\ \bar{s}_{B1}$	700
								800
								900
								1000
								1100
								1200
								1300
$\rho(1450)^0$					•• ••		$s_{B4}\ \bar{s}_{B4}$ $s_{B2}\ \bar{s}_{B2}$	1400
								1500
$\rho_3(1690)^0$				••	• •		$s_{B2}\ \bar{s}_{A1}$ $s_{B4}\ \bar{s}_{A1}$	1600
$\rho(1700)^0$				•••			$s_{A1}\ \bar{s}_{A1}$ $s_{A1}\ \bar{d}_{B1}$	1700
								1800
								1900
								2000

218

Quarks	u_{A1}	u_{B1}	d_{A1}	d_{B1}		s_{A1}		s_{B1}		(and their anti-quarks)	
Quarklets				d_{A2}	d_{B2}		s_{A4}		s_{B2} s_{B4}		Mass
Mesons										Pair Structure	Scale
(MeV Mass Names)											0
											100
											200
											300
											400
											500
											600
											700
				••							800
$a_0(980)^0$				••						$d_{B1}\bar{d}_{B1}$ $d_{B1}\bar{d}_{B1}$	900
											1000
											1100
$a_1(1260)^{+1}$	•							••	•	$u_{A1}\bar{s}_{A4}$ $s_{B2}\bar{s}_{A4}$	1200
$a_2(1320)^0$								•	•••	$s_{B4}\bar{s}_{B4}$ $s_{B4}\bar{s}_{A4}$	1300
$a_0(1450)^0$				•	•		•	•		$d_{B1}\bar{d}_{A2}$ $s_{B1}\bar{s}_{A1}$	1400
											1500
											1600
											1700
											1800
											1900
$a_4(2040)^0$	•	•••					•	•		$u_{A1}\bar{u}_{B1}$ $u_{B1}\bar{u}_{B1}$ $s_{B1}\bar{s}_{A1}$	2000
											2100
											2200
											2300
											2400

Quarks	u_{A1}	u_{B1}	d_{A1}	d_{B1}		s_{A1}		s_{B1}	(and their anti-quarks)	
Quarklets				d_{A2}	d_{B2}		s_{A4}	s_{B2} s_{B4}		Mass
Mesons									Pair Structure	Scale
(MeV Mass Names)										0
										100
										200
										300
										400
										500
										600
										700
										800
										900
										1000
										1100
$b_1(1235)^{+1}$	•			••		•			$u_{A1}\bar{s}_{A1}$ $d_{A2}\bar{d}_{A2}$	1200
										1300
										1400
										1500

This completes the four pi related series. The b (1235) meson has two other structural forms which may occur (not shown. It is not certain that alternate forms for other particles have not been overlooked.) Next there are the four PDG f/η (eta) related series. (These are followed by the four PDG f/η' related series.)

Quarks	u_{A1}	u_{B1}	d_{A1}	d_{B1}	s_{A1}	s_{B1}	(and their anti-quarks)	
Quarklets			d_{A2}	d_{B2}	s_{A4}	s_{B2} s_{B4}		Mass
Mesons							Pair Structure	Scale
(MeV Mass Names)								0
(eta)								100
								200
								300
								400
η 547.51⁰			•		•		$d_{A1} \bar{s}_{B1}$	500
								600
								700
								800
								900
								1000
								1100
								1200
								1300
η (1405)⁰				•	• •	•	$d_{B1} \bar{s}_{A4}$ $s_{B2} \bar{s}_{A1}$	1400
η (1475)⁰					• •	• •	$s_{A1} \bar{s}_{A4}$ $s_{B4} \bar{s}_{B4}$	1500
								1600
								1700
								1800
								1900
								2000

Quarks	u_{A1}	u_{B1}	d_{A1}	d_{B1}	s_{A1}	s_{B1}	(and their anti-quarks)	
Quarklets			d_{A2}	d_{B2}	s_{A4}	s_{B2} s_{B4}		Mass
Mesons							Pair Structure	Scale
MeV Mass Names)								0
								100
								200
								300
								400
								500
								600
								700
								800
								900
								1000
								1100
								1200
								1300
$f_1(1420)^0$		• •		• •			$u_{B1} \bar{u}_{B1}$ $s_{A1} \bar{s}_{A1}$	1400
$f'_2(1525)^0$					• •	• •	$s_{A4} \bar{s}_{B1}$ $s_{B4} \bar{s}_{B4}$	1500
								1600
$f_0(1710)^0$				•	•	• •	$s_{B1} \bar{d}_{A2}$ $s_{B1} \bar{s}_{A4}$	1700
								1800
								1900
$f_2(2010)^0$		• • • •			•	•	$u_{B1} \bar{u}_{B1}$ $u_{B1} \bar{u}_{B1}$ $s_{B1} \bar{s}_{A1}$	2000
								2100
								2200
$f_2(2340)^0$					• •	• • • •	$s_{B4} \bar{d}_{B2}$ $s_{B4} \bar{s}_{A4}$ $s_{B1} \bar{s}_{A1}$	2400

220

Quarks	u_{A1}	u_{B1}	d_{A1}	d_{B1}	s_{A1}	s_{B1}	(and their anti-quarks)		
Quarklets			d_{A2}	d_{B2}	s_{A4}	s_{B2} s_{B4}			Mass
Mesons							Pair Structure		Scale
(MeV Mass Names)									0
(phi)									100
									200
									300
									400
									500
									600
									700
									800
									900
φ(1020)⁰		••		•		•	$s_{A4}\bar{d}_{A2}$	$s_{B4}\bar{d}_{A2}$	1000
									1100
									1200
									1300
									1400
									1500
φ(1680)⁰		•		••		•	$s_{A1}\bar{d}_{B1}$	$s_{B2}\bar{s}_{A1}$	1600
									1700
φ₃(1850)⁰				••		••	$s_{B1}\bar{s}_{A1}$	$s_{B1}\bar{s}_{A1}$	1800
									1900
									2000
									2100
									2200
									2300

Quarks	u_{A1}	u_{B1}	d_{A1}	d_{B1}	s_{A1}	s_{B1}	(and their anti-quarks)		
Quarklets			d_{A2}	d_{B2}	s_{A4}	s_{B2} s_{B4}			Mass
Mesons							Pair Structure		Scale
(MeV Mass Names)									0
									100
									200
									300
									400
									500
									600
									700
									800
									900
									1000
h₁(1170)⁺¹ (if +1)	•			•		••	$u_{B1}\bar{s}_{A4}$	$s_{B4}\bar{s}_{B4}$	1100
h₁(1170)⁰ (if 0)				••		••	$d_{B1}\bar{d}_{B1}$	$s_{B4}\bar{s}_{B4}$	1200
(if considered in this group)									1300
									1400
									1500

(Note that the charge on meson h₁ (1170) is definitely uncertain in limited empirical data. Like the b₁(1235) this series has a single PDG accredited member, but it may not belong in this f/η group of series, rather only in the f/η' group of series below.)

This completes the PDG f/η (eta) related series. The four PDG f/η' related series are next.

Quarks	u_{A1}	u_{B1}	d_{A1}	d_{B1}	s_{A1}	s_{B1}	(and their anti-quarks)		
Quarklets			d_{A2}	d_{B2}	s_{A4}	s_{B2} s_{B4}			Mass
Mesons							Pair Structure		Scale
(MeV Mass Names)									0
									100
									200
									300
									400
									500
									600
									700
									800
$\eta'(958)^0$		••		••			$u_{B1}\bar{d}_{B1}$	$d_{B1}\bar{u}_{B1}$	900
									1000
									1100
$\eta(1295)^0$		•		•		••	$d_{A1}\bar{d}_{B1}$	$s_{B4}\bar{s}_{B2}$	1200
									1300
									1400
									1500
$\eta_2(1645)^0$		•		•	•	•	$s_{B2}\bar{s}_{A1}$	$d_{A1}\bar{d}_{B1}$	1600
									1700
									1800
									1900
									2000

Quarks	u_{A1}	u_{B1}	d_{A1}	d_{B1}	s_{A1}	s_{B1}	(and their anti-quarks)		
Quarklets			d_{A2}	d_{B2}	s_{A4}	s_{B2} s_{B4}			Mass
Mesons							Pair Structure		Scale
(MeV Mass Names)									0
									100
									200
									300
									400
									500
$f_0(600)^0$ (uncertain PDG mass, if 600)			•			•	$s_{A1}\bar{s}_{B2}$		600
(changed to $f_0(500)$)									700
									800
$f_0(980)^0$			•	•••			$d_{A1}\bar{d}_{B1}$	$d_{B1}\bar{d}_{B1}$	900
									1000
									1100
$f_2(1270)^0$			•	•		••	$s_{B2}\bar{d}_{B2}$	$s_{B4}\bar{d}_{A2}$	1200
$f_1(1285)^0$		••	•	•			$u_{B1}\bar{u}_{B1}$	$s_{A1}\bar{d}_{A2}$	
$f_0(1370)^0$ (uncertain)		••			••		$d_{A1}\bar{d}_{A1}$	$s_{B1}\bar{s}_{A4}$	1300
(if 1370)									1400
$f_0(1500)^0$					•••	•	$s_{B1}\bar{s}_{A4}$	$s_{B4}\bar{s}_{A4}$	1500
									1600
									1700
									1800
$f_2(1950)^0$	••••				•	•	$u_{A1}\bar{u}_{A1}$ $u_{A1}\bar{u}_{A1}$	$s_{B1}\bar{s}_{A1}$	
$f_4(2050)^0$		••		••	•	•	$d_{A1}\bar{d}_{B2}$ $d_{A1}\bar{d}_{B2}$	$s_{B1}\bar{s}_{A1}$	
									2100
									2200
$f_2(2300)^0$				••	•	•• •	$d_{B1}\bar{d}_{B1}$ $s_{B4}\bar{s}_{A4}$	$s_{B1}\bar{s}_{A1}$	
									2400

222

Quarks	u_{A1}	u_{B1}	d_{A1}		d_{B1}		s_{A1}		s_{B1}			(and their anti-quarks)	
Quarklets				d_{A2}		d_{B2}		s_{A4}		s_{B2}	s_{B4}		Mass
Mesons												Pair Structure	Scale
(MeV Mass Names)													0
													100
													200
													300
													400
													500
													600
$\omega(782)^0$	••••											$u_{A1}\bar{u}_{A1}$ $u_{A1}\bar{u}_{A1}$	700
													800
													900
													1000
													1100
													1200
													1300
$\omega(1420)^0$				•		•				•	•	$s_{B1}\bar{d}_{B2}$ $s_{B4}\bar{d}_{A2}$	1400
													1500
$\omega(1650)^0$					•	•				•	•	$s_{B1}\bar{d}_{B2}$ $s_{B4}\bar{s}_{A1}$	1600
$\omega_3(1670)^0$			•		•			•		•		$s_{A1}\bar{d}_{B1}$ $s_{B1}\bar{d}_{A1}$	1700
													1800
													1900
													2000
													2100
													2200
													2300
													2400

Quarks	u_{A1}	u_{B1}	d_{A1}		d_{B1}		s_{A1}		s_{B1}			(and their anti-quarks)	
Quarklets				d_{A2}		d_{B2}		s_{A4}		s_{B2}	s_{B4}		Mass
Mesons												Pair Structure	Scale
(MeV Mass Names)													0
													100
													200
													300
													400
													500
													600
													700
													800
(Nominally the PDG preferred Group for this series.)													900
													1000
$h_1(1170)^{+1}$ (if +1)		•			•						••	$u_{B1}\bar{s}_{A4}$ $s_{B4}\bar{s}_{B4}$	1100
$h_1(1170)^0$ (if 0)					••						••	$d_{B1}\bar{d}_{B1}$ $s_{B4}\bar{s}_{B4}$	1200
													1300
													1400
													1500

This completes the four PDG f/η' related series.

PARADIGM SUMMARY OF THE NUMBER OF 45 PDG 2006 ACCREDITED LU MESONS CONTAINING:

Quark & Anti-quark Types	u_{A1}	u_{B1}	d_{A1}		d_{B1}		s_{A1}		s_{B1}		
Quarklets				d_{A2}		d_{B2}		s_{A4}		s_{B2}	s_{B4}
# of Mesons	7	10	9	7	11	7	22	11	20	12	14
# Quark/lets	19	17	10	10	21	7	32	18	21	14	23
				84	(13 ave.)				108	(21 ave.)	
Has Some Up and/or Down Quarks: 35							**Only Up &/or Down Quarks: 7**				

Quarks			Quarklets			s_A and/or s_B			Quarks & Qklets		
All Qrks	Mostly Qrks	50% Qrks	Mostly Qlets	All Qlets	Some Qlets	Some S	Mostly S	All S	Two Qs	Four Qs	Six Qs
17	9	8	7	4	28	38	20	10	5	33	7
									Lightest	--	Heaviest

Table 8-1e (Cont.) PDG 2006 Accredited Light Unflavored Mesons by PDG Series and MQP Structures

From this table it is clear by the number of kinds of PDG LU Mesons that PDG accredited Baryons containing mostly the lightest up and down quarks, though they are much more numerous (98 to 28) by kinds of particles in the Baryon Series tables, are much more stable and much less likely to be involved in decays or impact breakages that yield Mesons than are the few Baryons that contain mostly the four heavier types of quarks. (Since there are always 3 quarks in each baryon, each one must necessarily contain either mostly u and d or mostly s, c, b, and t quarks.) There is systematic bias.

Furthemore, it is also indicated, by the prevalence in Table 8-1e of component quark/quarklets between the d_A & s_B types of more than minimal mass and the same charge in normal matter (or oppositely in anti-matter), that the close groups of clusters of meson masses, such as those between particles $b_1(1235)$ and $a_1(1320)$ in Table 8-1d, will eventually be proven to be excited-orbit isomers of one or two heavy isotons (baryon components) of only one or two meson series (such as one having s_{A1}) rather than in representing the large number of separate meson series now indicated by the PDG in the LUMs with only three quark types to provide true separate series below the Charm class. At this time, the presence of quarklets combined with the A and B mass levels of each quark is too complicated for a clear proof of this indicated condition, and it must therefore remain as a suggestion or a conjecture for further investigation as more LUM data accumulates in the future. However, this indication is a strong one. - - It can be further substantiated as an insight by graphing the break-outs of Table 8-1e on the mass sequences of Table 8-1d, and noting that s_{A1} (and heavy quarklets) is prominent in this cited cluster, but s_{B1} (or their very heavy combination) takes the dominant mass-building lead for clusters in the heavier half of Table 8-1d. That is an interesting follow-up for thoughtful readers with large layout tables. (Paper forms are better for actually seeing and grasping information.)- - An even more interesting follow-up is continuing the search for alternate pair structures for measured PDG mesons in Table 8-1d or even continuing on into a new similar table for the strange, charm, or bottom mesons separately listed by the PDG tables at pdg.lbl.gov. First, it will be necessary to use the power law to calculate masses for the possible strange, etc., quarklets formed by losing one or several conserved charged or neutral pairs of quantal vortices in particle collisions, either from cosmic rays or reactors or colliders. Then the trials of one, two, or three pairs of quarks/quarklets can begin by exponential law calculations for specific PDG meson masses as for Table 8-1d above. This Is always done in steps of 2 components, starting with the lightest pair of quarks/quarklets, and using the table layout as a general guide in the new range. Because of the greater number of quarklets that can be added to the increased number of quarks, it is certain that there will be a greater number of trial pair options for each specific PDG meson. Any clues from the PDG literature will be useful. For each candidate pair of quarks, its mass Σ is added up, and its initial exponent y is computed by the inverted logarithmic exponential law, and then the full significant figure mass is run up with the original law to obtain the rounded mass for the record and next step if needed. For more than one pair of quarks, repeat the sum and exponent process with the two next lightest pairs.   Then sum that mass with any third heaviest pair of quarks and find the sum's exponent. In each of these steps after the first calculation of the quark (and quarklet) masses the number N of items being summed in the equation is two, usually two previously combined pairs or doublet groups of quarks, not the individual quark themselves over and over. (Note that this takes account of the fact that in this MQP paradigm's structure of mesons with two or three pairs of quarks the quarks are not all interacting closely in direct contact with each other, as occurs with the three quarks in a baryon box corner, but each pair is separated somewhat with linkages only from one member of the pair to one member of another pair. This partial separateness reduces the degree of overall mass energy accumulation by interaction of each quark's quanta with all others in the meson. Unlike the baryon box corners in a nucleus, four quark spheres cannot find a way of synchronizing and summing all their quantal orbits in a pyramid of direct contact between all four quarks for closer linkage and larger interaction mass of a two-pair or four-quark meson.) Charges must be conserved in every case and be consistent with the rule that full meson (or other complete) particles can have only plus or minus one electron level of net charge or be neutral, though there are also a few PDG empirical baryons with no negative quarks and a +2 net charge (after the Ω prototype plan with all three quarks of the same charge.) In conserving charge note that half the quarks (and quarklets) in every meson are antiquarks or antiquarklets which have the opposite charge from the same type of quark. Quarklets retain the charge of the basic quark from which they are derived by loss of one or two or more neutrinos in destructive collision (or "decay") of an original baryon or closely bound pair of baryons (or also of very heavy initial mesons from collisions and decays.)   An ordinary shirt-pocket "Scientific" calculator, preferably with a "Solar" (or electric lamp light) power cell (not batteries), will do all the computing needed. Overall, where there are two or three pairs of quarks or quarklets in a meson, the mass law must be applied in steps in two-wall corners which together create the larger particle out of as many as three pairs with loose quark ends. The stepped calculations for this are generally consistent with empirical PDG masses as long as the first connection step is between the two lowest mass quarks and then with similar two component steps in ascending order to the largest quarks/quarklets involved.

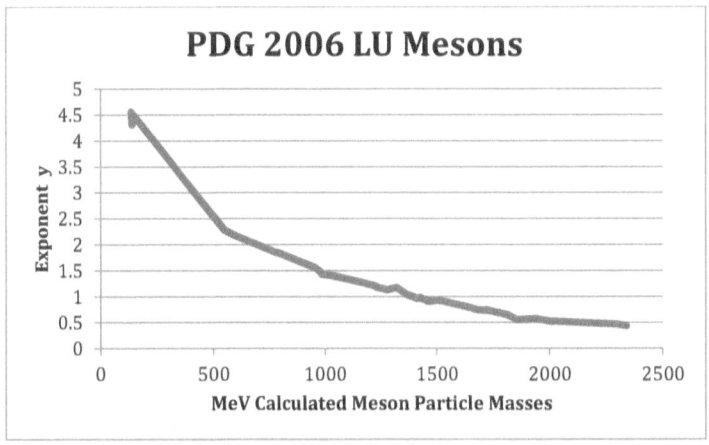

Figure 8-1c Exponent y for All 2008 PDG LU Mesons vs MeV Particle Masses

This curve of Exponent y for the entire PDG listing in Table 8-1c of the 2006-8 LU Mesons as analyzed by the MQP Paradigm shows that the overall curve is generally smooth except for discontinuities between series and for irregularities in the vicinity of cases in which structures were found only for charged mesons rather than for the neutral mesons found to the time of writing for the great majority of the LU mesons. Separate curves for the two main series with enough PDG members to check smoothness tend to show that the all-neutral f_0 meson series with two or four quarks are smooth and that the π meson series with charged mesons at the ends of the curve are consistent with the possibility that mesons with differences in charge should not be members of the same series because of the requirements of charge on structures. However, uncertainties in PDG masses are large enough to raise questions until more certain data are available.

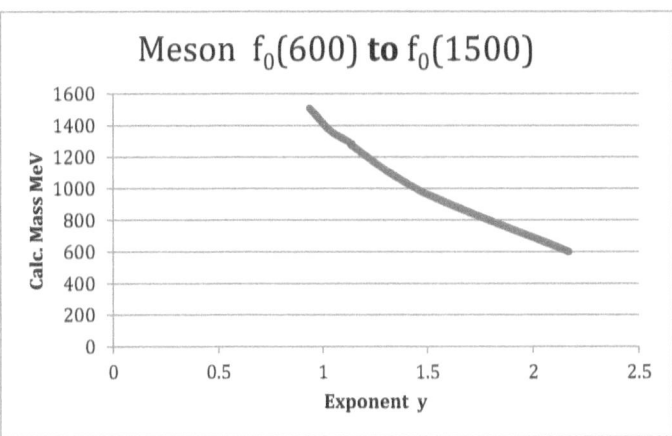

Figure 8-1d The Series of f_0 Mesons (with the coordinates reversed.) The smoothness of this curve for exponent y and particle mass is quite remarkable as well as unanticipated for two to four quark mesons, but six quark mesons are not as firm in that indication for this curve and do contribute to the lack of smoothness in the next curve.

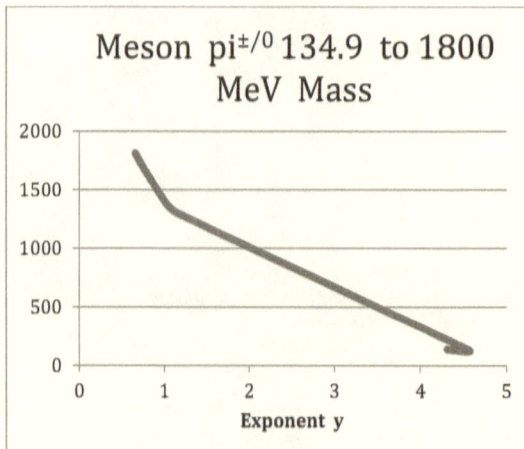

Figure 8-1e An improper π Meson Series Calculated Mass vs Exponent y. This shows very clearly the two initial and final irregularities for the definitely incorrect inclusion of the observed and structured charged π mesons in the same series with the five neutral π mesons. Again the definiteness of these curve breaks from the straight line of the other five members of this series is very distinct. The break at the smaller mass could be observed in the large curve of Figure 8-1c, but the large display at the high mass shows a possible error in a single series for charged and neutral mesons.

Since only two LU meson accreditations by the PDG have been recorded in eight years, it may be expected that few more will be designated in the near future, and that meson analysis to confirm or counter reclassification of mesons with separate series for charged and neutral particles in the LU class will be done predominantly in the heavier types of mesons. The larger quarks will permit generating larger numbers of quarklets, which will make such an extension much more laborious.

Though the meson series are not as definitely organizable at present as the baryon series, there are still some interesting group relations. If a heavy cosmic ray particle comes into the upper atmosphere, its most likely impact will be on a Nitrogen nucleus at atomic weight 14. Without including the content of the ray, the 7 protons and 7 neutrons in one destroyed nitrogen nucleus have 21 up quarks and 21 down quarks made up of a total of 42 each ++ quantal pairs, 21 each -- quantal pairs, and 126 each +- quantal pairs that are releasable into a Recursor re-organization process (to be discussed in more detail in a later chapter.) If these component pairs should re-arrange themselves under their own forces primarily into groups large enough and diverse enough to include the components for any of the uds quarks in three groups and for any of the anti-uds quarks in three adjacent groups (3 ++, 4 --, and 5 +- per group or their reverse charges), then any three of the two-pair LU mesons might result, with the residues being re-distributed in any suitable force arrangement. If all the groups are close enough, this situation could resolve into any one of the 6 pair LU mesons. In addition to the randomly re-assorted residues from these 6 groups, there could also be sufficient pair structural material in a larger single group to build 7 each π^+ mesons without further residues. -- Alternatively, the pair groups might arrange themselves into suitable structure for only 5 each π^0 mesons, with a great deal of residue that might make up any LU meson that can be constructed from a single d_A type of quark and up to 11 each u_A or 22 each anti-d_A types of quarks, or scatter into other parts of the cosmic ray crash debris with up to 126 Extreme class +- electron neutrinos possibly.

(The shortage above of negative quantal pairs compared to positive pairs brings this about.)

Is There an Ultimate Quirk of The Quarks?

A suitable mass for the form of two light up-quark/antiquarks, which would have distinctly lighter mass than the listed pi mesons, does not appear in the accredited PDG table of empiricly observed particles. To bring this lighter form to the accredited pi mass would require an exponential law y exponent of 5.14, significantly higher than the otherwise general asymptotic limit of 5.0. This light form does not fit (under the exponential range of the mass/charge law) to the empiricly observed mesons except as a part of the four and six quark mesons where its very low two-quark mass can exist in combinations with larger mass sums in which its y curve is always well below that critical limit. This definitely reduces the Precursor and Recursor options at self-assortment only for u_A quarks in the otherwise evenly symmetric self-activation of quarks, in an MQP Paradigm with no other such special peculiarities of asymmetry that have been found. The only apparent mechanism for enforcing this special limitation of the lighter up quark is that it may not have quite sufficient mass momentum to remain in an expanded linking orbit at the appropriate velocity between times of sufficiently powerful attraction into that special quark link position in the necessarily phased application of that force by the orbiting charged quanta of other potential quarks. If the original destroyed nucleus were antimatter, the equivalent antimatter bias would appear. Detailed estimates of this effect cannot presently be precise enough to prove this possibility for lack of refined scaled measurements on asymmetricly located vortices, as noted earlier. (Making such measurements will require a well funded follow-on experimental program.)

Thus, in very rough estimate, after a cosmic ray impact on usual or anti-baryons, when mesons are forming in Recursors, the average lightest u_A quark cannot appear positively under this empiric constraint in either the quark or the oppositely charged antiquark form simultaneously in the lightest range of meson components. Consequently, it can only be presumed that two usual up quarks are left to join up primarily with a single random down and make a proton hydrogen nucleus to keep the present routine cosmos going, while the excess anti-up pairs can only delay further and possibly dissipate because their matching anti-downs are still able to play at mesons a few micro-microseconds longer. If a temporary excess of either matter or antimatter appeared at a very active cosmic ray area, that excess could be rapidly amplified and become irreversibly dominant in abundance, leading to a wide-spread dominance of the initial accidental advantage. Any such hidden imperfection of symmetric particle activity could well be the ultimate structural bias toward the present matter/anti-matter system.

Special Reference for Chapter 8 on Mesons

LATTIMER, J. M, AND PRAKASH, M., (2004) , THE PHYSICS OF Neutron Stars, Science (2004) 304. #5670, pp 536-542, 23 Apr 2004, and references [plus citations of that paper in later updates on pulsars and magnetars up to the time of writing.]

228

Chapter 9

The Baryon "Decay" Necessities for Further Neutrinos - - Dark Matter

SUMMARY: Because they are so difficult to detect and appear so necessary in theory, neutrinos have received a large amount of empirical attention in attempts to determine their actual physical existence and define their mass ranges. It is now generally acknowledged that neutrinos must empirically exist in some form in the outputs of subatomic particle reactions such as baryon particle decays, but with a great deal of uncertainty about, not just the specific masses of designated types of neutrinos, but even the limits of mass ranges for their various types. This situation is evidenced in the large biennial variations in empirical neutrino listings of the recent international Particle Data Group (PDG) Reports and their evaluated references (Olive, 2014). In this MQP review of the situation it is clearly indicated that this scattered and piece-meal, but long-known, body of empirical data requires the existence and reaction inputs of abundant neutrinos of a wide variety of types, including several new apparent types with estimated mass ranges previously considered to be beyond the limits of neutrino masses in any locally observed reactions. These data effects have been disregarded by others through a conventional avoidance of accounting for conservation of each specific charge of the particles involved in a reaction on the assumption that a conserved neutral balance of over-simplified net charge in each reaction is sufficient. A new, re-classified order of neutrinos appears when such shortcuts are no longer assumed. That is made possible by the particle component accounting which arises necessarily wthin the MQP Paradigm of unified forces generated by unavoidable causal consequences of the simple Power and Exponential Laws of particle mass and charge structures.

Among lepton particles grouped with quarks (LQ) in the PDG "elementary" particles, the neutrinos take part in a significant number of baryon decays in the PDG Summary Tables of particles (Olive, 2014.) These listings show neutrinos as largely undetected outputs in such decay reactions to account for some of the empirical disappearances of mass energy in various quantities without the release of equivalent radiation and kinetic energy of observed particles in quantum mechanics experiments. (The types of these neutrinos were previously named for other participants, such as electron, muon, or tau neutrinos, in general classes of particle reactions. These titles are still conveniently useful in mass estimates as names for the distinctive portions of their prior PDG limits of mass ranges, but without initial restriction to the original three types of neutrinos whose masses may occur in those ranges. There are distinct tendencies, but perhaps not the firmly restrictive separations in definite types once used by the PDG.)

In prior PDG biennial reports there have also been wide neutrino uncertainties recognized by stating only upper limits of masses for various specific neutrino types. Recently the PDG has recognized inconsistencies with much smaller astrophysical estimates on neutrino masses from within exploding stars with very long transit times before earth observations. These long transit times permit high probabilities of changes of observed/estimated neutrino mass (oscillations in PDG terms) between the three mass classifications. This led to PDG removal of the charged lepton names for neutrinos, since the names were for original generations of neutrinos that do vary in apparent mass with time of flight. Accounting for neutrino mass from moment to moment is enforced in MQPP by conserving coaxial pairs of charged quanta under the widely confirmed principle of conserving electric charges. That conserves every charge through impact-changeable QM structures and requires actual reactions with balanced equations at every "oscillation of masses." (That is, a picking-up or dropping of sub-neutrinos, in effect, at every definite impact of the initial neutrino with other particles, usually other neutrinos, though such impacts do not permit actual before-and-after measurements. In MQPP every determination of any change in mass requires a full re-determination of true mass. This does not permit loose estimates of apparent mass with non-conserved charges to be a standard desired procedure.)

The PDG type of reporting of neutrino occurrences has been going on for about a century in the QM classic "beta" decay of whole atomic nuclei when electrons from various nuclei were called negative beta particles as distinct from the much bigger and more noticeable alpha decay particles (positive helium nuclei.) With present improved PDG accuracy, beta decay into release of an electron comes from disruption of one neutron in a nucleus. So it is an elderly prototypical example for present-day particle Quantum Mechanics, when the PDG Summary Tables show early in their baryon listings an abbreviated reaction equation of a neutron decaying to a proton, an electron, and an electron anti-neutrino with released energy. (This can be easily checked at pdg.lbl.gov on line for 2014 and recent years, with biennial updates, courtesy of Lawrence Berkeley Lab, funded by the US Government as an important function of the Energy Dept.)

However, the accredited listing equation takes a QM shortcut conserving only net charge (assuming that adjustments are created new from the QM pseudo-vacuum.) Even conserved PDG input and output charges of the quarks and electron in this decay cannot be equal. Only a collision by a large neutrino could put in the decay's missing charges for these PDG particles without putting out other observable debris.

In Classic Beta Decays, the input of a real PDG neutron with udd quarks has separate charges $+2/3$, $-1/3$, & $-1/3$. The output proton with uud quarks has charges $+2/3$, $+2/3$, & $-1/3$, while the electron adds $-3/3$ (and its antineutrino must carry at least a neutral $+-$ or $-+$ pair. presumably of minimal complexity and mass due to the relatively small difference between neutron and proton masses.) In these PDG accredited QM charges, there is an input deficit of $+2/3$ and $-2/3$ in real charges that could be supplied by an undetectable input MQP neutrino carrying the input charges $2++$ $2--$ in 4 pairs of quantal bits. The mass of that neutrino is in the tau neutrino range that was formally recognized by the PDG through at least 2006. Such a neutrino mass could have become involved by a collision that triggered the supposed decay. (Balancing this PDG accredited decay by MQP paradigm quantal bits comes to the same result, except that a collision could also provide the energy for 60° phase shifts of quark S Eq. orbits, and that there was also an original output deficit of another unseen e neutrino through 2006.)

This pattern of mismatches of input and output charges repeats in all but one of the 15 principal, 33% to 100% decay modes of major baryons with adequate PDG accredited data for checking the charges. This constant necessity for added neutral input shows that "decays" of unstable baryons do not necessarily occur spontaneously but because of impacts from real neutrinos or neutral hadrons. In 12 of the 14 cases there was no suitable neutral PDG baryon or meson for the impact result. In 9 of the 12 the necessary neutrino is heavier than the prior PDG 18 MeV upper limit for tau neutrinos, the nominal heaviest. (Mass compared to hadrons is irrelevant.) This is a distinct weight of PDG evidence of the existence of such neutrinos and of the necessity of a larger galactic reservoir than previously accredited of abundances of neutrinos with a wide variation of simple microquantal structures from the MQP paradigm. This constitutes a definite and significant, but unrecognized, contributor to dark matter for which the evidence continues to accumulate! There are many other decays (mesons, etc.) that could occur in the same way. (A large definite effort to estimate the dark matter impact of the neutrino reservoir needed for observed hadron lifetimes is required.)

It is also particularly interesting, that in every PDG decay case checked there is both some kind of neutral deficiency in the quanta pairs for the input empirical baryon to yield the quanta for the empirical output and also a net neutral deficiency of pairs of other conserved charge types in the PDG output to account for the structure that is listed in the PDG input. (The PDG accredited conservation of charge is only in net charge.) The neutral impact particle requirements found are compatible with both kinds of deficiency. The input deficiency requires each of these baryon decays to be triggered usually by an single impact from a neutrino defined herein or by a second neutral particle that merges into the decay. There must be floods of these impacting neutrinos. Mean lives of unstable baryons (all but protons and neutrons) are very short and very difficult to measure. (Only single digit per cent of mean lives are well known and PDG accredited.)

The required neutrinos are widely varied. In any case the input particle may also be required to provide any numbered SEq. orbit phase sites or the momentum for disruptive reorganization in a recursor of any unsuitable SEq. sites. Some of the required input neutrinos (as well as some of the added output neutrinos) could be much larger than the maximum accredited upper limit for a tau neutrino in the PDG summary tables of the 2004 and prior biennial reports (not included in 2006 and later.) There may indeed be a much heavier class of neutrinos than the nu tau range, as the PDG Notes and Comments continue to recognize by searches for very large special neutrinos.

When a baryon decays and mass is lost among the decay products, it is valid to expect that the energy which was creating the continuing emanation of mass effect for the mass difference lost is then available as kinetic energy of the products. But the microquantal pairs themselves, which carry the conserved charge, must also be conserved in any decay or ordinary collision reconstruction action. (The separation, rearrangement, creation, or any terminal destruction of individual pairs of quanta would be reserved to the interiors and jets of "black holes" or at least of "neutron" stars, both of which are not under consideration at this point.)

It is feasible then to construct a two dimensional Table of Required Additional Neutrino Quantal Pairs in Main PDG Accredited Decay Channels of Principal Baryons to

Other Baryons or Mesons and Mesons to Mesons or other fragments (using lightest mass pathways and leaving aside separate classical calculations of kinetic and radiation energy converted from interaction mass energy losses.) This gives the smallest usable impact neutrinos. They can be larger, with larger otputs. As noted, only adequately documented PDG accredited cases of main 30-100% decay channels or modes have been checked as yet, not the thousands of less outstanding accredited observations.

<u>Standard PDG Procedures in PDG Neutron Decay - - Plus MQP Quantal Pair Analysis</u>

$$n \quad \rightarrow \quad p \quad\quad e^- \quad \bar{\nu}_e + energy$$

```
Quarks         uA  dA  dA       uA   uA   dA
Quanta Pairs   2++ 1-- 1--      2++  2++  1--    3--    1+-
               2+- 4+- 4+-      2+-  2+-  4+-
```

Collect Terms 2++ 2-- 10+- → 4++ 4-- 9+- (Charge <u>NOT conserved</u>, only Net charge. Equation NOT balanced.)

DEFICITS In: small ν_{tau-} (2++ 2--)   **Out**: ν_e (1+-) Its own anti-particle
(That do appear in PDG output.)   by anti-spin along velocity direction.
(Inputs not present in PDG output.)

Restatement of PDG "decay" equation to balance it:

A <u>small</u> tau neutrino (<<1++,1--,4+-=18.170 MeV or <<18.2MeV, PDG 2004 limit) is necessary to enter the n decay (initiate decay and provide either the required initial S Eq. orbit sites or the energy to disruptively reorganize them) and balance the PDG output. Aside from release of reduced interaction mass energy in lower total mass decay products, then:

```
n                  Vtau-  →     p              e-    v̄e    νe + energy
uA  dA  dA                      uA   uA   dA
2++ 1-- 1--        2++          2++  2++  1--   3--   1+-   1+-
2+- 4+- 4+-        2--          2+-  2+-  4+-
```

Collect terms 4++ 4-- 10+- → 4++ 4-- 10+-

This determines a definitely balanced equation, not only in conserved charged quanta, but also in neutral pairs too. (The released energy is then separately calculated as usual.)

Table 9-1a Correction of PDG Neutron Decay

Table 9-1b Continuing Condensed Baryon 33%-100% "Decay" Findings:

PDG Particle Decay|Quark etc. Structures | Input Deficit | Output Deficit

$n \to p\ e^-\ \bar{v}_e$ $u_A d_A d_A \to u_A u_A d_A\ e^-\ \bar{v}_e$ small $v_{tau-}(2++,2--)$ add $v_e(1+-)$

$N(1440) \to N\ \pi$ $u_B d_A d_A \to u_A d_A d_A\ u_B \bar{u}_B$ $v_{tau++}(3++,3--,2+-)$ none
(if $N\ \pi = n\ \pi^0$.)

$N(1700) \to N\ \pi\pi$ $u_B d_B d_A \to u_A d_A d_A\ u_B \bar{u}_B u_B \bar{u}_B$ $v_{tau+++++}(6++,6--,4+-)$ none
(if $N\ \pi\ \pi = n\ \pi^0 \pi^0$.)   [or $\Lambda(1520)^0$ if in its short life, no change]

$\Delta(1232) \to N\ \pi$ $u_A u_B d_A \to u_A u_A d_A\ u_B \bar{u}_B$ $v_{tau++}(3++,3--,2+-)$ none
(if $\Delta(1232)^+$ & PDG $N\ \pi = p\ \pi^0$.)

$\Delta(1905) \to N\ \pi\pi$ $u_A u_A d_B \to u_A u_A d_A\ u_B \bar{u}_B u_B \bar{u}_B$ $v_{tau+++++}(7++,7--,2+-)$ none
[if $\Delta(1905)^+$ & PDG $N\ \pi\pi = p\ \pi^0\ \pi^0$.]   [or $\eta(958)$ if in its short life, then
                                (Cont.)   at limit $(7++,7--,4+-)$ & $2v_e(1+-)$]

$\Lambda^0 1116 \to p\ \pi^-$ (64%) $u_A d_A s_A \to u_A u_A d_A\ d_A \bar{u}_A$ $v_{tau+}(8+-)$ $v_{mu-}(1++,1--)$
           $\to n\ \pi^0$ (36%) $u_A d_A s_A \to u_A d_A d_A\ u_B \bar{u}_B$ $v_{tau}(1++,1--,4+-)$ none
                                                                    (near 2004 PDG limit)

$\Sigma^0 1193 \to \Lambda\ \gamma$(photon) $u_A d_A s_A \to u_A d_A s_A\ \gamma$ none   none
(If $\Lambda = \Lambda^0 1116$, an isomer of $\Sigma^0 1193$) There is only a mass interaction energy
drop for γ emission by quantal shift to less energetic orbit or spin sites in quarks.)

$\Lambda(1810) \to N$ antiK (20-50%) $u_A d_A s_B \to u_A d_A d_A\ s_B \bar{d}_A$ $v_{tau++}(1++,1--,8+-)$ none
[if $\Lambda(1810)^0 \to n$ antiK^0]

  $\to \Sigma\ \pi$ (10-40%) $u_A d_A s_B \to u_A d_A s_A\ u_B \bar{u}_B$ $v_{tau+++++}(6++,6--,4+-)$ $9v_e(9+-)$
  (if $\Lambda(1810)^0 \to \Sigma^0 1193\ \pi^0$)   [or $\Lambda(1520)^0$ if in its short life] [or $v_{tau++(9+-)}$]

$\Sigma^- 1197 \to n\ \pi^-$ $d_A d_A s_A \to u_A d_A d_A\ d_A \bar{u}_A$ $v_{tau+}(8+-)$ $v_{mu-}(1++,1--)$

$\Xi^0 1315 \to \Lambda\ \pi^0$ $u_A s_A s_A \to u_A d_A s_A\ u_B \bar{u}_B$ $v_{tau}(1++,1--,4+-)$ $v_{mu}(3+-)v_e(1+-)$
(If $\Lambda = \Lambda^0 1116$)

$\Omega^- 1672 \to \Lambda\ K^-$ $s_A s_A s_A \to u_A d_A s_A\ s_B \bar{u}_B$ $v_{tau+++}(11+-)$ $v_{tau-}(2++,2--)$
(If $\Lambda = \Lambda^0 1116$)

(Table Continued)

(Cont.) Table 9-1a Condensed 33%-100% Baryon Decay Findings:
PDG Particle Decay|Quark etc. Structures | Input Deficit | Output Deficit

$\Lambda_c(2593)^+ \rightarrow \Lambda_c^+ \pi^+ \pi^-$ $u_B c_A d_B \rightarrow u_A c_A d_A$ $u_A \bar{d}_A$ $d_A \bar{u}_A$ $\nu_{supertau}(1++,1--,16+-)$ none
(If $\Lambda_c^+ = \Lambda_c^+ 2286$) (May break up in stages.) (Est. at stability limit ±.)

$\Sigma_c(2520)^+ \rightarrow \Lambda_c^+ \pi$ $u_A c_A d_B \rightarrow u_A c_A d_A$ $u_B \bar{u}_B$ $\nu_{tau++}(3++,3--,2+-)$ none
(If $\Lambda_c^+ \pi = \Lambda_c^+ 2286\ \pi^0$)

$\Xi_c^+\ 2468 \rightarrow \Xi^0 \pi^+ \pi^0$ $u_A s_A c_A \rightarrow u_A s_A s_A$ $u_A \bar{d}_A$ $u_B \bar{u}_B$ $\nu_{tau++++}(4++,4--,6+-)$ $\nu_e(1+-)$
(If $\Xi^0 = \Xi^0\ 1315$)

$\Lambda_b^0 5624 \rightarrow \Lambda_c^+ l^- \bar{\nu}_e$ $u_A d_A b_A \rightarrow u_A c_A d_A$ $3-- 1+-$ $\nu_{tau}\ -(2++,2--)$ $\nu_{mu}(3+-)$
(If $\Lambda_c^+ l^- \bar{\nu}_e = \Lambda_c^+ 2286\ e^- \bar{\nu}_e$)

Wherein neutrinos are usually not observable except as input or output deficits in observed events. Where later PDG Summary Data changes conditional selections. or a selection is changed, the input deficit may change, but is still needed.

Table 9-1b Condensed 33%-100% Baryon "Decay" Findings (+Energy Out.)

Minimum Resultant Types in Local Neutrino Population

(Necessitates at least Quasi-Stable Neutrinos impacting Baryons
in Definable ≥ 33% PDG "Decay" Modes)

Minimum Limit Structures		Balance
small vtau-(2++,2--)	2 cases	Bare Balance
vtau((1++,1--,4+-) **PDG Limit**	18.17 MeV	Well Balanced
vtau+(8+-) stability back up	2 cases	Well Balanced
vtau++(3++,3--,2+-)	3 cases	Well Balanced
vtau+++(11+-)		Well Balanced
vtau++++(4++,4--,6+-)		Well Balanced
vtau+++++(6++,6--,4+-)	2 cases	Bare Balance
vtau+++++(7++,7--,2+-)	10.78 GeV	Well Balanced
$\nu_{supertau}(1++,1--,16+-)$	(approaching stability limit?)	Balanced

At least these (or larger?) Impacting Neutrinos
Can be Very Active on Unstable Quark Targets.

Table 9-1c Minimum Resultant Types Recorded (by PDG) in Local Neutrinos

Clearly, with a real empirical basis for being without doubt, particle physicists have been recording a high and consistent level of large specific neutrino impacts on unstable baryon quarks for many decades and have been mis-identifying the resultant breakages as simple baryon decay. The very brief mean life times of the unstable baryons makes it difficult to record full data on other effects of the break-up encounters sufficiently well to enable a full MQP reconstruction even for all the relatively few 33% to 100% PDG accredited "decay" modes. Enough of these were conditionally limited within the accredited break-up descriptions to give a reliable level of result. If the PDG accredited experiments are re-examined more broadly, it is predicted that the breadth and reliability of these reconstructions will grow steadily and will make a major contribution to the dark matter required by galactic gravity observations, at least in the local Milky Way galaxy.

This opening into the Neutrino structural requirements (especially when coupled with the astrophysical neutrino estimates from distant supernovae, etc., as other structural requirements) makes it necessary to go more deeply into the broader general neutrino structural consequences of the MQP quanta themselves. They are taken up in the next chapter.

Chapter 10

Expanded Neutrino Mass Ranges, from Pre-Quarks, Pre-Baryons,& "Higgs" Actions to Micro-Neutrino Photons, with Bare Black Holes in a Stable Universe

SUMMARY: Like the electron, neutrinos are not involved with the complications of quarks and their interferences between ordinary orbits, but while usual neutrinos, with masses in the range of electrons to medium quarks, are single spheres of orbits, other neutrinos may have more unusual MQP structures. Since they survive long passages of time in travel from distant space, from the time the early supernova stars exploded, to reach earth in this current era, on the whole neutrinos must be far more stable than every quark-based particle except the proton (- - the stabilizer of the galaxies, the stars, the comets, and the earth we live on.) The sources of neutrino stabilities must necessarily originate in their basic structures. First, the majority of the structures must be balanced, either as groups of interacting micro-quantal pairs orbiting in spheres, or with a proportion of others in staticly spinning sites from which quantal pairs interact with pairs orbiting around them. The almost universally conserved pairing of uniform right or left spinning quanta, of course, is the most fundamental necessity for the essential balance of larger forms. Second, after extending that type of balance to the high mass limit of stability in terms of the ratio of coherent forces to the inertial momentum of orbits, a low end limit of mass makes a transition from natural highest orbital momentum with their spinning conical bases of quanta outward to minimization of mass with conical bases inward, masking the sources of mass interactions.as well as minimizing the numbers of paired quanta to single pairs in both types of the structural variations. Third, the two extremes of structure give rise to the structural mechanisms of the previously mentioned Precursory and Recursory high mass transition forms and the Photons' radiating low mass transition form of wave-particles. Even the black hole may be basicly a very large neutrino based on the over-riding stability of cohesion by great mass gravity force, Thus neutrinos tie all the basic forms of Micro-Quantal Particle Paradigm matter together into a coherent unitary source of all the generally agreed particles and all the established forces of nature in a stably renewed Universe, to wbich the force of resonant entanglement is expanded in a final chapter.

In the previous chapter instead of finding causes from impact neutrinos, the last century's QM mis-reads the numerous break-ups of most baryons (except the very stable protons and their protected combinations with nearly stable neutrons) as unstable self-decays of the heavier baryons through their own internal actions. In that mis-reading, QM reaction equations are mis-stated from true balance of the opposite sides due to QM acceptance of beginning and ending net charges and dis-regarding losses and gains of unlike charged groups and quarks that must be accounted for exactly in any truly balanced equation. The MQP paradigm identifies all the real components of the equation so that they can be and must be computed. It develops that the elements missing from the QM net charge equations are the missing neutrinos, and they are both numerous and wide-spread in the usual particle mass range. This was explored in the previous chapter for the 15 cases of 33% to 100% supposed baryon decay modes for which adequate PDG documentation is available (or can be readily estimated.) All but one case found neutrino impacts required to balance the equations in terms of both quarks and true conservation of actual charges as necessary in both QM and MQPP, but not included in the PDG application of QM for the usual range of particle mass from electrons through the various quarks. The thousands of other baryon (and meson) decays in the PDG tables remain for re-analysis in this way. It will take decades of detailed work by many skilled analysts to accomplish this. From the limited range covered in the previous chapter it is clear that major neutrino contributions to the widely sought dark matter will be found when this is done. But it will take a major program.

The very large neutrino assemblies and the micro-neutrinos require exploration of the prior MQPP structures which cannot avoid natural spread into wider ranges. Starting again from the structural capabilities of the micro-vortices of Chapters 2 through 7, these structures can be indicated clearly by the relative synchronizing positions of the cones of the central drivers of the individual vortices in schematics of spheric orbits. The simplest of these are the single and triple neutral coaxial pairs of uniform symmetric turbulent conic vortices in the prototypical electron neutrino and mu neutrino (which are based on the charged electron spheric orbital structure of Chapter 5 and its relevant scaled symmetric force equations, as well as the scaled non-symmetric force estimates between turbulent conic vortices in pairs of separate electrons and of quarks in Chapters 5 through 7, as well as the Charge-Mass Law Equation 1-2. {This also includes the much weaker mass gravity and inertia effects of vortex interactions at other frequencies in those chapters.}) It is simplest to take up first the neutrinos that are most similar in structure to the prototypic electrons.

As was clear in Chapter 5 (and in Chapter 1) the neutrino most closely related in structure and mass to the electron is not the one called the electron neutrino (nu e, v_e), but the muon neutrino (v_μ). The names were given to them many decades earlier when the first indications of the existence of neutrinos became distinct, and it was apparent that need for the vaguely lightest neutrino was detectable when the electron appeared in a particle reaction, while the apparently mid-massive neutrino became needed when the mid-range charged lepton (CL) or muon appeared, and the probably most massive neutrino could be assigned when the most massive CL, the tauon, appeared. It was much later that the upper limit of mass of the mu neutrino was found to be probably about a third the mass of the electron, with indications that the mass of v_e is more likely to be at least five orders of magnitude (OM) less than the electron mass, while the tau neutrino upper limit should be about an OM below the muon mass. (And so they have continued to be regarded until quite recently, when, due to conflicts with cosmic astrophysics, all neutrino masses were (confusingly) made formally ambiguous in the PDG accreditations, though not in the continued practice of many particle physicists.)

Actually it was the adjacent PDG mass sequences (with almost exactly a factor of 3 difference) of the electron itself and the upper limit of the also prototypical mu neutrino (in Table 1-1, Page 35) that made visible the secret of Nature's inherent simplicity of conserved charged and neutral pairs of like and unlike-rotating conic quanta which underlies the MQP Paradigm of the particles and their interactions arising in Chapters 1 through 7 and shown in very similar Figures 6-1 (page 143) for the electron and 10-1a (next) for nu mu. This was made possible by the rare purity of having only charged like-rotating pairs in e and only neutrally counter-rotating pairs in v_μ. It is also easy to see with the fingers of anyone's two hands the simplicity of the reduced mass turbulence arising between the currents of a co-axial pair of conic counter-rotating vortices (or two adjacent non-co-axials) compared to the conflicting current turbulences that cannot be avoided frictionally between like-turning gyres, whether in co-axially bound pairs or in separated pairs. (It is also these different frictions between members of pairs that keep driving the concentric orbits of the pairs with various mutual forces between currents.)

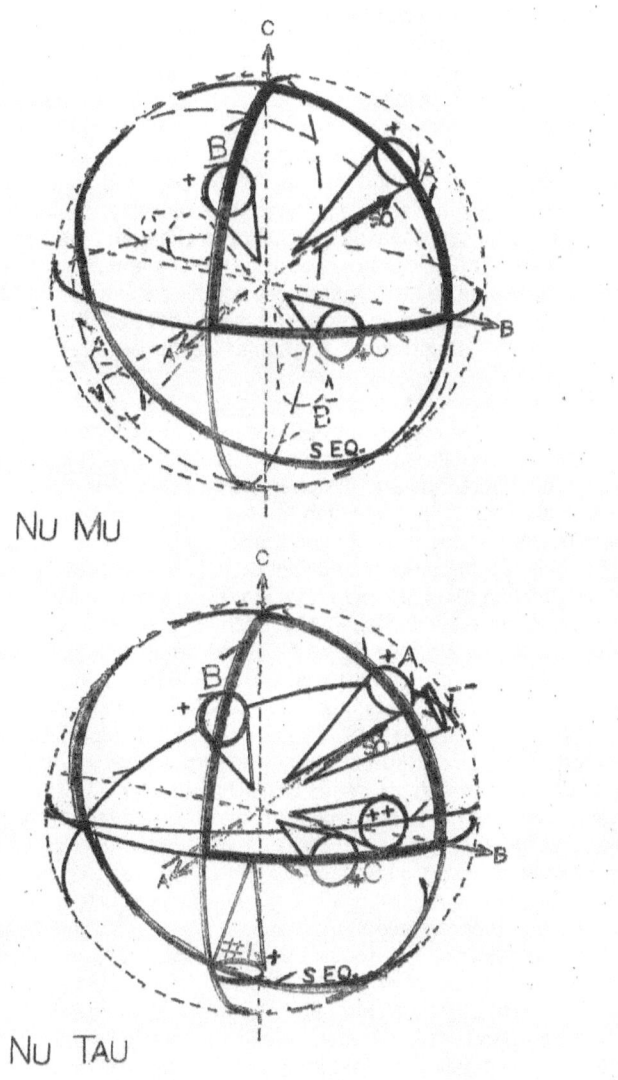

Fig. 10-1a MQP Orbital Structures of the Mu and Tau Neutrinos at Sync Times for Gyre Drive Cones. All the drive cones are shown for Nu Mu, but only those for the front face of Nu Tau. See text.

Figure 10-1a shows schematicly the most widely recognized larger QM and MQP neutrinos with definitive MQP sphericly orbital structure and mass near the prior PDG mass limits under the Mass-Charge Power Law Eqiation 1-2, as listed in Table 1-1. The v_μ structure in this figure is exactly like the electron structure of Fig. 6-1 except that one vortex of each pair of gyres is rotating in the opposite direction with the opposite negative charge, so that each pair is neutral and thus has 1/3 the mass, as does the mass of the entire particle. It is not feasible to show the six coaxial gyre cone drivers on the opposite face of the Nu Tau. (Since there is no predominant direction of rotation of the neutrino gyres, the orbits may be rotating in either direction around the primary octant, which may be a distinction between normal and anti-neutrinos if that is truly necessary, but at present in MQPP they are considered self-anti-particles with either of the two opposite principal octants as primary.) Any of this type of spherical orbit structures balances itself under its own forces around a sum axis (through an octant centroid, which becomes SO, as shown in the earlier structures herein and in Fig10-1b.)

That balancing starts a balanced set of static spinning sites at octant centroids, within a spherical orbit set or alone, as in the Extreme class options of Fig. 10-1b next, typically of neutral pairs. All of these occur often in the higher mass particles such as the large spherical orbital neutrinos and the charged particles. The cylinder-packed doubled pairs of E_{OB} are frequently present in the very large particles with full sets of orbiting pairs, wherein the cylinders must synchronize by rotating at sub-multiples of the angular rate of the orbits.- - If not interacting with orbital pairs, these configurations will have further reduced friction mass development that fit an increased whole number exponent x of 3 in Equation 1-2, the Mass-Charge Power Law. If sub-figure E_{OC} is operating alone to provide the electron neutrino, at x=5 due to loss of orbital rotation with reduction of mass eddies, it yields the mass of 2.8846 eV just below the QM upper limit of 3. Yet smaller neutrinos have higher values of x and transitional configurations of sub-figure E_{2C} in Fig.10-1c a page later. The E_2 configurations progressively reduce frictional mass interactions either in toroid merger or by gradually smothering the toroidal and spiral wave actions on the bases of the drive cones until they almost vanish in the base-to-base configuration with minimal friction of base toroid rings that sweep off at speed. At this point the exponent x of 3 in the Power Law is preliminarily valued at about 30, or more exactly at 31.11 if the neutrino $E_{2C'}$ in Fig. 10-1c is rounded in mass energy at 10^{-12} eV. Similarly, the photon is a true neutrino in Fig. 10-1c-d, with mass resolved at 2.2×10^{-8} eV when the exponent x of 3 for the mis- shaped base toroid and spiral wave systems increases to 22. This also accommodates astrophysics requirements of neutrinos scaled down to 10^{-6} eV mass (summarized in the Appendices to Howard, 2005, as is consistent with the MQP Paradigm taper of Extreme micro-neutrinos and other particles to fit PDG empirical accreditations where they have been posted and fitted to. More detailed fits that vary with photon wavelength band, spiral wave electret, or mass eddy waves may be developed when necessary or useful.)

Next in order would be an E_1 random non-symmetric group of possible neutrinos that do not seem to be observed, and so are not shown herein (but are shown in a web-site.) This leads again to the prior symmetric, but not necessarily spheric nor orbital, E_2 configurations, which include a broader aspect of the photon as neutrino (Fig. 10-1c-d.)

239

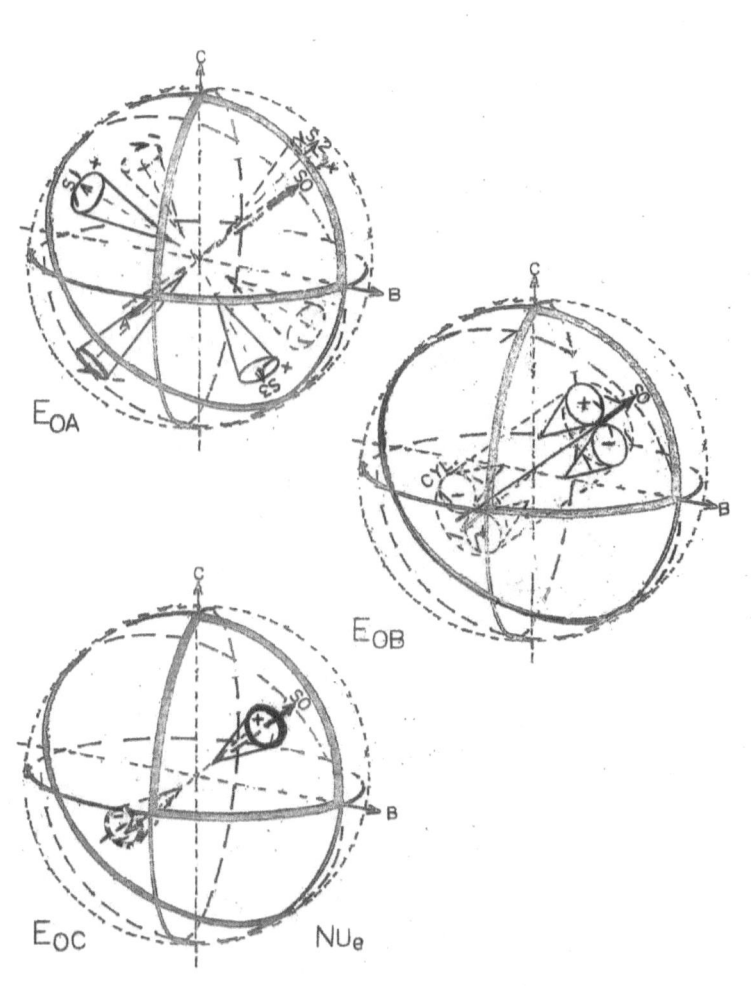

FIG 5

Fig. 10-1b Staticly Spinning Coaxial Pairs at the Octant Centroids, Including an Electron Neutrino on SO.

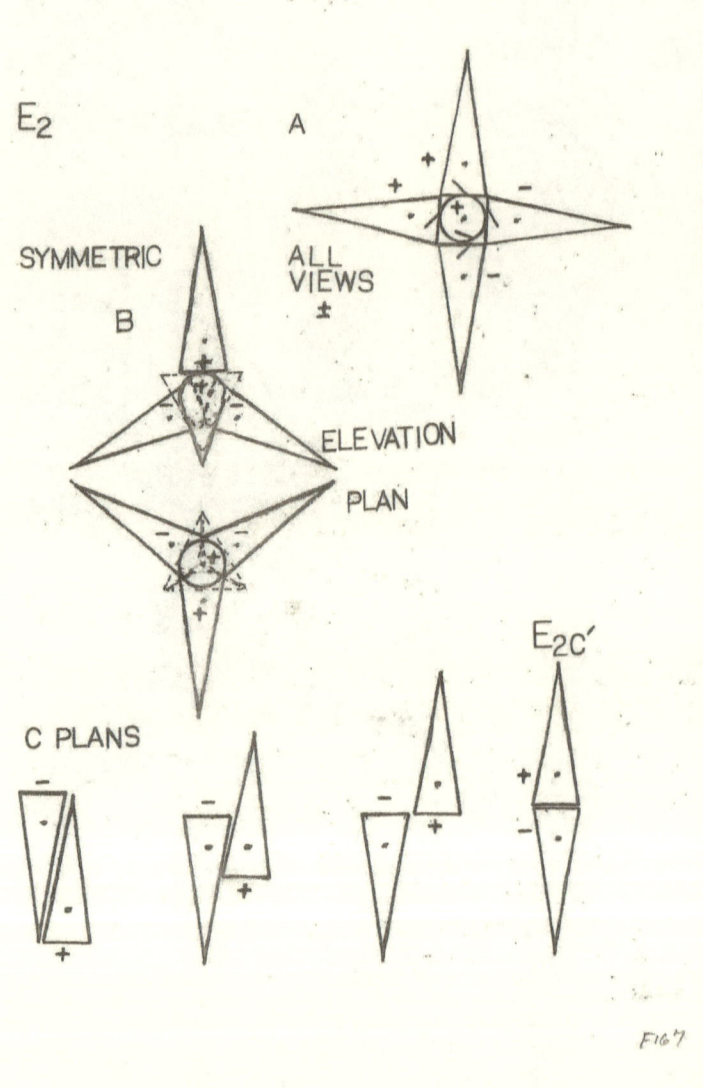

Fig. 10-1c Symmetric but Not Necessarily Spheric Nor Orbital Neutrinos. These may have one, two, or three pairs of gyres. It is possible that E_{OC} also converts directly to E_{2C}, but until such an energy change is observed, that is uncertain. The E2C' is the most adaptable Neutrino of all, becoming the Photon carrier of radiation energy as shown in following figures & text (or implied in many other publications.)

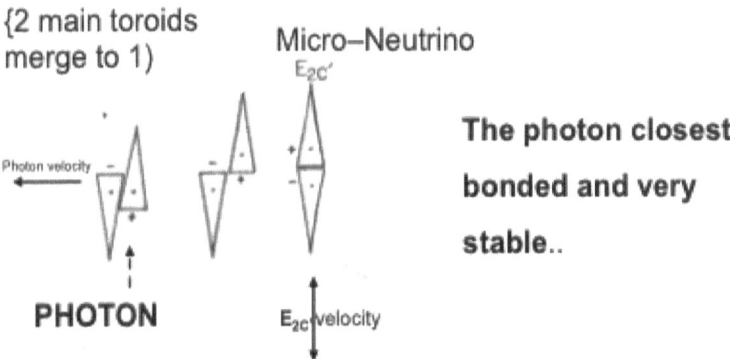

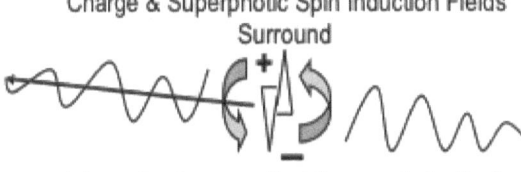

(shown in planar vertical linear polarization)

Fig. 10-1d The Photon Neutrino, Radiation Carrier of Electro-Magnetic Wave Energy (looking down a bit.)

The fingers of the two hands put thumb-to-thumb show how the currents and toroids of the oppositely rotating gyres of the smallest micro-neutrino merge and support each other in generating the wave actions of the photon, however they may be rotating as an assembly. The closeness or tightness of the bond between the two centroids of the conic vortex drivers side-by-side makes the photon <u>the most stable particle of all</u>, whether the stability is determined on the basis merely of the force to mass ratio or on the linked velocities of the inner currents around the drivers or on the closeness of cone centroids. This bond cannot be broken by another particle (except perhaps a photon of higher energy) without first absorbing away the carried energy of electromagnetic wave transport stored in its rotations other than the internal gyre currents. This is the peak of the conservation of coaxial and near coaxial pairs of the MQP centrally driven, turbulent conic vortices in computations by the scaled force equations of Chapters 4 and 5 on the assumption of Planckian values of the micro-diameter and spin frequency of the ultimate gyre drive cone as indicated by the near point radius of the electron (Mac Gregor, 1992) accelerated to approaching the velocity of light. [See requirements on internal forces of black holes (and possibly of neutron-type stars) in a renewable class of universe below.]

As shown here, the storage of energy in a photon is in the rotation of the body of the locked gyres in the vertical plane containing the velocity vector of the entire particle as one of many options. The plane of energy storage polarization could be rolled around that vector to any angle from the vertical plane. Then this storage plane containing the particle velocity vector could be either rolled continuously around the vector or rotated transversely to it in any of a complex combination of continuous orbital motions or fixed polarizations while the particle is moving along the velocity vector at the speed of light (c). The freedom of energy storage orientations has a special QM/optics designation of Orbital Angular Momentum (OAM) for the amount of storage as angular momentum at any orientation and phase or frequency of energy cycle. [It can be complex. OAM is a major optics topic which must be studied separately. Its freedom of storage is part of the process of its generation in storage in the photon, which is summarized briefly here as it occurs in the MQPP couplings. These couplings may necessarily act primarily at the velocity of entanglement (an undetermined number of OM faster than c) where that is applicable, and so may function here with non-relativistic classical forms of equations.]

First, the energy to be stored in a photon is typically accumulated by an electron in an excited orbit around the nucleus of an atom. (It is important to finding the high total dark-mass density of the suitable neutrinos that the mean life of storage of the energy in the higher electron velocity may be very short.) MQPP presumes that, at an appropriate point in time and electron position in orbit as well as electron spin orientation and electron phase of gyre orbits in the SEq plane, an $E_{2c'}$ neutrino moving axially (+ first) at close to the speed of light enters the electron (moving at about c/100) approximately along its axis and disturbs it just enough that the electron instantly accelerates with its stored energy in classical jerk toward its next position in a less energetic orbit. (This occurs when the linear neutrino's pointed axial end is only about half imbedded in the electron sphere at the next approach of the electron gyre bases to the electron SO axis. The neutrino inertia is microscopicly small compared to the electron's inertia.) The jerk action gives a snap in re-orientation and acceleration of both the electron and the neutrino sufficient that the linear neutrino is converted to the parallel-gyre photon configuration and ejected away at the speed of light again, while spinning with OAM in any suitable plane to carry away the electron's excess of atomic external energy between its initial and next orbits (which are generally known to be well defined.) The OAM quickly stores a fraction of its energy in the preparatory fore and aft induction fields of the photon by acting like a nearly lossless long-line antenna (with an entanglement type of induction field inducer) until a tuned receptor is approached. The photon thus proceeds until it meets typically a low energy electron with orbits that provide exactly the necessary absorption and vectors of movement to take up the OAM and wave field storage energy and jerk the photon back to its prior neutrino state. The delay may take a cosmic time interval, unless there is an impact or field that alters the OAM and wave field situation in the photon. - - This variable neutrino becomes orbital and/or spheric in nature when it converts to its photon form, but then goes back to a linear neutrino when the radiative photon energy is absorbed from it. As a photon it has many striking variations of BI (bases inward) orbital rotation depending on the exact spherical angle (with respect to its initial coaxial velocity) at which radiation energy is fed into it in similar processes.

Note also the microscopic mass energy difference (between the two linear and photon forms of the neutrino) which must be taken up from the OAM and returned to it at the times of its storage in the photon and its delivery to the next electron typically. As pointed out on page 237 the exponent x of 3 in the Power Law is about 30 with neutrino mass energy at 10^{-12} eV when the two base spiral wave systems that generate mass eddies eliminate most of their friction interactions, and photon mass is 2.2×10^{-8} eV with exponent 22 for interactions of the two exposed base toroid & spiral wave systems.

While the photon continues to move at the speed of light in the local medium, an extremely slow red shift is observed in the OAM frequency (and energy) due to the small losses of maintaining the entanglement induction field fore and aft along the photon velocity vector, and any other similar very long-term losses. This light red shift makes the stored OAM appear to indicate that the Universe is expanding at a significant rate and/or accelerating in whatever other expansion or local velocity process it actually has, (Bahcall, 2015) though it is not expanding or accelerating in general or overall as far as can be told from this type of long-term red shift of internally stored energetic frequency. The red shift is evidence requiring and confirming the MQPP structures. This causal process requires that cosmic astrophysicists re-evaluate our relation to a "Big Bang" (?), putting it at a different time in the past from the present widely accepted calculation of about 13 billion years, if not eliminating it again in favor of the prior steady-state renewed universe. (In another subsection of this chapter below an additional necessary MQP action is found to interact with and accentuate this correlation with quantal structure.)

The Larger Neutrino-Like Precursor/Recursor/"Higgs" States

The Particle Data Group takes note (in separate review sections of its reports) of an evident prevalence of Dark Matter over visible matter in the universe due to the large effects of gravity of galaxies on otherwise inexplicably high velocities of galaxy rotations. Such PDG reviews are very detailed in attempting to isolate specific types of neutrinos and/or other possible dark matter that might provide such unaccounted gravity force. Attempting to correlate with the PDG reviews of dark matter in detail is beyond the current scope of this empirical MQP paradigm. That would take lengthy additional work. Due to the MQP indication in the last prior chapter of the presence in the local galaxy of a large number of both heavy and light neutrinos beyond PDG accreditation, and to the PDG discussion in its 2014 edition of the "Higgs" neutral particle discovery at about 125 GeV by CERN, the structures, structural variations, and structural actions of neutrino-like particles that are available through the MQP paradigm in the larger mass ranges are summarized without full recognition of all the numerous PDG equivalents.

One of the larger structures found necessary in the MQPP is a Precursor mechanism for generating such PDG "observed" and formally accredited particles as the bottom quark (or the top quark) within the overall structure of a hadron, especially in a constructive type of hadron such as a baryon. In view of the complexity of these very largest quark structures, as well as their instability and short mean lives, and the difficulty of comprehending their structures readily, a similar model of one of the very simplest and moderately stable baryons, a neutral one, the neutron, with its simplest two kinds of quarks, is put through a simplest, minimal possible, but quite important, Precursor generation process that can be understood quickly, as a sample of them all.

PDG Limit Neutrino Impacts Create Baryon **PreCursor** by Quark Pairs

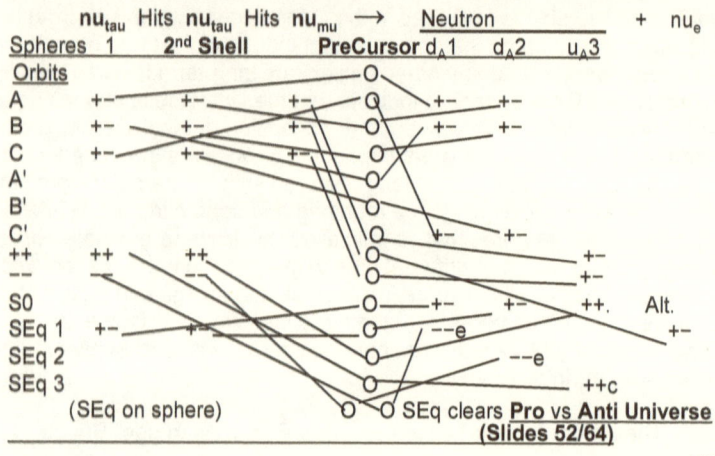

Fig. 10-2a Succession of Possible Simplest Precursor Orbit Structures for a Neutron in One Orbit Cycle. Three nominal PDG neutrinos at their PDG upper limit sizes come together in a random cloud of dark neutrinos to generate a neutron with three minimal quarks and one electron neutrino excess at process end. The input particle groups of gyre pairs go through roughly three cloud-like approximations of a larger particle before settling out finally in the last formation as a pro- versus an anti-matter universe particle. (Slide numbers in the lower corner refer to Power Point slide numbers in a briefing on the MQPP Concept included herein as the last overview chapter, which has a slightly different perceptive viewpoint.)

This minimal Precursor example begins spontaneously in a random location in a large cloud of independent neutrinos of unknown density when two tau neutrinos in their separate orbital spheres happen to come together with sufficient energy to break them apart in a small dense cloud of interacting pairs that takes the shape of an expanded sphere of orbitally moving coaxial pairs of gyres with a second shell of moving coaxial pairs around it. This enlarged cross section enables a quick tuned-to-fit capture of a mu neutrino. The three sets of moving and forcefully interactive coaxial pairs of gyres expand a bit more with their impact energy to make room for moves between pairs, but rapidly assort themselves into three cooperating quark spheres with expanded SEq links and a single neutral pair that does not fit in anywhere shrinks in interaction mass energy and exits the triplet of neutron quarks, all within one orbit timeframe. The closer combination of all the pairs but one builds up much larger interactive mass eddy energy from its quasi-unlimited spinning energy inputs. Other inertial momentum energy moves it away with whatever energy is left over from the new structural formation.

The time it takes to progress through the Precursor process of Fig. 10-2 from the left input side of the figure to the right output side can only be estimated from the necessary velocity estimates concerning the electron in Chapter 5 and from the required motions of the input neutrinos, which must be energetic enough to break them apart. It is generally accepted in physics and in the MQP paradigm that the velocity of whole particles will be not greater than the speed of light. It is obvious in Chapter 5 that in order to maintain a moving structure the internal velocity of the components must be significantly greater than the external velocity, or not less than c. It is clear from the PDG measurements of the charge radius of a typical baryon such as the proton, which is made of three quarks, that a typical particle sphere must be smaller than Mac Gregor's comparison of the usual electron with the prior Compton radius and greater than his estimate of the maximally accelerated point electron at less than about four orders of magnitude smaller. Therefore, the time of one approximately required orbit of the parts of a small neutrino at impact or neutron at output must be given wide bounds, but must be understood as probably well within the inverse frequency bounds of the orbits of an electron x2πr in Chapter 5. It will probably not be fruitful at present to attempt to resolve it more finely.

A detailed simulation of this interactive process, using the smoothed and scaled radial forces of Chapters 3 to 5, plus trial estimates of the asymmetric and lateral forces, should make it possible to smooth these force estimates sufficiently for an informative slow motion simulation of the Precursor process through an orbit cycle or two, if not longer. Finding smoother transition paths of input gyre pairs to output gyre sites is a valid simulation objective. Such initial simulations would be very useful in planning and designing equipment for the previously identified further lab measurements of forces. Early preparation for most critical angles of relative gyre positions and force directions at relative gyre separations and spin frequencies is especially important, as is reduction of instrumental and model vibration over a wide range of frequencies. Relative strengths of typical gyre assembly paths between sites under measured forces with 3D records would be an ultimate goal. - - But again un-accounted <u>Dark Matter Neutrinos</u> play a lead part.

A similar process might be called a Recursor when it is followed by a part or all of a particle which had already grown beyond a simple sphere but is reorganized by an energetic impact. An example simple enough to follow on a single page would be the classic neutron "decay" with beta electron emission (page 231.) As shown next in Fig. 10-2b, this could expand through three shells of component partial orbits before again collapsing within less than a full orbit of most components into a new structure formation with altered phase of at least one gyre in an SEq plane, but with actually changing all three. Again, the particular changes made in orbits are not pre-ordained, but are worked out by the forces and inertias between gyres as they initiate circles around each

246

other on momentum. A full simulation might readily follow other paths of re-combination, and a second simulation might take a slightly different impetus from the initial impact and follow yet another set of paths with equivalent final result. It might even be difficult to find the particular set of initial conditions which reliably results in this indicated final outcome. It is only a possible outcome. [But there may be a proton plus electron output bias rather than the anti-world bias or equal options of an anti-proton plus positron output. If so, the Pro-world versus Anti-world bias may arise through necessities of the neutron up quark structure with a slight, uncertain energy variation from the directions of the near interferences between static SO sites and ++ or − − orbits in Table 6-5, page 148. (Small energy variations at near interferences must eventually be very closely checked against refined lab force data in every case in which they occur. They may also be systematicly pro-world in trend.)] (The "Higgs" also appears to be a variable Recursor.)

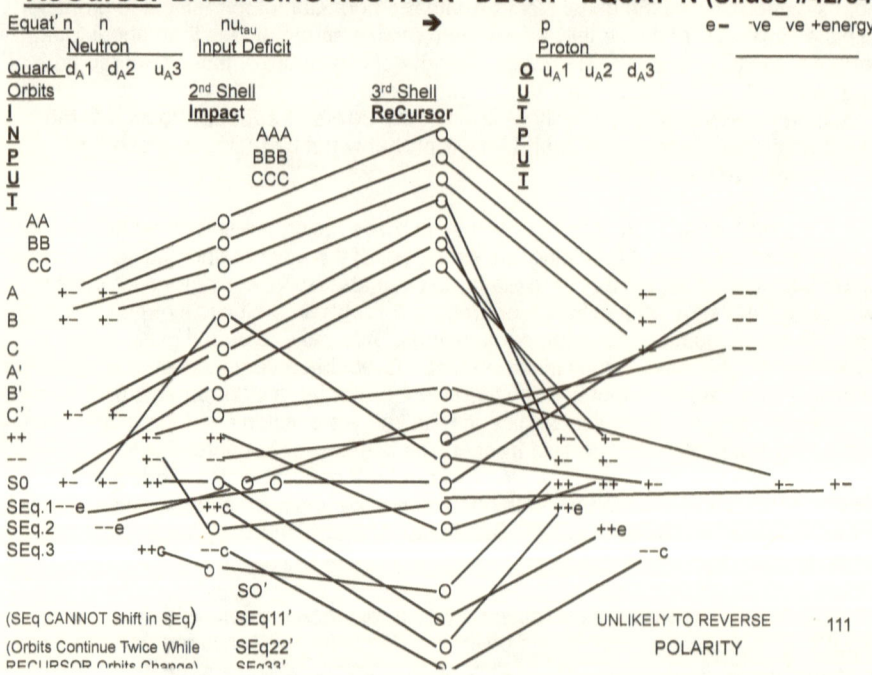

Fig.10-2b A Possible Recursor Process of Reorganization After an Energetic Neutron-Neutrino Impact. This figure makes it clear essentially what has to happen for this classic transition to occur (plus a C'C'C' lable in Shell 3.) The neutron is not a simplistic combination of a proton and an electron with a bit of energy and an insignificant output neutrino, as it had appeared to be for well over a half century. Neither are other supposed negative beta ray (electron) "decays" simplistic fallings apart of a couple of continuously pre-existing

components. There is no electron hidden in a neutron. It is unbalanced, and most of the components required to make a stable proton and an electron are present in a neutron, until the rest happen to come by. Many such reactions may further recombine into active hydrogen atoms. (This is one of dozens of similar cases, brilliantly, imaginatively, and extremely complexly described in pre-2000 QM theories of "decay." There are other beta "decays", alpha "decays" giving off helium nuclei, and gamma "decays" emitting radiation. Many absorb impacting neutrinos to yield more stable, re-balanced particles.)

A much broader importance of the Precursor/Recursor processes (described above in their simplest and most readily displayed cases) develops when they are worked out for the heaviest or largest, published particle measurements in the current 2014 to 2015 (or prior) PDG literature. These very large observed and broadly accepted particles have a close common relation in deriving most readily and exactly from the same large particulate Precursor as their main source of input mass. Another key factor that occurs in some of these cases is that there is first a single reasonably random impact between two particles, one large and one small, with their well understood "cross sections" for such impacts, followed by a rapid expansion of the components cloud into an estimated at least 2 times larger radius outer shells around inner shells of orbiting components with 2+ squared ratio of effective cross-section for any second impact event. This larger cross section is also aided by higher component velocity of excitation from the first impact before possible third neutrino impact events, which do appear to occur as serial "decays" in the 2014 PDG Top Quark Review section. The impact of a second and third neutrino is greatly increased in probability by the first rapid expansion. (This is not a simultaneous impact of three particles, which would be highly improbable.) Due to the large number of tangled crossing paths in these larger Precursor processes they cannot be shown usefully in the same path detail as the two introductory cases, but appear in the same types of group summaries as those in prior chapters (see p. 231 or -3.) The comparative broader importances in the total wide range of masses that take part in such Precursor and Recursor developments of many particles are also kept clear by the condensed summaries for quick comparisons and ready review.

The most common possible source for all these very large particles is the neutral combination (not qute like a big neutrino) of the most stable and universally predominant small particles (except the photon and perhaps other tiny neutrinos) that appear on ordinary planets and stars in many variations. They are the two protons (including their six quarks) and the two electrons of the neutral hydrogen molecule. Distributed in eight tiny spheres as the source is, it is amazing how very light it is, and yet how very massive it can become under the influence of the exponential Charge-Mass Power Law with the impact of only one relatively large, but not necessarily greatest, added tau neutrino input deficit. After that Precursor expansion in up to three (or more) concentric spheres of orbital and static sites, there can be condensation (with residues) into a top quark in a top neutron; a "Higgs" equivalent neutrino; a Z boson; a W^-, a W^+, or the two together; or other suitable QM "decay" outputs that may occur, possibly after more neutrino impacts:

Table 10-1 Typical Very Large PRECURSOR INPUT NEUTRAL & OUTPUT Structures in Power Law
2 Protons & Electrons (8++,2- -,16+-)+(6- -)=(8++,8- -,16+-), 2x 938.27 + 2x 0.511 = 1877.562 MeV
+SuperTau Neutrino   + (3++,3- -,12+-), 36^5x36x10.9525 [(6/18)+{12/(3x18)}] = 13245.116MeV
For Precursor Inputs   = (11++,11- -,28+-) {at Mass Energy Sum 15.12268GeV} = 6863.559GeV
    POSSIBLE OUTPUTS - - - by MQP Power Law
Top Quark (in baryon)   (10++,8- -,8+-)   172.119 GeV Mass   (PDG 173.21 ±0.51 ±0.71 GeV)
Top Neutron (series base) (10++,10- -,16+-)   + neutrino & energy
"Higgs"("Boson")   (11++,11- -,2+-)   126.51 GeV   + " (CERN ATLAS 126±0.5657GeV)
                                                                     (PDG 125.7±0.4{late CERN ave})
Z^0 ("Boson")   (0,0,27+-)   90.522 GeV   + "(PDG 91.1876 ±0.0021{excited?})
W^+ ("Boson")   (9++,6- -,8+-)   79.705 GeV   (PDG 80.385 ±0.015 { "   ?})
W^- ("Boson")   (1++,4- -,20+-)   79.862 GeV   (PDG 80.585 ±0.615 ?("   ?})
W^+ & W^- ("Bosons") (10++,10- -,28+-)   sum   + "

    In Table 10-1 it is remarkable to see what a Power Law mass (6,863.6 GeV) the required two separated precursor constituents with an initial mass sum of 15.12 GeV run up to under the Mass-Charge Power Law when they are combined into a single-particle expanded Precursor of 50 interacting pairs or 100 total interacting component vortices for their single orbit period of full interaction. This rapidly strips down into the largest observed PDG particle of a Top Quark in a smallest suitable baryon (Neutron subseries base with two down quarks.) There is an escaping residue tau (++++) neutrino of (1++,1- -,12+-) [which could possibly have gone into a heavy bottom quark with 1 down quark and the top quark in a much heavier mass {with residue neutrino of (1++,1- -, 1+-)}. However, the heavy bottom quarks have been observed very rarely, if they have been observed and PDG accredited in baryons at all. (See bottom baryons above.) For the present, there is insufficient evidence to include the heavier bottom quark in a baryon in a precursor, though it is clearly not impossible here.]

    This Precursor discussion is not consistent with the PDG statement in the Top Quark Review Section of its 2014 report cited earlier that top quarks form and decay before inclusion in a hadron can occur. The MQP paradigm finds that the quarks and hadrons necessarily form together to the same extent simultaneously. - - (It is possibly significant that the stated review finds that top quark mass was well measured at CERN before 2014 in lepton decay channels in a bottom quark stage of decay with decay jets and a single neutrino residue.) - - Even at CERN, where the intent has been to impact Protons alone (which could not possibly yield a sufficiently large precursor input for any of these heavy outputs observed there), it is most likely in the rare level of very large observations that a high speed proton's quarks first impacted a suitable large neutrino that was passing by. Such a combination would generate a double shell expanded cloud of components which would then have a much higher probability of quickly capturing another accelerated proton impact. The further expanded + charged cloud would even more quickly gather two stray electrons (which are very difficult to eliminate) and would be sensitized or tuned toward any further deficit in small penetrant neutrinos (which from the short lifetimes of unstable baryons can always be assumed to be passing nearby.) All of this action might require an extended lifetime of the precursor cloud or the loss of a few fairly complete clouds within the delays between observed "Higgs" occurrences in CERN. - - Additional neutrino impacts would permit two large outputs from such a precursor. Large added neutrino inputs to precursors are also required to enable reportedly observed decays to some massive particles by "Higgs" particles, as reported by CERN and reviewed by PDG. (A few of such smaller input deficits might be obtained from a Recursor cloud generated by two colliding protons, particularly if the two neutralizing electrons are also unexpectedly picked up.) - - All the PDG "Higgs", W, Z, and nominal "gluon" bosons may also be described as Recursor substates or small Precursors from possibly undefined sources larger than the boson value with or without added neutrinos captured because of enlargement. Beyond such decay chain functions, the W and Z bosons, as appearing and disappearing particles, are not yet found to be definitely necessary in MQPP. (The definite photon "boson" is taken up here as a neutrino.) - - Top quark plus top anti-quark production simultaneously would require much larger precursor structure in ++ & - - inputs than the common one shown above, at least by the capture of a second large neutrino before the combined production. - -

In this aspect of Precursors or Recursors, as complete evacuation of any collider experimental space as possible is very important. In the event that one molecule of the largest fraction (about 80%) of the earth's atmosphere, or Nitrogen, remains in a collider and takes part in a collision that distributes its parts in a Precursor cloud of components, it can contribute as much as (84++,42- -,252+-) to the mix at some point. Unless this is recognized in some way and isolated, that is enough to confuse thoroughly any conclusions that involve the particular point discussed. In the CERN reports reviewed and summarized in the cited PDG report above, such random interferences are well minimized by requiring hundred and thousands of repeated measurements to establish the probable reliability of all marginal observations before they can be reported. That is separate from the probability that the events connected with the measurement occurred structurally and causally as described, and does definitely permit alternative variations of serial chains of events behind the measurements. It is in this region of reconstruction of events that the MQPP differs on a limited number of points (though they may be very important points) from the SM of the last century of QM. The measured Data accredited by the PDG is the basic starting point in effect for both QM and MQP paradigms of natural serial action. - - Beyond that correlation the discussion of comparative "decay" versus impact reactions and Precursor or Recursor actions can go into many volumes of detail which primarily demand retabulation and weighting of the alternative reactions that fit experimental hard data on hand. That will take years of work for the future. - - Here this introductory MQP Paradigm must return to its yet broader consequences.

Necessities About the Continuity to Black Holes

Without being prepared for it, the micro-necessities of the MQPP have, at this point, led to the inevitability of quantal structures (a Precursor neutrino) at least a full order and a half of magnitude (OM) and more, at its smallest, greater than the mass of the largest PDG observed and accredited ordinary particle. The associated PDG review includes the requirement for further enlargement to generate the <u>combined</u> top and anti-top quarks as well. This adds a necessity for at least 0.6 OM increase in precursor mass (if not more in actual particles), bringing that neutrino-like entity to about 2 OM greater than the largest widely recognized particle, in order to be able to generate any one out of all the PDG-listed large particle combinations observed arising out of quite similar initial CERN conditions of very high energy particle collision.

Such a necessary particle type opens the door to a wider necessity for particulate structures that fit the requirements within astronomical bodies, including the progression into and beyond the previously noted "neutron" stars. These are situations in which great gravity and density become the most important factors rather than being only associated factors or approaching negligible side issues. Temperatures or combined forces and energetic velocities also increase greatly or may be subordinated by density and the compressions of gravity. As an eventuality, even a black hole, if not obviously charged, might have aspects very like a precursor in being similar to an over-large re-collecting and re-processing neutrino. - - - [That general view could also apply in their ways to star clusters and galaxies during their observed passages into each other and apparent merging or ripping each other apart, much like particles in a collider. It is notable in this view that isolated <u>barred</u> spiral galaxies, with outer spirals from the tips of non-spiraling central bars of star and gas luminosity (Longair, 1996, simplified summary volume, Our Evolving Universe) have exactly the same shape and flow pattern as the centrally driven vortices of Chapter 3 for a brief period just after the drive cone is switched off while the cone is slowing down or after its being taken out of the fluid. The bars and attached spirals of both kinds of barred spirals are direct vortex model evidence that black holes at galaxy centers evidently can and do decrease in effective maintenance of central galaxy drive by star-to-star gravity and frictional coupling, etc. (as well as at other times gathering and accumulating mass, momentum, and galaxy drive.) For the galaxies this bar shape must occur through loss of their black hole's mass and/or accumulated angular momentum whenever such loss exceeds in-falling mass and momentum. This is also evidence that can best be explained jointly with further direct evidence of the observed highly relativistic axial jets of ejected black hole mass discussed below. Putting these two black hole type observations together establishes a necessity to consider that the most central and fully relativistic axial jets can actually be formed as mass-loss mechanisms within the central body of the black holes (rather than as part of the incoming disk of galaxy mass which moves axially to form the much less relativistic broad cone of more slowly ejected rotating matter in the central bulge of the galaxy without its going into the event horizon of the black hole itself.) This requirement is one of a series of progressive necessities within astronomic bodies which provide the series of particulate structures that can generate the jet, as follows next.] - - -

It is not useful here to try to reclassify stars beyond their evolving face values nor to match all their internal conditions. That will require extensive follow-up analysis briefly noted below. This paradigm can only assemble a number of adaptations of particle structure that do fit many aspects of star conditions. The ordinary small to mid-range star (like the solar system sun) appears to have core and outer mantle conditions that are compatible with the Usual particle structures of the MQPP discussed herein, and also with the Extreme tiny neutrinos beyond the electron neutrino, including the MQP photon. The solar nuclear reactions and magnetic turbulence too can be assumed as usually described in the literature (though in MQPP at present magnetism is not required to be more than a special summary of electric force effects under relative motion of gyres, as suggested by interpretation of the electron fields of Lorentz, 1952.) Overall the normal particle-to-particle bonding forces are predominant within larger particles and nuclei. Temperature energies and local entanglements are predominant between most particles and nuclei. Density effects and gravity are sufficient to keep the entire solar system and its members together. Minimal packing densities accommodate usual bonding-force-dominated structures between larger particles. Since many small baryons exist and emit their spectral evidence from the sun, it is beginning to approach conditions which deep in the core of somewhat larger stars (at or beyond the neutron star level, with wider ranges of reactions) would from baryon disruptions also create traces of ordinary meson chain residues up to about six quarks, which then definitely appear suitable under larger star conditions without further detailed accounting for their assembly.

Thus, as stars and planets begin to get larger by orders-of-magnitude accumulation of matter and gravity, atoms must necessarily begin to adopt structures that are modified from the usual bonding force balances with temperature compensations. In cooler surface spots of such larger bodies, if not hot enough to maintain full ionization of all elements, atomic electron orbits may be twisted out of the usual planes and/or otherwise compressed closer to nuclei. Then nuclei must be modified with bond variation to occupy less volume than usual by addition of exterior forces to the attractive forces between the members of each pair. Initially this would further stabilize the usual bound neutron-proton pairs in nuclei (noted earlier.) Eventually neutrons in nuclei could, however, become separated by being forced laterally away from the protons which stabilize them, and they would then become vulnerable to penetrating neutrinos which can convert the neutrons to scrambled mesons that could link in continuous neutral chains (noted under mesons earlier.) Such long chains could stabilize in reduced volume by multiple satisfaction of end-bonds, with the dissociated protons and electrons clustered neutrally around them (as sometimes discussed for neutron stars.) But, the natural meson structures in right-angle coupled spheres (Fig. 8-1a, Page 205), while more densely packable in overlapping chains than equivalent independent baryon triplets of spheres, are not at maximum packing density in the usual right angle chains. Further densification in the largest stars would then occur by break up of the meson quark spheres and loss of the right angle bonds to straight-line bonds assisted by gravity compaction forces until gyre-to-gyre cohesion within chains of effective pairs is supported mainly by alternating point-to-point thrusts and base-to-base couplings of individual vortices, which take over the chains. In these circumstances of increasing density and gravity forces, protons and electrons would likewise be disrupted and joined in the great increase in density and paralleled chain structure.

While all this is occurring, the mass interaction energy for particle content of effectively larger and larger particles would be increasing exponentially under the Charge-Mass Power Law, further compounding the chain mass and density increases. The conversions of kinetic momentum to relatively static locations would be raising temperatures further. The various neutron star, pulsar, magnetar, quasar, and supernova steps to the formation of black holes described in the (very large and widely known) cosmology literature (not specificly referenced in detail here due to its length and availability) would necessarily occur. The long chain structures would maintain the neutrino-like coherence up to the black hole by simple entrainment at its event horizon of added compressed structures from the equatorial rotating masses surrounding black holes generally. Within the event horizons, increasing internal gravity forces would accrue as the adapted meson-like particle chains progress in spirals deeper into the hole interior. Forces between condensing chains would ultimately condense individual gyres sufficiently that the point-to-point and base-to-base load surfaces of pairs become unstable, both by sliding together point-to-point and by breaking apart base-to-base until singles are effectively separated. (The forces at which this conversion occurs can be estimated from the point thrust of a condensed electron gyre during acceleration of the electron to velocities approximating the speed of light in prior chapters.) This process necessarily completes disruption of particle structures by gravity forces in the axial centers of the black holes. Freed vortex quanta of both rotary charges in disorganized outer portions of that deep axial volume (with toroidal rings and their stores of gravity eddies stripped away by effective drag) then are able to accelerate on point thrust against the frictions to super-photonic velocities of escape from the black holes along whatever axial magnetic lines are locally established. The small angle jets of these superluminal particles along black hole axes have been observed by their deceleration in collision excitation of sequential bodies of ordinary particles outside of the holes with emission of highly excited spectra. This is the evidence (cited below) of the existence of the jets, of the invisible superphotic quanta within the jets, and of the provision of those real quanta by essentially the break-up process briefly described here above.

General acceptance of the axial jet evidence that Einstein was incorrect in the position (or assumption from the prior evidence) that nothing could go faster than the speed of light has not been uniform. In 1996 a prestigious simplified summary of cosmology by an eminent astronomer published by the Cambridge University Press (Longair, 1996) cited data on a black hole jet with an apparent speed of about eight times the speed of light, then cited Einstein for its impossibility, and sketched a geometry that was supposed to show that the actual speed of the jet was about 98% of c. Equally prestigious publications from mid-2015 continue to follow this tradition with special classic justifications of velocity (or lack of publication.) [An apparent collision shock jet velocity measures 7c ±0.8c (Meyer, E.T, et al., 2015).] Typically, a report on quasar superluminal radio jets in our galaxy with a velocity >2c (Yadav, 2006) was found only in a workshop report and in arXiv. In a search of publications that found about 60 related reports only the following definitely stated jet speeds greater than c. Eleven definite findings of jets from active galactic nuclei (AGN, obscured black holes) over five to 12 year satellite measurement periods have velocities from 1.7 to 6.9c (Lister, M. L., et al., 2013). Optical measure of other galaxy jet knot velocity up to 4.5c (Meyer, E.T, et al., 2013). Apparent AGN jet feature velocities range from 3.9 to 13.5c (Cohen, M. H., et al., 2015). Other apparent superluminal motion of jets or jet features reported include: (Mirabel and Rodriguez, 1994); (Dhawan, V., et al., 2000); and (Cheung, C. C., et al., 2007). Thus there is a small body of recent definite observational commitments to at least apparent black hole and quasar jet velocities well above c, including a few that do not call them apparent velocities.

It is clesr that Einstein's upper limit of all velocities to the speed of light in a vacuum was based on the previously observed (early 1900s) <u>external</u> limit of waves and of entire particles made of at least two or more MQP quanta that are bound together in close interaction which builds up external features such as toroids with close coupling to the local medium. To clarify, <u>external</u> superphotic velocity is found in MQPP only in the mechanism of wave entanglement for entire particles associated with internal quantal structure (see next chapter) and for the slim centrsl conic structure of a single isolated quantum when it is separated to accelerate with unrestricted point thrust, which can initiate only by separations within black holes and their approximate equivalents of great density, energy, and gravity, such as those of quasars, pulsars, etc., noted earlier. <u>Internal</u> superphotic velocities are necessarily required within all MQPP leptons and quarks and also in the internal and external coupling mechanisms of the quantal components of these particles herein, both above and below. (Spin of LQ particles is essentially internally generated, though externally measured.) There is no direct true disagreement with established Einsteinian physics at velocities of multi-quantal whole particles below c, except that there is in MQPP a very generally definable medium in the vacuum (Appendix 5B to Chapter 5 on the Electron), and there necessarily exist superphotic physical mechanisms for the observed phenomena of entanglement (next chapter.) (The future medium may continue from the present property of having its own abstract or real 3D orthogonal straight lines of space dimension that do not curve with gravity nor with entanglement pressure-type waves in it, as well as a usually invisible straight line 4th time dimension line on which the 3D geometry translates as redefined at will, most often orthogonally through the general origin in a plane from the crossing plane of two planes on the 3D axes.) MQPP also necessarily covers actions at extremely high velocity (and frequency) above c and beyond the scope of Einsteinian physics such that relativity at and below c does not apply to these actions without special redefinition (which MQPP does not presently include.)

These structures and mechanisms also become evidence of a particle restorative process in black holes (and in the progressive star gradations of black hole generators) which appear to separate with internal forces the otherwise conserved gyre pairs. Each gyre is then able to accelerate on its inherent scaled point thrust of Chapters 3 to 5 to a velocity of escape along the black hole (or progenitor) axis in the observed small angle jet process, which is definite evidence of such an MQPP consequence. The jet ends in eventually unavoidable deceleration by combined external impacts and field interactions, with recombination as restored simple pairs for further recombination in stars and various impacts as cyclically restored particles. There is variation of the single gyre's ability to accumulate and retain mass eddies as the toroid bearing most eddies is blown away in the process of acceleration, dispersed in the jet exhaust, later recaptured, with Power Law exponent x for the jet or jet pair particled varying continuously between >30 and >5. These processes yield net simplifications of observed phenomena. Otherwise the elaborate mathematics of QM, optics, spectroscopy, astrophysics, fluid mechanics, relativity, and gravity continue to describe black hole exteriors necessarily.

By a remarkable coincidence, between the completion of the prior evening's first draft (only slightly revised below) and the next day's final revision of the summation paragraphs of the end of this chapter, two jointly published but separately observed, analysed, and authored reports appeared (in a top quality journal, with celebratory news releases) which are completely consistent with and unavoidably confirmative of the jet overall MQPP particle process and necessary radiation output herein (Lee et al., 2016) and (Bartels et al., 2016.) These two new reports specifically find millisecond (rotation frequency) pulsar stars to be the sharply clumped or point-like (not diffusely scattered) sources of the local area's portion of the universe's most energetic (gamma ray) radiation from the general vicinity of the center of the Milky Way galaxy. (This is close enough to earth for separate sharply focused jet sources to be resolved distinctively and well differentiated from diffusely scattered radiation released from randomized "dark matter" "annihilations" previously hypothesized from or predicted by theories about dark matter.)

During the above described gyre deceleration that would yield the now well-observed gamma ray very high energy radiations by impacting locally available classic particles along the jet path, the gyre quanta of de-energized velocity are available for rapid local recombination into pairs of quanta, either with points inward in tumbling random orbits suitable for further recombinations into usual particles, or with bases inward in neutrinos suitable for activation as photons. This sequence of actions (or more broadly, this type of evidenced action sequences, however varied in steps or organized) establishes the black holes and their special star relatives and progenitors (such as pulsars, and now most particularly pulsars with millisecond rotation frequencies) as MQPP particle regenerators in a continuously renewable universe that has never required a "Big Bang." - - The ultimate MQPP simplicity! - -

- - - - - - - - -

(MQPP began as an entirely empirical and completely non-theoretical paradigm on observed actions and processes, plus only distinctively required structure under specifications of necessity as restricted to carrying out the observed functions on the simplest and most minimal basis. It has continued in that mode, with the possible lapse of considering additional functions that the most minimal structure also unavoidably performs. This inclusion of additional unavoidable capabilities is also a fully empirical necessity, which may interfere with full performance of functions or add to their completeness. Theorists may object that this is all that theory does in fact, especially successful theory that broadens the application of observed functions, or increases efficiency, or facilitates practical design, even when calculated from first principles. Perhaps that is right. Perhaps successful theory is only a perfection of the empirical observation. For this borderline between the empirical and the theoretical, including where it may lead in a further empirical way that perhaps should not be considered part of MQPP, or preferably should continue to be, also see Appendix 10A below.)

APPENDIX 10A
A Seeming Contradiction in Quantal Actions (re Expansion of the Universe)

Primary empirical observations over earlier chapters herein are:

That fully turbulent conic vortex gyres (centrally driven in a specially defined, very low density fluid medium by viscous surface friction of right regular, flat-based, rigid cones, of ≤ 30° total angle, which are naturally formed from the same medium) are necessarily required in particle interactions as the active particle components.

That the drive cones are also necessarily axially driven by an internal source of angular momentum which is an effectively semi-infinitely dense core, of interchangeably equivalent and not necessarily precisely known size.

That this observation is due to a necessary characteristic of the visco-elastic medium being spun by the drive cones axially at a Planckian extremely high frequency > 10^{30} cps and without usual perceptible mass other than the external medium eddies of frictional mass energy waves.

That this is a small part of turbulently interactive vortex mutual force measurements scaled from lab data to match electron particle data from the PDG.

That this observation is necessitated (at length) by the further general observation that electrons as light photon emitters (especially of light spectra from hydrogen atoms) have not varied detectably in the ratios of emitted photon frequency patterns from the earliest observed 13 to 14 billion year old data recently received from distant space to current lab data corrected for local conditions (as summarized by Bahcall, 2015, in a US National Academy of Sciences report.)

That photon velocity losses as an entire particle are constantly replaced by the drive cone momentum energy storage mechanism over quasi-infinite observed time periods without apparent change on arrival from velocities of locally emitted photons.

That photon frequencies, on the other hand, are emitted by the necessarily separate rotation of the entire photon structure with its separate mechanism of loss of rotationsl angular momentum and frequency.

Exactly similar frequency energy loss in observed red shift of light has previously been attributed generally to source motion relative to receiver because of expansion and accelerating expansion of the Universe after its initiation in a "Big Bang." These two descriptions of the natural process of light energy transmission from an extremely distant source are apparently mutually contradictory.

As discussed above in connection with Fig. 10-1d, the MQPP empirical photon is a modified stable neutrino of two vortices. It is observed only in flight at local speed of light with induced wave trains at the proper frequency tapering out effectively before and after within the cylinder of the approximately quarter to half wave length near vicinity of the particle flight path in free space (very much like a long-line antenna.) The photon particle, made of two oppositely pointed, paraxial bound gyres with exposed turbulent spiral wave disks which induce the alternating electric photon field, is also rotating on a separate axis through its center of mass at the flight path with either a fixed or a systematicly rotating 3D orientation compared to the flight path. (For simplicity the Fig. 10-1d axis is fixed at a right angle to a fixed vertical plane containing the roughly horizontal flight. The figure is viewed from a little above the path and offset from the plane. The circular broad double arrow indicates the rotation in the vertical plane on the separate axis for linear polarization.) The rotation of the particle shown by the double arrow is the rotation given to it by the interaction with the emitting electron at the instant of emission (while the photon is arriving as a neutrino at substantially the speed of light and passing through the vicinity of the electron with only a nearly infinitesimal increase in speed and beginning of increased emission of mass eddies internally.) That separate rotation carries the energy of the photon to the receiving electron. The rotation frequency is the frequency of the photon rotation with respect to its electron emitter at the instant of emission. If, and only if, at that instant the emitting electron were actually moving away from or opposite to the direction of the flight path to the receiver, the measured received rotation frequency would carry a Doppler red shift of that rotation frequency in proportion to the velocity of movement in that opposite direction. Such variation of shifts within a beam of such light are very small compared to the broad offset of the shift with measured distance of the emitter from the eventual receiver.

The prior general assumption that such a supposed mechanism is the primary or sole reason for the observed red shift of long-lived photons is no longer necessary and should not be applied as if in demonstration of the expansion or acceleration of expansion of the universe. As noted with the figure discussion, the rotation of the photon particle which carries the photon frequency and energy is due entirely to the emission energy donated by the emitting electron. It is not due to the spinning of the gyres, which is powered by the gyre drivers, whatever they are. There is a separation of two sources of continuing energy in addition to the added energy transmitted from emitter to receiver by the photon. This particular emission energy rotation of the photon donated by the emitting electron does necessarily power both the transmitted frequency and the constant small but continued loss of energy of the fore and aft wave train set up along the flight path (which may be varied by fields that are along the path.) The loss of such energy at a given frequency is proportional to the time of flight, and does decrease the specific whole body rotation involved in the red shift of the photon. It is the primary measure of the time of flight. (The gyre spinning does maintain variable replacements of the speed of light in the space vacuum variations without detectable loss overall because it is provided through the jet drive of the two cone's point thrusts. That thrust is necessarily deflected by a pressure wave front on the entire particle when pointed forward from each cone point, and then exhausted aftward freely with increased momentum when pointed in that direction in a pitching rotation scaled model of jet aircraft engine exhaust jets.) - - This separation of energy sources (for quanta and entire particle) removes the apparent contradiction here between the empirical MQPP action and prior astrophysics theory of the same gross observational data. - - - - Again, this undermines the justification for the supposed expansion or acceleration of expansion of the Universe. Consequently, there is no concrete and real necessity for the "Big Bang." If it occurred, it would have been at a much more distant time than previously estimated. MQPP must take the position that it did not occur, until a true necessity for it is established. Such a necessity is not really in sight. There are empirically necessary and not largely theoretical requirements in MQPP to be met for any re-establishment of the Big Bang theoretical hypothesis. - - - In this situation (in which we are taken as actually existing) the only necessary Universe is the empirical, potentially stable, continuously self-renewable Universe with no physically established beginning (and no conjectural parallels or quasi-logical offsets.)

SPECIAL REFERENCES FOR CHAPTER 10

Bahcall, N. A. (2015) Hubble's Law and the Expanding Universe, Proceedings National Academy of Sciences, Vol. 112, #11, 3173-3175, (doi: 10.1073/pnas.1424299112) and included References (Riess, 1998) & (Betoule, 2014)

Riess, A. G., et al. (1998) Observational Evidence from Supernovae for an Accelerating Universe and a Cosmological Constant, Astron. J., Vol 116, #3, 1009-1038

Betoule, M., et al. (2014) Improved Cosmological Constraints from a Joint Analysis of the SDSS-II and SNLS Supernova Samples, Astron. Astrophys., Vol. 568 (A22), 22-53

Longair, M. S. (1996) Our Evolving Universe, Cambridge University Press, Cambridge UK, CB2 1RP, & New York City, NY, USA 10011-4211

Lorentz, H. A.(1952) The Theory of Electrons, Dover Books on Physics, Dover Publications. Inc., 920 Broadway, New York City, NY, USA

Cohen, M. H., et al., (2015) Studies of the Jet in BL Lacertae, II, Superluminal Alfven Waves, Astrophys. Jour., 6 April 2015, Vol 803, (1), Article #3, ISSN 0004-637X, (CALTech Repository)

Lister, M. L., et al., (2013) Mojave. X. Parsec-Scale Jet Orientation and Superluminal Motion in AGN, Astronomical Jour., Nov. 2013, Vol. 146, (5), 120-142 (and arXiv: 1308.2713v1 [astro-ph.CO] 12 Aug 2013)

Meyer, E. T., et al., (2013) Optical Proper Motion Measurements of the M87 Jet: New Results from the Hubble Space Telescope, Astrophys. Jour. Letters, Vol 774, (2), doi:10.1088/2041-8205/774/2/L21

Meyer, E. T., et al., (2015) A Kiloparsec-scale Internal Shock Collision in the Jet of a Nearby Radio Galaxy, Nature (Letter), Vol 521, 495-497 (28 May 2015), doi:10.1038/nature14481

Mirabel, I. F., & Rodriguez, L. F., (1994) A Superluminal Source in the Galaxy, Nature (Letter), Vol 371, 46-48 (01 Sept 1994) doi: 10-1038/371046a0

Dhawan, V., Mirabel, I. F., & Rodriguez, L. F., (2000) AU-Scale Synchrotron Jets and Superluminal Ejecta in GRS 1915+105, Astrophys. Jour., Vol 543, (1), 373, doi:10-1086/317088

Cheung, C. C., Harris, D. E., Stawatz, L., (2007) Superluminal Radio Features in the M87 Jet and the Site of Flaring TeV Gamma-Ray Emission, Astrophys. Jour. Letters, Vol 663, (2), L65, doi: 10.1086/520510

Yadav, J. S., (2006) Relativistic Superluminal Radio Jets in Microquasars in Our Galaxy, Proceedings of Science, VI Microquasar Workshop: Microquasars and Beyond, Sept. 18, 2006, Como, Italy

Nakamura, M., Garofalo, D., & Meier, D. L., (2010) A Magnetohydrodynamic Model of the M87 Jet. I. Superluminal Knot Ejections from HST-1 as Trails of Quad Relativistic MHD Shocks, Astrophys. Jour., Vol 721, (2), 1783, doi: 10.1088/0004-637X/721/2/1783

Bartels, R., Krishnamurthy, S., & Weniger, C., (2016) Strong Support for the Millisecond Pulsar Origin of the Galactic Center GeV Exess (Gamma Ray), Phys. Rev. Lett., Vol 116, 051102, 4 Feb 2016

Lee, S. K., Lisanti, M., Safdi, B. R., Slatyer, T. R., & Wei Xue, (2016) Evidence for Unresolved Gamma Ray Point Sources in the Inner Galaxy, Phys. Rev. Lett., Vol 116, 051103, 4 Feb 2016

Chapter 11

Extension of the MQP Paradigm in Entanglement Throughout the Cosmos

SUMMARY: Entanglement of whole particles was introduced here in the chapter on electrons. That is the main ordinary particle around which entanglement practice has built up, except that it has not dealt with the electrons themselves so much as the photon, a highly unusual particle, which the electrons very regularly and usually emit in enormous numbers. That, of course, is because the electrons in ordinary nature are always present in enormous numbers themselves. Gradually over the entire book, down to the very mysterious neutrinos in the just prior (next to last) chapter, the background has been built up on this, to the description of just how it is that the tumultuously always intrusive electron, mixed into so many natural actions, can unobtrusively emit all the photon modifications of special neutrinos, which we can never really observe except as the basicly and very simply modified, highly interactive, and most evidently obtrusive, attention-capturing photons that flood around us from every star that is not obscured in some definite way and from every heated or highly excited object which has some excitable electrons in many clouds of orbits around its central massive cores or nuclei.- - - It is a most mysterious process, this electronic emission of photonic light, which has been very difficult to get into and grasp. It happens in the briefest of usual actions which is over before the superacute (imaginary) observer would think it had begun. Every step in the action is a necessary consequence of the MQP vortex quantum as observed at scale in the lab. Humans will never really be able to observe it as a step-by-step action. Yet it is at the heart of what is probably the sharpest and most essential human sensory capability.- - - And the most extreme aspect of the electron-photon-light action is that it is also necessarily at the heart of the very highest velocity natural entanglement interaction between particles. This velocity is necessarily present in a real action because the interaction is observed reliably over and over again, though all the experts have denied its presence over and over again (because they cannot see or feel it.) For a half-century there has been an imaginary logic that describes an unembodied necessity of its action which has made the entanglement process an engineering tool that large numbers of researchers use repeatedly while continuing to deny its existence as a real action.- - - It is far past the time that this should be dealt with empiricly as a real particle action at a very high new velocity. Actually that was introduced faintly and as unobtrusively as possible in the last prior chapter without fully admitting it, but here is the point at which to add some readily available MQP details which eventually can be and must be measured and reduced to tables of hard dats. - - - [There are beginning to be repeated empirical indications that entanglement occurs in many other types of particle interactions, but the bulk of work has been done with photons and electrons, and that is concentrated on, with obvious applicability in more limited ways to all particles as constructs of MQP quanta assemblies. <u>The greatest of these missing elements is the concentrated non-relativistic entanglement forces between vortex drivers in black hole densities and closest proximities with the most minimized mass eddy beginnings and peel-offs plus dwarfed spiral waves, the missing elements in cosmic analysis to date.</u> Also herein the advanced technology of entanglements requires a scaled, quasi-Planckian frequency of entanglement coupling at the Planckian ratio of about $f_{Plk}/Vp/2\pi$ times the gravity force (at f_{Plk} times the frequency of gravity coupling f_G) at the same standard coupling distance of GD times an entanglement coupling constant C_e which is to be determined eventually.]

In the just previous chapter on neutrino-like interactions there was a gross first description of the inherently necessary process of converting an Extreme minimal linear neutrino that is initially traveling very close to the speed of light into an emitted photon of many OM greater, though still sub-microscopic, mass traveling at exactly the speed of light. That description in words was very fragmentary. Any kind of typical mathematical description of the process would require many more specific actions, whether given in words alone or in math symbols or in both. These are actions that change speeds and directions of travel, that initiate or end emission waves, that begin or end or change rotations and especially the creation of new rotating objects, such as added or (correctly) <u>dis</u>-continuously replaced mass eddies. These would involve new or ending displacements, velocities, accelerations, and the <u>jerks</u> that cause these changes.

Something has to act to cause the jerks and their consequences as accelerations and velocities of actions beginning and ending at definable times. This is especially true of complete beginnings and endings of vigorous actions that do not exist at all before these beginnings and after these endings (such as the generation and detachment of mass eddies, however microscopic they are.)

There are a number of these sudden beginnings and endings of vigorous special actions with brief and necessarily strong jerk in the turbulent centrally driven conic vortex. (These were not being looked for, searched out, earlier. Now that new causes are necessary, these jerk wave actions must be established and clearly defined.) The most obviously visible starting and stopping action in a turbulent conic vortex is in the wall of turbulent and rapid circular currents that suddenly stop and subside into smooth outer waves at the GD of the vortex. (Fig 2-10, page 61.) The initiation of these multiple small bodies of turbulent momentum may occur at random and build fairly slowly over perceptible periods of time with low jerk that cycles at a wide frequency band, but the endings are sudden in more vigorous jerk frequency bands that should be readily measurable in differences as counter-pressure waves that stop the spread of the turbulence waves and collapse them at the GD wall. As waves of pressure change in water they would be distinctly different from the predominantly viscous waves that build the turulences at much slower relative velocities. While the massive turbulence body waves that can be readily seen travel within the GD at a few centimeters per second, these jerk pressure waves reverberate back and forth between the bodies within the GD at not less than the speed of sound in water or about 330 meters per second, depending on temperatures, etc., with at least 4 OM difference in speed. With scaling in size over at least 11 OM to the electron and further to the photon vortex, the difference there in speeds would grow proportionately. (GD & toroids are stripped off in black holes, etc.)

> - - - [The idea of these separable types of waves at different natural velocities in a medium may be difficult to get used to. (This is not the place to develop this concept fully since it is a part of a separate voluminous discipline of fluid mechanics, and many waves are mixed viscous and pressure waves.) The extremes of largely viscous waves from light breezes on the surface of water move relatively slowly at a few centimeters per second but speed up as they get larger with pressure components, while the high pressure sound wave components of the wake of a bullet fired into water will travel at hundreds of meters per second once they are separated from the bullet. - - - Actually, since air is readilty compressible, the tornado in air is probably the situation with the most widely varied and at the same time most easily understood differences between viscous and pressure waves. These range from light pressure airs at the edge of the disturbed volume to central mixed vortex waves at hundreds of meters per second. Yet a lightning bolt may propagate at thousands of meters per second while creating a pure pressure thunder wave that starts very fast but quickly settles to a sound wave at a few hundred meters per second in air and several times that in any solid chrystalline rock below.] - - -

Back to the jerk wave situations in the quantal turbulent vortex: The next most prominent jerk process is the initiation and breaking away of multiple mass eddies at the edge of the base of the drive cone where each spiral wave breaks away from the base and extends in a lagging curve to the GD. The spiral wave lag itself beyond the base appears to be the next. The initiation of the spirals near the axis in the driven base is next. Within the base the spiral wave is wrapped in sub-spirals on the free spiral surface, but this subspiral does not make a complete pass around the spiral wave, since each subspiral begins where the spiral tube revolves away from the surface of the cone base and ends nearer the edge of the base where the spiral tube comes back to the frictional drive of the cone base again. By looking more closely more jerks can be seen, but these are the principal ones. They are impulse pressure actions to exceed some starting limit on initiations of new gross actions that continue with lower frequency cycles until similarly shut down sharply without a decreasing wave train. There are multiple, predictably odd, harmonics of the knife-like action step.

A large amount of drive cone energy and power is going into these start-and-stop jerk operations, which act through sharp pressure waves that reverberate resonantly back and forth in the vortex and couple out into the medium with detached pressure waves at velocities evidently higher than the observed laminar shear circular waves departing from the edge of the GD and than the spiral waves or the frictional mass eddy waves or the large laminar flow waves on the sides of the drive cones and in the laminar toroidal currents. As scalable model velocities, the viscous waves travel at one lower speed of sound which scales to the speed of light, and the more vigorously driven impulse pressure waves that operate the very brief and intense jerks travel at OM higher velocity for which the difference in OM must be necessarily magnified in the extreme scaling to the velocities and dimensions of the quantal vortices. As noted numericly in the appendix to Chapter 5, the numbers of frequency bands, the widths of the resonances, the numbers of sharper resonances possible within the individual frequency band resonances, and the sum and difference frequencies that multiply the resonances, provide an enormous number of separately tunable transmission and reception operating resonances of great tuned selectivity for the natural coupling of entanglements of each type of particle. This is especially true of the electron and the photon with their combinations of simplicity and great stability noted repeatedly throughout this MQP paradigm of particle structures and actions.

In the medium of the appendices to Chapter 5, entanglement is the resonance linking force that is faster than the medium's ultimate worm (string-type) knots can disentangle, approaching the velocity at which gyre drivers become rigid and quasi-infinitely dense, the Planckian 10^{40} cps at V_P 10^{40} mps. - - - It is much higher than the 10^{20} scales (beyond 10^{17} mps) where the electron's gyres get to the effective point size electron at acceleration to very close to the speed of light. In the water model it is equivalent to the hyperspeed evacuation bubbles that appear and collapse at the trailing edge of wedge propellers. In the space medium this would occur as a very tense stretching or compression of the interlinked writhing (string-like) worms of the medium faster than they could fully disengage or further engage, well beyond 10^{30} scales. (Estimated measureable in ultimately refined lab experiments as tuned average sums only in the water model at the median between the spiral wave electric and the point thrust forces.) Baseline frequencies would be high enough that there is very little area under each curve. - - - Entanglement is the most ultimate empirical simplification, a simplicity beyond the ordinary continuing superluminal internal actions of MQPP.

It is predicted from multiply repeated observation of the initial lab models of the quantal turbulent vortices that refined separate measurements of the summed entanglement pressure wave force vectors and resonant frequency patterns (as distinguished from the lower frequency viscosity waves of the nominal strong, point thrust, electric, and gravity eddy waves) can and will be accomplished when lab resources are allocated. This is a necessary step in full realization of the empirical MQPP simplification of the last century's heroicly complex model of the sub-nuclear particles and in providing the working auxiliary and supplementary correlations and benefits outlined or briefly indicated herein.

See the Foreword for an abbreviated initial summary (which does not include additional progress in the final chapters above) of The Many Striking Correspondences of the Micro-Quantal Particle Paradigm with the Quantum Mechanics of the Sub-Nuclear Particles. A similar but yet earlier briefing in overall summary is attached as a final chapter below. An Afterword, Acknowledgements, and References in addition to those attached to chapters follow the briefing.

In conclusion of the main text, the ultimate requirement of the PDG empirical data, and of the MQP paradigm of sub-nuclear particles developed from it, is that micro-quantal conic drivers of centrally driven turbulent conic vortices are the single source of all the self-renewing universe's energy and forces, which may die out eventually or be replaced. But, as its bio-chemical physics constructs or parts of its end items, we humans are unable to observe directly either this background universe or its medium. We can only hypothesize that these sole spin-momentum energy sources must necessarily be present, each with an effectively extreme density of another class than ordinary mass density. This apparent momentum density must be greater than the observable converted mass energy losses since some very distant beginning of the present cosmos, much more distant in time than is now considered proper from 20th Century Quantum Mechanics. We can go on with our observable consequences in physics of the result, and herein begin to do so on the basis of the empirical Mass-Charge Laws observed in the PDG data tables and their consequence the Micro-Quantal Particle Paradigm.

381

AFTER-WORD

    This book summarizes well over a decade of private research. If accepted and utilized, these findings must lead to a wide-spread follow-up over many decades. New research must include: Broader empirical applications, funded experimental refinements and extensions (especially for empirical measurement of asymmetric mutual forces between vortices and of force couplings in separated frequency bands.) Eventually there must be theoretical grounding in fluid mechanics (if not also in a new type of knotted ether forms), as well as testing of inherently necessary predictions.

    In this we achieve an opening to a new synthesis of science, not only of all the forces of Nature, but also of all the particles and waves in a single interplay of actions. This will slowly clarify for our understanding all the remaining major mysteries of which we are now aware, those between ordinary life, light, black holes, exploding supernovae, a continuing cosmos, and some unclear aspects of relativity beyond light speed, with further probability mechanics for another twilight of things then to be observed dimly but not yet within human grasp. Still, it can take cycles of reviving civilizations to approach such conclusions, just as it has taken a civilization of Rome, and now a long expansion of European civilization to find the real meaning of the Greeks' initial concept of atomic Nature.

Acknowledgements: I thank Fred E. Howard, III, for many constructive comments on general physics involved; Katherine M. (Howard) Douglas for significant data acquisition, figure preparation, and endless assistance; D. Bailey Howard for distinct improvements in internet data acquisition; and H. Blevins Howard for much assistance with figures and computing equipment, for shooting video data, and for keeping watch on internet data postings without which many correlations to current literature would not have been observed. These contributors constitute the rest of the Howard Particle Physics Group which provided organized, continuing, research support wherever it was needed and also many influences on research and publication direction. I owe much appreciation to Cheryl Mack and Christi Rountree of the US Air Force Armament Laboratory Technical Library for patient and long-continued assistance with searches of the background literature after my retirement emeritus. Particular appreciation is due to Matthew Clark, whose woodworking students at the Okaloosa County Vocational and Technical School (in Fort Walton Beach, Florida) turned the large 30 degree cones in maplewood for exploratory experiments in conic vortices (with Blevins Howard.) This research would not have been started and could not have been attempted without access to the biennial voluminous reports of the international Particle Data Group as published on-line by the US Lawrence Berkeley National Laboratory.

    It was these empirical trials, which the cones and video camera enabled, that discovered the full wave-particle structure of the micro-quanta and made the structure both quantitative and specific in fitting the necessary requirements from the PDG data and the equation laws. The fit to the necessities is both precise and informative with much unexpected meaning, particularly as to the nature of mass with a low level of forceful gravity coupling and inertia, as well as to a real basis for entanglement. This congruence of structures with functional consequences, like that of the initial mass and charge laws, is too complete and thorough to be anything other than Nature's doing, observed here only fragmentarily, with much more to reveal in time.

LITERATURE CITED IN MAIN VOLUME
(in addition to Special Lists at Chapter ends.)
CHAPTER 1 AND FOREWORD

ALBRIGHT, C. H., 2004. Normal vs. inverted hierarchy in type I seesaw models, Phys. Lett. B, 599, 285-293.

BABU, K. S., AND S. M. BARR, 2000. Mass relation for neutrinos, Phys. Rev. Lett. 85, 6, 1170-1173.

BAHCALL, J. N., P. I. KRASTEV, AND A. Y. SMIRNOV, 1998. Where do we stand with solar neutrino oscillations? Phys. Rev. D, 58, p. 096016-1-22.

BANDOS, I. A., J. A. AZCARRAGA, J. M. IZQUIERDO, AND J. LUKIERSKI, 2001. BPS states in M theory and twistorial constituents, Phys. Rev. Lett., 86, 20, 4451-4454.

BARGER, V., S. L. GLASHOW, D. MARFATIA, AND K. WHISNANT, 2002. Neutrinoless double beta decay can constrain neutrino dark matter, Phys. Lett. B, 532, 15-18.

BERGSHOEFF, E., A. SALAM, E. SEZGIN, AND Y. TANII, 1988. Singletons, Phys. Lett., B, 205, 2-3, 237-244.

BOYER, T. H., 1982. Classical model of the electron and the definition of electromagnetic field momentum, Phys. Rev. D, 25, 12, 3246-3250.

BRANCO, G. C., M. N. REBELO, AND J. I. SILVA-MARCOS, 2004. Universality of Yukawa couplings confronts recent neutrino data, Nucl. Phys. B, 686, 188-204.

CALDWELL, D. O., AND R. N. MOHAPATRA, 1993. Neutrino mass explanations of solar and atmospheric neutrino deficits and hot dark matter, Phys. Rev. D, 48, 7, 3259-3263.

CATANI, S., D. FLORIAN, G. RODRIGO, AND W. VOGELSANG, 2004. Perturbative generation of a strange-quark asymmetry in the Nucleon, Phys. Rev. Lett., 93, 15, 152003-1-4.

CHACKO, Z., L. J. HALL, T. OKUI, AND S. J. OLIVER, 2004. CMB signals of neutrino mass generation, Phys. Rev. D, 70, 085008-1-18.

CLOSE, F., M. MARTEN, and C. SUTTON, 1987. The Particle Explosion, Oxford University Press, New York, NY. 239 pp.

DAVIES, C., and LEPAGE, G. P., et al., 2010, Determination of the masses of the common up and down quarks, Phys. Rev. Lett. **104** (2 Apr. issue)

DOWLING, J. P., (ed.), 1997. Electron theory and quantum electrodynamics 100 years later, Plenum Press, New York, NY. 356 pp.

DUGNE, J. J., S. FREDRICKSSON, and J. HANSSON, 2002. Preon trinity---A schematic model of leptons, quarks, and heavy vector bosons, Europhys. Lett., 60 (2),188-194

DURR, S., et al., 2008, Ab initio determination of light hadron masses, Science **322**, 1224

EIDELMAN, S., et al., 2004, Particle Data Group biennial report, Phys. Lett. B 592, 1,

(and at pdg.lbl.gov) including:
AMSLER, C., and O. G. WOHL, 2004, Note on the quark model.
GROOM, D. E., 2004, Note on understanding two-flavor oscillation parameters
HOEHLER, G., and R. L. WORKMAN, 2004, Note on N and Delta resonances.
KAYSER, B., 2004, Note on neutrino mass, mixing, and flavor change
MANOHAR, A. V., AND C. T. SACHRAJDA, 2004, Note on quark masses.
NAKAMURA, K., 2004, Note on solar neutrinos
OLIVE, K. A., 2004, Note on low mass neutrinos.
VOGEL, P., AND A. PIEPKE, 2004, Note on electron, muon, and tau neutrino listings.

ELGAROY, O, AND O. LAHAV, 2003. Upper limits on neutrino masses from the 2dFGRS and WMAP, J. Cos. Astropart. Phys., 1475-7516/03/62120-1-24.

ELLIS, J., M. RAIDA, AND T. YANAGIDA, 2002. Observable consequences of partially degenerate leptogenesis, Phys. Lett. B, 5456, 228-236

FUKUDA, S., AND 116 OTHERS of the international SUPER-KAMIOKANDE COLLABORATION, 2000. Tau neutrinos favored over sterile neutrinos in atmospheric muon neutrino oscillations, Phys. Rev. Lett. 85, 3999-4003.

GOUVEA, A., AND J. W. F. VALLE, 2001. Minimalistic neutrino mass model, Phys. Lett. B, 501, 115-127.

GSPONER, A., AND J. HURNI.,1996. Non-linear field theory for lepton and quark masses. Hadr. Jour. 19; 367-373.

HAISCH, B., A. RUEDA, AND H. E. PUTHOFF, 1994. Inertia as a zero-point-field Lorentz force, Phys.Rev. A 49, 2; 678-694.

HANNESTAD, S., 2003. Neutrino masses and the number of neutrino species from WMAP and 2dFGRS, J. Cos. Astropart. Phys. 1475-7516/03/62004-5-19.

HESTENES, D., AND A. WEINGARTSHOFER, (eds.), 1991. The electron, new theory and experiment, Kluwer Academic Publishers, Boston, MA. 405 pp.

HOWARD, F. E., JR. 2005. Elementary Particle Mass Sub-Structure Power Law, Florida Scientist, 65, #3, 175-205

JIMENEZ, J.L., AND L. CAMPOS, 1999. Models of the classical electron after a century, Found. Phys. Lett.,12, 2, 127-146.

JOSHIPURA, A. S., 1995. Degenerate neutrinos in left-right symmetric theory, Phys. Rev. D, 51, 3, 1321-1325.

KAUS, P., AND S. MESHKOV, 2004. Neutrino masses, mixing, and hierarchy, Phys. Lett.B, 580, 236-242.

KIM, J., 1998. Explanation of the masses of quarks and leptons in a supersymmetric preon model, J. Phys. G, Nucl. Part. Phys. 24; 1881-1902.

KING, S. F., 2003. Phenomenological and cosmological implications of neutrino oscillations, J. Phys. G, Nucl. Part. Phys. 29, 1551-1559.

KRONFELD, A. S., 2008, The weight of the world is Quantum Chromodynamics, Science **322**, 1198

LIPMANOV, E. M., 2003. Quasi-degenerate neutrino masses, Phys. Lett. B, 567, 3-4,268-272.

LUNARDINI, C., AND A. Y. SMIRNOV, 2003. Probing the neutrino mass hierarchy and the 13-mixing with supernovae, J. Cos. Astropart. Phys. 1475-7516/03/63726-2-48.

LUTY, M. A. AND P. H. MOHAPATRA, 1997. A supersymmetric composite model of

quarks and leptons, Phys. Lett,. B, 396, pp. 161-166.
MAC GREGOR, M. H., 1992. The enigmatic electron, Kluwer Academic Publishers, Boston, MA. 165 pp.
PARTICLE DATA GROUP (PDG) BIENNIAL REPORTS
See Eidelman and also Yao, and pdg.lbl.gov
PATI, J. C. AND A. SALAM, 1983. Supersymmetry at the preonic or pre-preonic level, Nucl. Phys. B, Part. Phys. 214, 1, 109-135.
PATI, J. C., A. SALAM, AND J. STRATHDEE, 1981. A preon model with hidden electric and magnetic type charges, Nucl. Phys. B, 185, 4516-428.
RODEJOHANN, W.,2002. Phenomenological aspects of light and heavy Majorana neutrinos, J. Phys. G, Nucl. Part. Phys. 28, pp. 1477-1498.
ROHRLICH, F., 1982. Comment on the preceding paper by T. H. Boyer, Phys. Rev. D, 25, 12, 3251-3255.
ROHRLICH, F., 1997. The dynamics of a charged sphere and the electron, Am. J. Phys. 65, 11, 1051-1056
ROSS, G. G., 2003. Theoretical implications of neutrino oscillations, J. Phys. G, Nucl. Part. Phys. 29, 1541-1550.
SALAM, A., 2000. Overview of Particle Physics. pp. 481-492 in DAVIES, P., (ed.), The New Physics, Cambridge University Press, Cambridge, UK. 516 pp.
SPRINGFORD, M., (ed.), 1997. Electron, a centenary volume, Cambridge University Press, Cambridge, UK. 330 pp.
TREIMAN, S., 1999, The Odd Quantum, Princeton University Press, Princeton, NJ. 262 pp.
VALLE, J. W. F., 2003a. Status of neutrino oscillations and non-standard properties, Nucl. Phys. B (Proc. Suppl.) 114, 159-175.
VALLE, J. W. F., 2003b. Standard and non-standard neutrino properties, Nucl. Phys. B (Proc. Suppl.) 118, 255-266.
WILCZEK, F., 2008, Mass by numbers, Nature **456,** 449
XING, Z., 2002. Model-independent constraint on the neutrino mass spectrum from the neutrinoless double beta decay, Phys. Rev. D, 65, 077302-1-4.
YAO, W.-M., et al., 2006, Biennial PDG Data Report, Jour. Phys. G, **33**, 1 (2006), pdg.lbl.gov, including Drees, M., and G. Gerbier, 22. Dark Matter (2006)

CHAPTER 2

EIDELMAN, S., et al., 2004, Particle Data Group biennial report, Phys. Lett. B 592, 1, (and at pdg.lbl.gov) including:
AMSLER, C., and O. G. WOHL, 2004, Note on the quark model.
HOWARD, FRED E., JR., 2010, Striking Matches of the MQP and QM, www.particlephysics.info
YAO, W.-M., et al., 2006, Biennial PDG Data Report, Jour. Phys. G, **33**, 1 (2006), (and at pdg.lbl.gov)

CHAPTER 3

REFERENCES TO VOL. 1, CHAPTER 4

Blackwell, K.G., 2000: The evolution of Hurricane Danny (1997). Mon. Weather Rev., 128, 4002-4016

Forbes, G. S., and Bluestein, H. B., 2001: Tornadoes: Contributions by T. T. Fujita. Bul. Am. Met. Soc., 82, 73-9

Holmes, J., Blackwell, K.G., and Wade, R., 2005: Collapsing precipitation cores in open- eyewall hurricanes at landfall. Pre-print, SE Coastal Atmospheric Processes Symposium (SECAPS), Univ. So. Ala., Mobile, AL

Fujita, T. T., 1976a: Graphic examples of tornadoes. Bull. Am. Met. Soc., 57, 401-412.

----, 1976b: History of suction vortices. Proc. Symp. on Tornadoes, Lubbock, TX, Texas Tech University, 78-88

----, 1992: The mystery of severe storms. WRL Research Paper 239 (A bound monograph). Wind Res. Lab, Dept. of Geophys. Sci., Univ. of Chicago, IL, USA. 298

Fujita, T. T. and Arnold, B., 1963: Preliminary result of analysis of the cumulonimbus cloud of April 21, 1961. MRP Res. Paper 16, Univ. Chicago, IL, USA. 16 pp.

Howard, F. E., Jr., 2009: Turbulent conic vortices: Part 1, Lab gyres define tornado band in Hurricane Ivan.

Lyons, W. A., and Armstrong, R. A., 2004:A review of electrical and turbulence effects of convective storms on the overlying stratosphere and mesosphere. Am. Met. Soc. Symposium on Space Weather, AMS Annual Meeting, Seattle, WA. Conference CD, 6 p

Lyons, W. A., Nelson, T. E, and Fossum, J., 2000: Results from the SPRITES '99 and STEPS 2000 field programs. Preprints, 20th Conf. on Severe Local Storms, Am. Met. Soc., Orlando, 4 pp.

Markowski, P. M., Straka, J. M., and Rasmussen, E. N., 2002: Direct surface thermodynamic observations within the rear-flank downdrafts of nontornadic and tornadic supercells. Mon. Weather Rev., 130, 1692-1721

McWilliams, J. C., 1984: The emergence of isolated, coherent vortices in turbulent flow. J. Fluid Mech., 146, 21-43

Melander, M. V., Zabusky, N. J., and McWilliams, J. C., 1988: Symmetric vortex merger in two dimensions: causes and conditions. J. Fluid Mech. 195, 303-340

Prandtl, L., and Tietjens, O.G., 1957: Fundamentals of Hydro- and Aeromechanics. Engineering Societies Monographs. Dover Publications, New York, USA

Rasmussen, E. N., Straka, J. M., Davies-Jones, R., Doswell, C. A. III, Carr, F. H., Eilts, M. D., and MacGorman, D. R., 1994: Verification of the origins of rotation in tornadoes experiment: VORTEX. Bull. Amer. Met. Soc., 75 (6), 995-1005.

Rasmussen, E. N., Straka, J. M., Richardson, S., Markowski, P. M., and Blanchard, D. O., 2000: The association of significant tornadoes with a baroclinic boundary on 2 June 1995. Mon. Weather Rev., 128, 174-191.

Rasmussen, E. N., and Collaborators, 2001: 29 June 2000 STEPS tornado near Goodland, KS. Severe Storms Research Papers, Cooperative Institute for Mesoscale Meteorological Studies, Oklahoma Univ., Norman, OK, USA.

Riccardi, G., and Piva, R., 1998: Motion of an elliptical vortex under rotating strain: conditions for asymmetric merging. Fluid Dynam. Res., 23, 63-88

Sentman, D. D., Wescott, E. M., Picard, R.H., Winick, J. R., Stenbaek-Nielsen, H. C., Dewan, E. M., Moudry, D. R., Sao Sabbas, F. S., and Heavner, M. J.2003:Simultaneous

observation of mesospheric gravity waves and sprites generated by a midwestern thunderstorm. J. Atmos. Solar/Terr 65(5), 537-550 (& 2002 preprint.)
Smyth, W. D., and Peltier, W. R., 1994: Three-dimensionalization of barotropic vortices on the f-plane. J. Fluid Mech., 265, 25-64
van Delden, A., 2003: Adjustment to heating, potential vorticity and cyclogenesis. Q. J. R. Meteorol. Soc., 129, 3305-3322
Ziegler, C. L., Rasmussen, E. N., Shepherd, T. R., Watson, A. I., and Straka, J. M., 2001: The evolution of low-level rotation in the 29 May 1994 Newcastle-Graham, Texas, storm complex during VORTEX. Mon. Weather Rev., 129 (6), 1339-1368.

References for Chapter 5 of Volume 1

1. Feynman, R. P., Leighton, R. B., & Sands, M., *The Feynman Lectures on Physics*, Vol. 2 (Addison-Wesley, Reading, MA, 1964).

2. Hestenes, D., Weingartshofer, A., Eds., *The Electron, New Theory and Experiment* (Kluwer Acad., Dordrecht, 1991).

3. Springford, M., Ed., *Electron, a centenary volume* (Cambridge Univ. Press, Cambridge, UK, 1997).

4. Dowling, J.P., Ed., *Electron Theory and Quantum Electrodynamics 100 Years Later* (Plenum, New York, NY, 1997).

5. Salam, A., in *The New Physics* (Ed. Davies, P.) 481-492 (Cambridge Univ. Press, Cambridge, UK, 1989).

6. Haisch, B., Rueda, A., & Puthoff, H. E. Inertia as a zero-point-field Lorenz force. *Phys. Rev.* **A 49**, 2, 678-694, (1994).

7. Slominski, W., Szwed, J., Phenomenology of the electron structure function, *Eur. Phys. J.* **C 22**, 123-127 (2001).

8. Muryn, B., Szumlak, T., Slominski, W., & Szwed, J., Measurement of the electron structure function. *Nucl. Phys* **B 126**, (Proc. Suppl.) 11 (2004).

9. Waite, T., Barut, A. O., & Zeni, J. R., in *Electron Theory and Quantum Electrodynamics 100 Years Later*, (ed. Dowling, J. P.) 223-240 (Plenum, New York, 1997).

10. Pan, J.-W., Bouwmeester, D., Daniell, M., Weinfurter, H., & Zeilinger, A., Experimental test of quantum nonlocality in three-photon Greenberger-Horne-Zeilinger entanglement. *Nature* 403, 515-519 (2000).

11. Mair, A., Vaziri, A., Weihs, G., & Zeilinger, A. Entanglement of the orbital angular momentum states of photons. *Nature* **412**, 313-316 (2001).

12. Mac Gregor, M. H., *The Enigmatic Electron*. (Kluwer Acad., Boston, 1992).

13. Howard, F. E., Jr., Elementary particle mass sub-structure power law. *Florida Scient.*, **68** (3), 175-205 (2005); & Erratum, Appendix Table C3, **69**, 2, 148 (2006). See www.electron-particlephysics.org for correction in place.

14. Howard, F. E., Jr., Sub-structure laws of particle masses and charges---a new systematic classification of subatomic particles. *Florida Scient.* **69**, 3, 192-215 (2006). (Also at www.electron-particlephysics.org.)

15. Amsler, C. *et al*. (Particle Data Group) Biennial Report of Particle Data Group for 2008., *PL* B **667**, 1 (2008). Download by Sections, Signed Notes, Summary Tables (accredited data), Particle Listing Tables (detailed & not yet accredited data), etc., at www.pdg.lbl.gov; as well as close equivalents below and biennially for subsequent years:

Yao, W.M., *et al*. PDG Report for 2006. *J. Phys*., G **33**, 1 (2006).

Eidelman, S., *et al*. PDG Report for 2004. *Phys. Lett. B* **592**, 1 (2004)

Hagiwara, K., *et al*. PDG Report for 2002. *Phys. Rev. D* **66**,010001 (2002)

16. Hestenes, D., in *The Electron, New Theory and Experiment*. (eds. Hestenes, D., & Weingartshofer, A.) 21-36 (Kluwer Academic, Dordrecht, 1991).
17. Hu, Q.-H., The nature of the electron. *Phys. Essays* **17**, 4 (2004)
18. Barut, A. O., in *The Electron, New Theory and Experiment*. (eds. Hestenes, D., & Weingartshofer, A.) 105-148 (Kluwer Academic, Dordrecht, 1991).
19. Goldhaber, M., A closer look at the elementary fermions. *Proc. Natl. Acad. Sci.* **99**, 33-36 (2002).
20. Pati, J. C., Salam, A., & Strathdee, J., A preon model with hidden electric and magnetic type charges. *Nuc. Phys.* **B 185**, 416-428 (1981).
21. Barut, A. O., Lepton mass formula, *Phys. Rev. Lett.* **42**, 19, 1251 (1979).
22. Boyer, T. H., Classical model of the electron and the definition of electromagnetic field momentum. *Phys. Rev.* **D 25**, 12, 3246-3250 (1982).
23. Rohrlich, F., Comment on the preceding paper by T. H. Boyer. *Phys. Rev.* **D 25**, 12, 3251-3255 (1982).
24. Rohrlich, F., The dynamics of a charged sphere and the electron. *Am. J. Phys.* **65**, 11, 1051-1056 (1997).
25. Campos, I., & Jimenez, J. L., Comment on the 4/3 problem in the electromagnetic mass and the Boyer-Rohrlich controversy. *Phys. Rev.* **D 33**, 2, 607-610 (1986).
26. Jimenez, J. L., & Campos, I., Models of the classical electron after a century. *Found. Phys. Lett.* **12**, 2, 127-146 (1999).
27. Bialynicki-Birula, I., Classical model of the electron. Exactly soluble example. *Phys. Rev. D* **28**, 8, 2114-2117 (1982).
28. Bergshoeff, E., Salam, A., Sezgin, E., & Tanii, Y., Singletons, higher spin massless states and the supermembrane. *Phys. Lett.* **B 205**, 2-3, 237-244 (1988).
29. Pati, J. C., & Salam, A., Supersymmetry at the preonic or pre-preonic level and composite supergravity. *Nuc. Phys.* **B 214**, 1, 109-135 (1983).
30. Gsponer, A., & Hurni, J.-P., Non-linear field theory for lepton and quark masses. *Hadronic Jour.* **19**, 367-373 (1996).
31. Rosen, G., Semi-empirical operator for the self-interaction masses of finite-size leptons and quarks. *Europhys. Lett.* **62**, 473-476 (2003).
32. Kim, J., Explanation of the masses of quarks and leptons in a supersymmetric preon model. *Jour. Phys.* **G 24**, 10, 1881-1902 (1998).
33. Luty, M. A., & Mohapatra, R. N., A supersymmetric composite model of quarks and leptons. *Phys. Lett.* **B 396**, 161-166 (1997).

34. Hayakawa, M., Mass hierarchy from compositeness hierarchy in supersymmetric gauge theory. *Phys. Lett.* **B 408**, *1-4* 207-212 (1997).
35. Bohun, C. S., & Cooperstock, F. I., Dirac-Maxwell solitons. *Phys. Rev.* **A 60**, 4291 (1999).
36. Slominski, W., & Szwed, J., Spin effects in the electron structure functions. *Eur. Phys. J.* **C 39**, 55-59 (2005).
37. Lanciani, P., The rest mass of (charged) leptons in 6-D spacetime. *Found. Phys. Lett.* **14**, 6, 541-551 (2001).
38. Dugne, J.-J., Fredricksson, S., & Hansson, J., Preon trinity---A schematic model of leptons, quarks, and heavy vector bosons. *Europhys. Lett.* **60**, 2, 188-194 (2002).
39. Sackett, C. A., Kielpinski, D., King, B. E., Langer, C., Meyer, V., *et al.*, Experimental entanglement of four particles. *Nature* **404**, 256-259, (2000).
40. Rarity, J. G., Getting entangled in free space. *Science* **301**, 604-605 (2003).
41. Aspelmeyer, M., Bohm, H. R., Gyatso, T., Jennewein, T., Kaltenbaek, R., *et al.*, Long-distance free-space distribution of quantum entanglement. *Science* **301**, 621-623 (2003).
42. Jones, R. T., & Cohen, D. *High Speed Wing Theory* (Princeton Univ. Press, Princeton, NJ, 1960).

Disc Title

A Hidden Simplicity
of
NUCLEAR PARTICLES
-

that <u>UNDERGRADS</u> to <u>PHDs</u> WILL <u>UNDERSTAND</u>
by
Fred Howard
<u>+- CHARGE SPIN INITIATES MASS</u>
<u>SINGLE</u> <u>SOURCE</u> <u>UNIFIES</u> <u>ALL</u> <u>FORCES</u>
<u>ELECTRONS BECOME REAL</u>

The VERY SIMPLE CONCLUSIONS, the
STRIKING MATCHES with Quantum Mechanics, &
the CONTINUED FURTHER CONSEQUENCES of
the Unnoticed 2005 Paper
(on PDG data) which
ANTICIPATED by 5 YEARS the 2010 first accurate
Quantum Mechanics (QM) Masses of the lightest Up
and Down Quarks of the Standard Model (SM), &
FORESAW by 3 YEARS the widely celebrated 2008
first Fairly Accurate QM computed Masses for the
SM Proton & Neutron (the Every Day Visible World),
SEPARATING those 2 ESSENTIAL MASSES, which
the QM/SM Computing Teams & Super-Computers
COULD NOT DO - - **NOW UNIFIES ALL FORCES**
in *Turbulent, Co-axial Pair, Conic Micro-Quanta* - -
A Simpler Basis for New 21st Century Physics
Underlies QM, QED, QCD, SM, & **GRAVITY**
due to MASS initiated by Single Source of +- Charge-Spin-Forces
(the TRUE "Higgs" effect phenomenon)
What Einstein & Feynman Were Looking For
That EVERYONE who wants to Can Understand
Even the Math now takes only a Bright Teen-Ager
and a Pocket Calculator

Now Reveals THE Hidden Variables
that cause MOST QM Random Probabilities
that had to wait for the decades of PDG Particles Summary Data to accumulate

www.ingramcontent.com/pod-product-compliance
Lightning Source LLC
Chambersburg PA
CBHW020633220526
45464CB00001B/133